MW00975104

370
Topics in Current Chemistry

Editorial Board

H. Bayley, Oxford, UK
K.N. Houk, Los Angeles, CA, USA
G. Hughes, CA, USA
C.A. Hunter, Sheffield, UK
K. Ishihara, Chikusa, Japan
M.J. Krische, Austin, TX, USA
J.-M. Lehn, Strasbourg Cedex, France
R. Luque, Córdoba, Spain
M. Olivucci, Siena, Italy
J.S. Siegel, Tianjin, China
J. Thiem, Hamburg, Germany
M. Venturi, Bologna, Italy
C.-H. Wong, Taipei, Taiwan
H.N.C. Wong, Shatin, Hong Kong
V.W.-W. Yam, Hong Kong, China
C. Yan, Beijing, China
S.-L. You, Shanghai, China

Aims and Scope

The series Topics in Current Chemistry presents critical reviews of the present and future trends in modern chemical research. The scope of coverage includes all areas of chemical science including the interfaces with related disciplines such as biology, medicine and materials science.

The goal of each thematic volume is to give the non-specialist reader, whether at the university or in industry, a comprehensive overview of an area where new insights are emerging that are of interest to larger scientific audience.

Thus each review within the volume critically surveys one aspect of that topic and places it within the context of the volume as a whole. The most significant developments of the last 5 to 10 years should be presented. A description of the laboratory procedures involved is often useful to the reader. The coverage should not be exhaustive in data, but should rather be conceptual, concentrating on the methodological thinking that will allow the non-specialist reader to understand the information presented.

Discussion of possible future research directions in the area is welcome.

Review articles for the individual volumes are invited by the volume editors.

Readership: research chemists at universities or in industry, graduate students.

More information about this series at http://www.springer.com/series/128

Salvatore Sortino
Editor

Light-Responsive Nanostructured Systems for Applications in Nanomedicine

With contributions by

S. Acherar · P. Arnoux · S. Ashraf · J. Atchison · F. Baros · J.F. Callan · M. Carril · C. Carrillo-Carrion · L. Colombeau · C. Conte · D. Costley · A. Escudero · C. Fowley · A. Fraix · C. Frochot · J. Garcia-Amorós · G.R.C. Hamilton · S. Kamila · P. Kubát · K. Lang · S. Maiolino · N. Marino · C. McEwan · A. Miro · A.M. Gazzali · J. Mosinger · W.J. Parak · B. Pelaz · D.S. Pellosi · L. Petrizza · P. del Pino · L. Prodi · F. Quaglia · E. Rampazzo · F.M. Raymo · Y. Sheng · M.G. Soliman · S. Sortino · S. Tang · E.R. Thapaliya · M. Toussaint · F. Ungaro · R. Vanderesse · N. Zaccheroni · K. Zaghdoudi · Q. Zhang · Y. Zhang

Springer

Editor
Salvatore Sortino
University of Catania
Catania, Italy

ISSN 0340-1022 ISSN 1436-5049 (electronic)
Topics in Current Chemistry
ISBN 978-3-319-22941-6 ISBN 978-3-319-22942-3 (eBook)
DOI 10.1007/978-3-319-22942-3

Library of Congress Control Number: 2015955259

Springer Cham Heidelberg New York Dordrecht London
© Springer International Publishing Switzerland 2016
This work is subject to copyright. All rights are reserved by the Publisher, whether the whole or part of the material is concerned, specifically the rights of translation, reprinting, reuse of illustrations, recitation, broadcasting, reproduction on microfilms or in any other physical way, and transmission or information storage and retrieval, electronic adaptation, computer software, or by similar or dissimilar methodology now known or hereafter developed.
The use of general descriptive names, registered names, trademarks, service marks, etc. in this publication does not imply, even in the absence of a specific statement, that such names are exempt from the relevant protective laws and regulations and therefore free for general use.
The publisher, the authors and the editors are safe to assume that the advice and information in this book are believed to be true and accurate at the date of publication. Neither the publisher nor the authors or the editors give a warranty, express or implied, with respect to the material contained herein or for any errors or omissions that may have been made.

Printed on acid-free paper

Springer International Publishing AG Switzerland is part of Springer Science+Business Media (www.springer.com)

Preface

Nanomedicine is a cutting-edge area of biomedical research and exploits the application of nanotechnology to medical science. It involves the design and development of novel nanostructured materials that, once engineered, promise a profound impact in prevention, diagnosis, and treatment of several diseases. Light is a very attractive trigger for activating specific diagnostic and therapeutic functions. In view of the easy manipulation, in terms of intensity, wavelength, duration, and localization, light represents a minimally invasive and finely tunable external stimulus for optical imaging and the introduction of therapeutic agents in a desired bio-environment, mimicking an "optical microsyringe" with superb spatiotemporal control. Furthermore, light triggering offers the additional advantage of not affecting physiological parameters such as temperature, pH, and ionic strength that is an important requisite for bio-applications. Therefore, it is not surprising that the marriage of photochemistry and nanomaterials has been attracting the interest of many researchers in recent years.

Fabrication of photoresponsive nanostructured materials implies collective cross-disciplinary efforts because of synthetic methodologies and the physical characterization techniques. However, photochemistry plays a dominant role. In fact, design and synthesis of photoactive materials certainly depends on the way atomic or molecular units are assembled into nanostructures with specific characteristics such as size, shape, coordination environment, and on the ability to predict the response of the obtained nanomaterials to light stimuli. The convergence of consolidated knowledge in photochemistry and the significant breakthroughs in materials science have recently led to outstanding achievements with excellent diagnostic, therapeutic, and theranostic performances. The technological advances in fiber optics and the possibility of exploiting unconventional photochemical strategies, which enable activation of these nanoconstructs with highly penetrating near infrared light, are pushing such materials towards practical applications, facilitating an entirely new category of clinical solutions.

This themed issue includes eight authoritative reviews written by experts in the field and covering a wide range of nanostructured systems engineered in such a way as to perform imaging, therapeutic, or both functions under the exclusive control of light input. The first two contributions focus on the design and fabrication of nanoparticles for in vitro and in vivo optical imaging, demonstrating the advantages offered by fluorescence spectroscopy in terms of versatility, sensitivity, and resolution. Prodi and co-workers provide a state-of-the-art summary in the field of silica luminescent nanoparticles, exploiting collective processes to obtain ultra-bright units suitable as contrast agents in optical imaging, optical sensing, and other high sensitivity applications in nanomedicine. With many fascinating examples, Garcia-Amorós, Raymo, and co-workers elegantly illustrate how photoinduced energy transfer between multiple guests co-encapsulated within polymeric nanoparticles may help to elucidate the transport properties of these biocompatible nanocarriers in a diversity of biological media. Discussion of nanomaterials for phototherapeutic applications focuses on photodynamic therapy (PDT), photothermal therapy (PTT), and emerging treatments based on the nitric oxide (NO) radical. The contribution by Quaglia and co-workers is entirely devoted to polymeric nanoparticles for PDT in cancer applications, illustrating the advantages offered by nanotechnological approaches and describing the pillars of rational design of these nanocarriers and their potential in preclinical models. Frochot and co-workers discuss a wide range of inorganic nanoparticles for PDT and show how they can be used to improve the PDT efficiency. Nanostructured bidimensional surfaces and their potential as photoantimicrobial materials, mainly based on PDT, are discussed by Mosinger and coworkers. Specific contributions are devoted to gold nanoparticles and quantum dots that represent fascinating multifunctional platforms for many biomedical applications. A detailed state-of-the-art review on gold nanoparticles and their wide use in nanomedicine as multitasking platform for optical imaging and sensing, PDT, and PTT applications is provided by Ashraf, Carrillo-Carrion, and co-workers. Callan and co-workers illustrate how quantum dots can be used as fluorescent tags with superb performances for the direct visualization or quantification of particular cell types or biomolecules and how they can serve as effective energy donors in sensing and therapeutic applications. The final contribution comes from Sortino's group and describe a variety of engineered nanoconstructs for the therapeutic release of NO combining multiple imaging and phototherapeutic modalities, and is mainly devoted to anticancer and antibacterial applications

I would like to express my gratitude to the colleagues and friends who enthusiastically agreed to contribute to this volume with their reviews. The contents of this issue is certainly far from being as comprehensive as this vast topic deserves. However, I really hope the readers – experts and newcomers – enjoy these chapters and find exciting stimuli and fruitful sources of inspiration for further developments of "smart" nanomaterials. In his famous talk in 1956 the Nobel Laureate Richard P. Feynman stated *"There is Plenty of Room at the Bottom,"* the cornerstone of nanotechnology. I can safely conclude that at the bottom there is plenty of room for photochemists, whose creativity enslaved to nanomedicine ensures a "bright" future for innovative research in this field. The International Year of Light, 2015, is definitely a very good omen!

Catania, Italy Salvatore Sortino

Contents

Top Curr Chem (2016) 370: 1–28
DOI: 10.1007/978-3-319-22942-3_1
© Springer International Publishing Switzerland 2016

Luminescent Silica Nanoparticles Featuring Collective Processes for Optical Imaging

Enrico Rampazzo, Luca Prodi, Luca Petrizza, and Nelsi Zaccheroni

Abstract The field of nanoparticles has successfully merged with imaging to optimize contrast agents for many detection techniques. This combination has yielded highly positive results, especially in optical and magnetic imaging, leading to diagnostic methods that are now close to clinical use. Biological sciences have been taking advantage of luminescent labels for many years and the development of luminescent nanoprobes has helped definitively in making the crucial step forward in in vivo applications. To this end, suitable probes should present excitation and emission within the NIR region where tissues have minimal absorbance. Among several nanomaterials engineered with this aim, including noble metal, lanthanide, and carbon nanoparticles and quantum dots, we have focused our attention here on luminescent silica nanoparticles. Many interesting results have already been obtained with nanoparticles containing only one kind of photophysically active moiety. However, the presence of different emitting species in a single nanoparticle can lead to diverse properties including cooperative behaviours. We present here the state of the art in the field of silica luminescent nanoparticles exploiting collective processes to obtain ultra-bright units suitable as contrast agents in optical imaging and optical sensing and for other high sensitivity applications.

Keywords Luminescent contrast agents nanomedicine • Luminescent probes • Optical imaging • Silica nanoparticles

Contents

E. Rampazzo, L. Prodi (✉), L. Petrizza, and N. Zaccheroni
Dipartimento di Chimica, G. Ciamician, Via Selmi 2, 40126 Bologna, Italy
e-mail: luca.prodi@unibo.it

Abbreviations

Φ	Quantum yield
AFM	Atomic force microscopy
AOT	Bis(2-ethylhexyl)sulfosuccinate sodium salt
APTES	3-Aminopropyltriethoxysilane
BHQ	Black hole quenchers
CT	Computed tomography
CTAB	Hexadecyltrimethylammonium bromide
Cy5	Cyanine 5
Cy5.5	Cyanine 5
Cy7	Cyanine 7
DDSNP	Dye doped silica nanoparticle
DLS	Dynamic light scattering
EPR	Enhanced permeability and retention effect
ET	Energy transfer
FAM	Fluorescein amidite
FRET	Förster resonance energy transfer
IR780	I2-[2-[2-Chloro-3-[(1,3-dihydro-3,3-dimethyl-1-propyl-2*H*-indol-2-ylidene)ethylidene]-1-cyclohexen-1-yl]ethenyl]-3,3-dimethyl-1-propylindolium (Cy7 dye)
MRI	Magnetic resonance imaging
NIR	Near infrared region
nm	Nanometre
NP	Fluorescent silica nanoparticles
ORMOSIL	Organic modified silica
PDT	Photodynamic therapy
PEG	Polyethyleneglycol, ethylene oxide, ethylene glycol
Plus NP	NP Pluronic silica nanoparticle
PPO	Poly(propylene oxide)
PS	Phosphatidylserine
PTT	Photothermal therapy
QD	Quantum dot
R6G	Rhodamine 6G
ROX	Rhodamine-X, rhodamine 101
$Ru(bpy)_3^{2+}$	Tris(2,2′-bipyridine)ruthenium(II)

TEM	Transmission electron microscopy
TEOS	Tetraethylorthosilicate
TGA	Thermo gravimetric analysis
VTES	Triethoxyvinylsilane

1 Introduction

Despite the great multidisciplinary research effort worldwide and the great steps forward in the understanding and treatment of cancer and other diseases such as diabetes and Alzheimer's, there is still much to be solved. Besides improved and personalized therapies, early diagnosis is fundamental to reduce morbidity and consequently mortality. With this aim, many different imaging techniques [1] are used to spot these diseases and monitor anatomical and functional changes [2–7]. The various imaging modalities already in clinical use differ significantly in basic features such as resolution, time of analysis, penetration depth, and costs, but, in general, they still suffer from low sensitivity and low contrast. The development of other imaging methods to speed their clinical translation becomes essential in this scenario, an aim that can be obtained by optimizing exogenous contrast agents. These imaging tools need to be stable in physiological conditions to enable selective targeting, high signal-to-noise ratio, and optimized time of circulation and clearance [8, 9].

Optical imaging is one of the most promising diagnostic techniques and it is already close to clinical application being habitually applied in preclinical studies on small animals [10, 11]. It presents the great advantages of high sensitivity, short time of analysis, and low cost, with a few drawbacks linked to small penetration depth (1–20 mm) and to relatively limited resolution. Because of the scattering of blood, tissues, and other biological components, deeper penetration in in vivo imaging can be obtained only by working in the NIR range, in the so-called biological optical transparency window (650–1800 nm) [12]. The use of NIR light absorption and emission ensures a low auto-fluorescence and tissue damage—a possible concern when long-term UV excitation is required—and reduced light scattering. It should be mentioned, however, that two water absorption peaks (centred around 980 and 1450 nm) also fall in this spectral range, and therefore the possible non-negligible and harmful local heating of biological samples when exciting via laser light in this window [13] has to be carefully evaluated. On the other hand, optical imaging presents exceptional versatility: (1) the luminescence process can start with excitation different from photo-excitation, obtaining luminescence in dark conditions as in the case of chemiluminescence [14], electro-chemiluminescence [15, 16], and thermochemiluminescence [17]; (2) light can be exploited not only for diagnostic purposes but also for therapeutic treatments (for example in the photodynamic and photothermal therapies, PDT and PTT respectively) [18, 19].

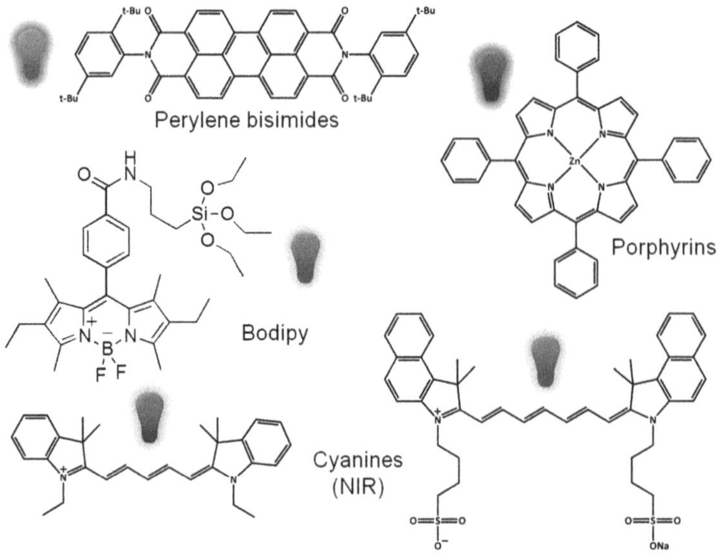

Fig. 1 Some families of organic luminophores presenting NIR absorption and/or emission

The most studied and common contrast agents for this kind of imaging are fluorescent molecules, in particular for in vivo applications as NIR absorbing and emitting luminophores [20]. There are some deeply investigated families of organic dyes with suitable features such as cyanines, phthalocyanines, and bodipys, to name a few (Fig. 1), but they also exhibit very poor water solubility and, generally, decreasing quantum yield, moving the emission towards the red part of the spectrum. Many different derivatizations of these molecules have been proposed to overcome these limitations and also to introduce targeting functions, usually at the expense of demanding synthetic procedures. Moreover, low weight (molecular) contrast agents suffer from fast clearance which decreases their ability to reach all organs because of the short available time of measurement for image acquisition.

In this context, the use of nanostructured contrast agents can provide interesting solutions [21, 22]. Lots of nanomaterials have been proposed which include dye doped nanoparticles [23] (for example, luminescent silica nanoparticles [8, 24]) or intrinsically luminescent materials (for example, quantum dots) [25, 26]. The use of non-toxic materials such as silica have inherent advantages, although dimension - and surface - related hazards need to be investigated for each system. The silica matrix can also be seen as a robust container able to ensure an almost constant chemical environment, and to improve chemical stability and resistance to photobleaching of the doping dyes. It is also transparent to visible light and therefore inert towards photophysical processes that involve the photoactive species embedded in the matrix [24, 27].

It has to be emphasized, however, that on passing from molecules to nanoparticles the molecular weight is increased by few orders of magnitude and

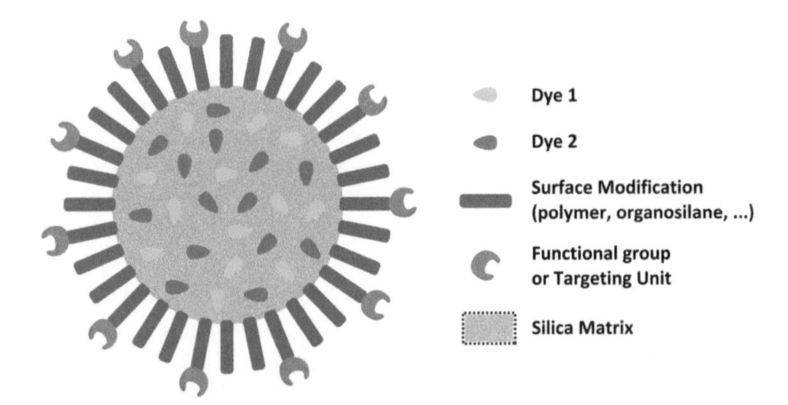

Fig. 2 Schematization of a dye doped silica nanoparticle with a surface derivatization and presenting targeting functions

this could cause possible viscosity problems with the solution being injected for in vivo applications in the case of low sensitivity techniques such as MRI and CT [28]. This is not the case with optical imaging where highly luminescent materials lead to very high sensitivity, and low concentrations of contrast agents are required. In the case of luminescent silica nanoparticles, each nanoparticle contains many dyes and this can lead to an increase in the overall signal-to-noise ratio. This goal can be reached thanks to the great versatility of these nanotemplates, which is largely because of the many different, typically mild, and non-cost effective synthetic strategies which allow the efficient derivatization of the inner core, the outer surface, or the mixing of different materials and the production of core-shell and multi-shell architectures [29]. All this opens up possibilities for the preparation of customized species for a large variety of applications (Fig. 2).

Bearing in mind the preparation of contrast agents for optical imaging in vivo, the ability to introduce antibodies, peptides, or proteins to guide localization in the target area is of fundamental interest, even if the biochemical difference existing among healthy and unhealthy tissues gives rise to the well-known *enhanced permeability and retention* effect (EPR). EPR consists in the preferential accumulation in the malignant areas of nanoobjects exceeding around 5 nm in diameter, an effect, however, which is not very linear, homogeneous, or predictable, and which can be considered really effective only in areas where the lesion is quite big and vascularised; it is not, therefore, reliable for the specific detection of early stage (sub-millimetre) metastasis. The few nano-preparations exploiting the EPR effect are already approved for clinical use and they consist of some nano-therapeutics mostly based on liposomal systems. However, for early diagnosis a targeted approach is necessary to design an effective in vivo contrast agent, and the versatility of dye doped silica nanoparticles is of particular relevance to this goal [30].

This multifluorophoric structure can present complex photophysical properties—multiple emissions with multiexponential decays—which can sometimes also be affected by the environment. On the other hand, a proper design can induce cooperative behaviours among some or all fluorophores in each nanoparticle, interpreting amplification effects (the status of one dye affects the properties of all the neighbouring ones) and other advantages as an increase of brightness and a decrease of noise. All these beneficial features arise thanks to the interconnection of the fluorophores within one nanoparticle which allows efficient energy transfer phenomena such as Förster Resonance Energy Transfer—FRET—which involves the deactivation of the donor excited state and the excitation of that of the acceptor [31, 32]. Homo-FRET can indeed be used to distribute the excitation energy among a variety of dyes. Competitive energy transfer pathways can be created and the fastest can induce a predictable direction of energy migration. All this is possible only after proper design of the interfluorophoric distances (FRET is proportional to r^{-6}) and a suitable choice of luminescent species to be mixed which should have appropriate photophysical properties (FRET is proportional to the overlap integral J which describes the matching between emission of the donor and absorption of the acceptor in terms of energy and probability) [33] (Fig. 3). One of the great strengths of FRET is that in a well-designed system many different consecutive, cascade, or independent FRET events can take place, both in organized and non-organized nanoobjects [34]. This multiple energy migration involves a very directional event or, on the other hand, the photophysical properties of the whole device. A very important point to bear in mind is that FRET can allow the creation of a species with very high pseudo-Stokes shift, a powerful tool to obtain NIR emitting probes [35].

Therefore the presence of a single kind of dye or of different emitting species as neighbours in a single nanoparticle can lead to diverse properties and also to collective behaviours which can be exploited to obtain ultra-bright species suitable for high sensitivity applications such as in vivo optical imaging.

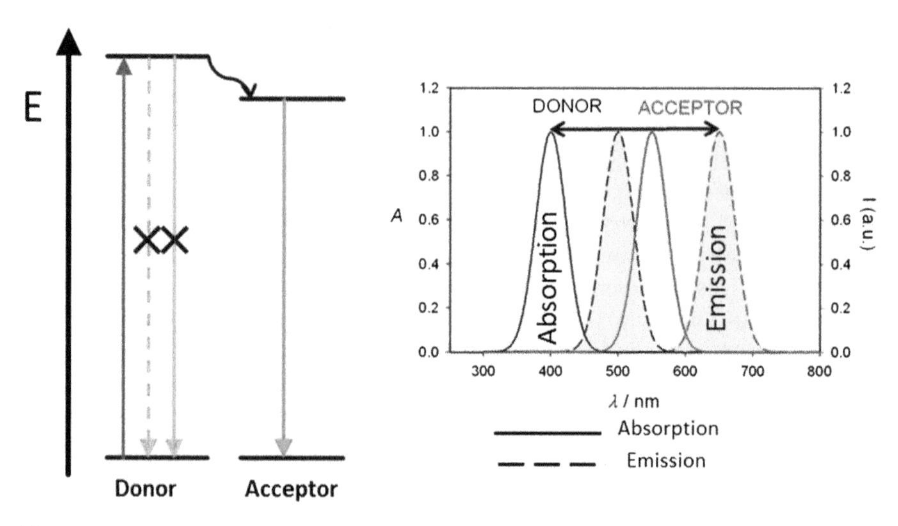

Fig. 3 Energetic scheme of FRET and of the induced pseudo-Stokes shift

Many interesting results, in fact, have already been obtained with nanoparticles containing only one kind of photophysically active species [36], and elegant systems have been prepared to address crucial problems advantageously, such as the mapping of sentinel lymph nodes [37, 38]. However, we strongly believe that luminescent silica nanoparticles doped with different dyes featuring collective processes make these nanostructures even more attractive. In this review we intend to underline these aspects, presenting selected examples to describe the state of the art of these materials as contrast agents in optical imaging and optical sensing.

We start with a very summarized discussion of the synthetic methods where the reader mainly finds a sketch of the rationale and goals of the preparation and relevant references for a wider dissertation of the subject. The reported examples are gathered in two chapters focussing on (1) nanoparticles containing different kinds of dyes and (2) luminescent nanosensors. As already mentioned, this work has no ambition to be exhaustive but we aim to give the interested readers a feeling of the progress in the area from a viewpoint that we hope stimulates the creativity and curiosity of researchers in the field.

2 Synthetic Strategies

Fluorescent silica nanoparticles (NPs) are powerful tools able to work as labels [8, 39] and sensors [40], usually obtained by means of wet sol-gel synthetic approaches characterized by simplicity, low costs, versatility and control over NP dimensions.

The first synthesis of highly monodisperse silica NPs was described by Stöber [41], who used tetraorthosilicate (TEOS) hydrolysis and condensation catalyzed by ammonia in a water-alcohol mixture to form silica NPs. Lately this strategy has been implemented by Van Blaaderen [41, 42] who obtained fluorescent silica NPs covalently doped with fluorescent dyes endowed with alkoxysilane groups, and more recently, by other groups who developed some interesting variations based on biphasic reaction media [43] and on heterogeneous nucleation processes [44].

The Stöber–Van Blaaderen method is straightforward for its simplicity, it supplies highly and stable monodisperse NPs in a wide dimensional range (~20 to 800 nm—with one pot procedures), and is applicable to many ethanol soluble organic dyes. These colloidal particles are stabilized by the electrostatic repulsion between the siloxane groups of the NPs surface, and the presence of these groups can be used for the introduction of functionality for biomolecule targeting. NPs surface functionalization is the major weak point of this synthetic approach because surface modification can be hampered by aggregation and loss of the pristine colloidal stability of the system, especially when tiny NPs are concerned.

The reverse microemulsions strategy (water-in-oil) [45] exploits the same hydrolysis and condensation processes of the previous method, but with chemical reactions that are confined in the water nano-pools of a reverse microemulsion, made of an isotropic mixture of a surfactant, water, and a hydrocarbon (usually cyclohexane). This approach is extremely versatile for obtaining functionalized NPs, because the addition of the reagents to produce the core and a functionalized shell can be made in sequence without intermediate purification procedures, reducing the possibility of aggregation. With surface functionality introduced to reduce non-specific binding and aggregation, NPs can be efficiently recovered and purified from the microemulsion by repetitive precipitation and washing steps. The strongest limitation to water-in-oil strategies is that they are exclusive to water soluble and, preferentially, positively charged dyes: this is why the most used dye with this approach is $Ru(bpy)_3^{2+}$.

An alternative and more recent strategy employs micellar aggregates or co-aggregates in water as self-organized templates to confine the silica NPs growth: such approaches are reported as direct micelles-assisted methods [46–48]. Because silica precursors such as TEOS or other organo-alkoxysilanes, and indeed the majority of the organic dyes, have a lipophilic nature, they can be efficiently entrapped in self-organized templates with a lipophilic core. When an organosilane (for example VTES, triethoxyvinylsilane, APTES, 3-aminopropyltriethoxysilane) are used together with a low molecular weight surfactant (AOT, bis(2-ethylhexyl) sulfosuccinate sodium salt, Tween 80) in the presence of ammonia, these NPs systems have often been referred to as ORMOSIL (ORganic–MOdified–SILica).

Prasad and coworkers made the greatest contributions in the field. These NPs, with a diameter spanning in the 20–30 nm range and a mesoporous silica matrix, were suitable for PDT applications [49], with dyes that usually need to be linked covalently to the nanoparticles to avoid leaching [50, 51]. Targeting and labelling capabilities for conjugation with bioactive molecules [52] were obtained with functionalization with alkoxysilane bearing groups such as $-NH_2$, $-COOH$, and $-SH$ by means of contiguous synthetic strategies, NPs were obtained functionalized with PEG [53], or by incorporating QDs [54] and Fe_3O_4 nanoparticles [55].

Hexadecyltrimethylammonium bromide (CTAB) micelles were used by Wiesner and coworkers to develop tiny mesoporous PEGylated NPs (6–15 nm): this approach allows control of the NP morphology, which depends on the number of micellar aggregates entrapped in each nanoparticle during the silica core formation [56].

A different synthetic possibility is to synthesize cross-linked micellar NPs featuring core-shell structures [46, 57]. This direct micelle-assisted strategy employs high molecular weight surfactants, such as Pluronic F127, F108, or Brij700 [58], or bridging organo-silane precursors [59]. The presence of self-organized templates and of condensation processes which are promoted in acidic environment makes these synthetic approaches quite different from the Stöber method. TEOS condensation is promoted by acids such as HCl [59] or HOAc [60] in a set of conditions ($pH < 4$) in which the hydrolysis process is faster than condensation—a situation that promote the formation of Si–O–Si chains which finally undergo cross-linking [61, 62]. During the NP formation, silica is confined in the micellar environment, and these surfactants undergo entrapment–adsorption processes, with the valuable consequence of NP surface modification in a one-pot procedure.

Pluronic F127, characterized by a tri-block PEG-PPO-PEG (poly (ethyleneglycol)-poly(propylene oxide)-poly(ethylene glycol), MW 12.6 KDa) structure and quite large micellar aggregates in water (~22 to 25 nm), is probably the most versatile surfactant for this synthetic approach and our group had a leading part in the development of fluorescent NPs obtained by Pluronic F127 micelles-assisted method (PluS NPs, Pluronic Silica NanoParticles) [39, 63]. The PPO inner core of these aggregates enables the use of TEOS as silica precursor, contributing to a more stable and dense silica network compared to ORMOSIL NPs. Great versatility is conferred by the possibility of using many alkoxysilane derivatized dyes—from polar to very lipophilic—which can be entrapped in the silica network under mild acidic conditions.

Their core-shell structure presents a hard diameter of 10 nm and an average hydrodynamic diameter of ~25 nm: the morphological properties were verified by TEM and DLS (Dynamic Light Scattering) [57], AFM measurements [46], and ^{1}H-NMR, showing the orientation of the Pluronic F127 PEG chains. TGA (Thermo Gravimetric Analysis) analysis performed on samples subjected to ultrafiltered samples also showed the stable linking between the surfactant and silica core [64]. PEGylated core-shell NPs are very monodisperse in water and physiological conditions, where they show low adhesion properties towards proteins and prolonged colloidal stability. Their versatility has been exploited in many fields: photophysical studies [57, 63], sensors [40, 65], biological imaging [60], fluorescent-photoswitchable nanoparticles [66], self-quenching recovery [34], and ECL [46, 64].

For applications involving labelling [67] or non-specific targeting [68], the functionalization of these NPs was achieved acting on the Pluronic F127 hydroxy end groups. We found that cellular uptake was influenced by PluS NPs functionalization, with cytotoxicity that was negligible towards several normal or cancer cell types, grown either in suspension or in adherence [68], for PEGylated NPs presenting external amino or carboxy groups. We recently exploited the core-shell nature of these NPs to develop NIR fluorescent NPs for regional lymph nodes mapping [37]. We synthesized negatively charged NPs having programmed charges positioning schemes to influence their biodistribution for lymph nodes targeting in vivo. We found that NPs having negative charges in the mesoporous silica core—hidden by a PEG shell—demonstrated a more efficient and dynamic behaviour during lymph nodes mapping. This last example showed once again the versatility of the core-shell silica-PEG structure, where the modifications of both core and shell parts play a synergistic effect in defining the efficiency of the fluorescent probe.

We have found that for in vivo imaging a fundamental requisite is the absence of dye leaching from the NP towards the biological environment, because this compromises the overall signal-to-noise ratiomaking fruitless the brightness increase produced by a nanoprobe. Trialkoxysilane derivatized dyes [69] can be synthesized efficiently to ensure condensation within the silica matrix, a critical issue, especially for NPs of dimensions lower than ~50 nm, characterized by a very high surface to volume ratio.

3 Multifluorophoric Doping of Silica Nanoparticles

As already anticipated in the introduction, even though dye doped silica nanoparticles are increasingly studied as contrast agents for fluorescent imaging, their clinical application is still only theoretical and further steps forward are needed in that direction. In our opinion, this process could be accelerated and facilitated by the optimization of FRET in multiple-dye-doped nanoparticles [33] because this could facilitate the preparation of materials with two very valuable characteristics: (1) NIR absorption and emission, and therefore suitable for biomedical application; (2) nanoparticles suitable for multiplexed analysis.

NIR excitation and emission are, in fact, extremely beneficial in the biomedical field but NP doping with a single NIR emitting dye has many drawbacks intrinsically relating to the nature of this class of emitters, typically characterized by very small Stokes shifts and quite low luminescence quantum yields. In such systems, high background signals are common, mostly caused by interference of the excitation light.

One of the most versatile and convincing strategies to overcome this problem and to obtain novel and customized photophysical properties is the design and activation of proper energy transfer events which take place only when a high local density of dyes is reached. The distance between the donor and the acceptor needs to be typically <10 nm for fast and efficient energy transfer events to occur, and therefore the segregation of many and different dyes in the same nanoparticle is a straightforward way to obtain suitable separation between them. This strategy can also lead to the formation of excimers and aggregates but homo- and hetero-FRET can have a synergistic role as shown in the following discussion.

Moreover, the need for cheaper and more rapid diagnostic protocols is a well-established implementation goal for current diagnostic strategies which are not always able to fulfil the time constraint imposed by some kinds of pathologies, and the use of NPs for multiplexed analysis, that is to say for simultaneous multi-analytes detection, can be an efficient answer.

A recognized advantage of silica nanoparticles is that they can be easily modified in different ways accordingly to the strategy in mind. In this context, Tan and co-workers have explored and proposed some very interesting nanoarchitectures. In a first example [70] they have developed aptamer-conjugated nanoparticles to detect (and extract with the help of magnetic nanoparticles) three different lymphoma cells in a single analysis. With this aim they used three diverse high-affinity DNA aptamers for recognition, each bound to a different fluorescent silica nanoparticle ($d = 50 \pm 5$ nm) doped with $Ru(bpy)_3^{2+}$, tetramethylrhodamine, or Cy5, respectively. The same authors later proposed a more elegant approach, allowing a higher multiplexing ability, based on FRET for multiplexed bacteria monitoring and imaging [71] (Fig. 4). In this work they prepared silica nanoparticles doped with a combination of up to three kinds of dyes (fluorescein, rhodamine 6G, and 6-carboxy-X-rhodamine) in different doping concentrations. In these conditions only a partial energy transfer occurred between the dyes and this allowed for the

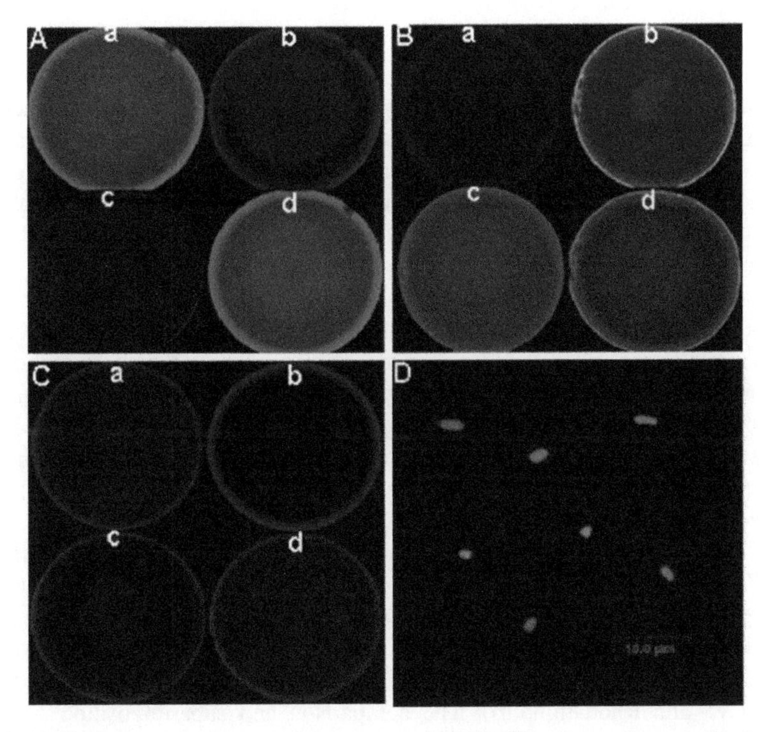

Fig. 4 (A-C) Fluorescence images of three types of FRET NPs taken under a confocal microscope: (**a**) FAM emission channel; (**b**) R6G emission channel; (**c**) ROX emission channel. (**d**) Combinatorial colour of the three channels. The three types of NPs exhibit *blue*, *orange* and *purple colours*. (D) Confocal image of three bacteria species specifically covered with the corresponding antibody labelled NPs (copyright Wang et al. [71])

detection, under a single excitation wavelength, of a number of bands equal to the number of doping fluorophores [72]. Their intensities could be controlled by varying the doping degree and ratio of all the luminescent species to obtain a large library of bar-coding NPs, thereby allowing the simultaneous detection of many bacterial pathogens or cancer cells [73]. The data collected in these papers show how this approach brings important advantages: (1) the simplification of the experimental setup needed for their detection because a single laser source was needed for their excitation; (2) the high specificity provided by the aptamers and the nanoparticles brightness could potentially remove the need for sample enrichment and amplification steps, allowing a fast diagnosis at the earliest stage of a disease.

K. Wang and co-workers [74] have also studied similar systems with silica nanoparticles (50 nm diameter) obtained via the microemulsion method and non-covalently doped with a tris-bipyridyl ruthenium complex and methylene blue units. Their main goal was to obtain a very large (around 200 nm) Stokes shift to prepare suitable fluorescent NPs for small animal in vivo imaging.

In this same framework, we have proposed Plus NPs that can be used as a scaffold to organize and to connect molecular components electronically

[57]. These very monodisperse water-soluble silica core-surfactant PEG shell nanoparticles (Plus NPs) were covalently doped in the core with a Rhodamine B derivative although, exploiting simple hydrophobic effects, a second luminescent species (Cy5) was adsorbed in the shell. These NPs could hence simultaneously behave as fluorescent labels (for imaging applications) and as carriers for the delivery of hydrophobic compounds (drugs). Moreover, the efficient communication via a partial ET between the active molecules trapped in the core and the hosted species in the shell was demonstrated experimentally.

Starting from these very promising results we have proposed two systems based on co-doping with two different fluorophores of the silica core of nanoparticles to obtain modular nanosystems for in vivo optical imaging with the goal of future use in clinics [60, 67]. The Plus NPs of both studies were doped with a cyanine and Rhodamine B in the core, the main difference being that one system was decorated on the outer shell with peptides able to target the hepatic cancer cells specifically [67] whereas the other non-targeted one was used to investigate the non-specific distribution and circulation of these materials in vivo [60]. The injection of the first nano-contrast agent in a mouse proved that the functionalization of the NPs with metastasis-specific peptides took to their homing and accumulation in the target tissues, resulting in specific visualization even of sub-millimetric metastases. In this system the energy transfer between the two doping fluorophores is partial, allowing the detection and acquisition of the emission signals of both of them (Cy5 and Rhodamine B). The comparison and superimposition of these images allowed for background subtraction and signal amplification, resulting in the enhancement of the imaging sensitivity. It should be noted that this is a modular approach, providing great versatility in achieving cancer cell-targeted delivery of a number of therapeutic, diagnostic, or teragnostic compounds.

In the second system [60] we wanted to use a single bio-label for both in vivo imaging and ex vivo microscopy on the same animal. A partial energy transfer between the two dyes maintains both signals, for in vivo experiments (NIR region, Cy7 derivative) and tissue investigations with optical microscopes equipped with light sources and detectors operating in the visible region (rhodamine B). In our nanoparticles it was possible to observe an efficient sensitization of the cyanine but also a residual 25% of luminescence of the rhodamine, sufficient for a high signal at the optical microscope, allowing the proposed goal of performing in vivo and ex vivo experiments with the same NP on the same animal to be achieved. It should be emphasised that this cannot be achieved with QDs or metallic NPs, because typically their one colour emission is tuned by their shape and dimensions.

Taking advantage of these results, we have then tried to exploit this modular and very versatile approach to optimize the system further for in vivo imaging [35]. Neighbouring Cy5.5 and Cy7 dyes in different ratios—both covalently linked in the 10 nm diameter core of the Plus NPs—led to a very sensitive contrast agent with high brightness in the NIR. Interestingly, the mixing of two different dyes advantageously led to a decrease the efficiency of the self quenching processes observed in single dye doped NPs, further increasing the brightness of the system. Small-animal in vivo imaging experiments have shown very promising results, indicating the great potential of nanoparticles realized with this approach (Fig. 5).

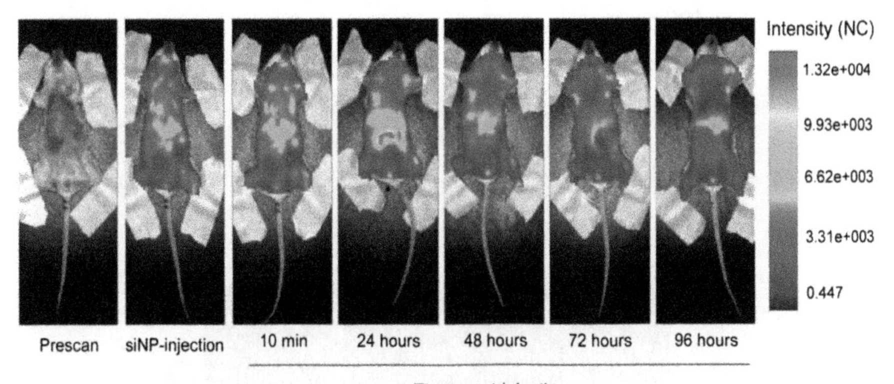

Fig. 5 Whole-body scan of a representative mouse in supine position; fluorescence intensity images were acquired at the indicated time post-injection and are displayed in normalized counts (NC). BALB/c mice were injected intravenously via the tail vein with 1 nmol of NPs and analyzed by optical imaging (copyright Biffi et al. [35])

We have discussed so far silica nanoparticles with multi-fluorophoric doping with two or more kinds of dyes able to give only partial FRET; it is important now to highlight that complete energy transfer in similar systems could represent a very valuable feature, allowing an increase in the number of analytical approaches towards the detection of many species via very bright, NIR emitting species. In spite of the brilliant and pioneering work of Tan and co-workers, an almost complete energy transfer in multi-doped silica nanoparticles had not been achieved until very recently, when we decided to exploit the particularly favourable characteristic of the Plus NPs to optimize these processes and prepare brighter contrast agents for medical imaging.

The first problem to overcome to enhance brightness in nanoparticles is to contrast the detrimental effects of self-quenching on the luminescence quantum yield, a very common feature when large amounts of dyes are present in a small volume. In these conditions, in fact, when the absorption cross section is additive, the luminescence can be negatively affected by their proximity, imposing a limit on doping degree to a relatively low number of luminescent units. We have proposed an alternative approach exploiting the efficient homo-FRET to boost the hetero-FRET towards highly luminescent energy acceptors [34]. We prepared PluS NPs doped in the core with approximately 40 coumarin dyes per NP, and with a bodipy that was chosen as energy acceptor because its quantum yield and spectral properties are not affected by neighbouring coumarin units. We found that a few molecules of acceptor (around 5) could very efficiently collect the light absorbed by all the coumarins via such a highly efficient energy transfer. Importantly, because of the high rate constant associated with this process, the self-quenching was almost completely prevented, and almost no loss of the excitation energy could be detected. The straightforward consequence is the linear increase of the brightness as a function of the coumarin molecules present in the NPs, allowing an unprecedented brightness (Fig. 6).

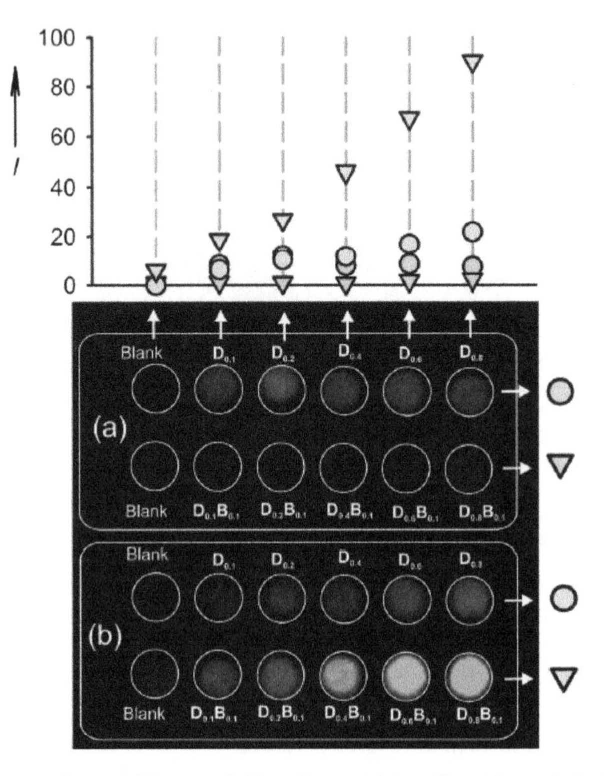

Fig. 6 Fluorescence images of a set of 12 wells containing dispersions of the NPs in water ($c = 1 \times 10^{-7}$ M), as detected with either a *blue* (**a**, $\lambda_{em} = 460$ nm) or a *green* (**b**, $\lambda_{em} = 530$ nm) *filter*. The total intensity determined by the integration of each spot is plotted in the *top graph* (copyright Genovese et al. [34])

Moreover, this system also presents the valuable advantage of a greater pseudo-Stokes shift which is produced by the wavelength difference between the maxima of the donor absorption and of the acceptor emission.

After experimentally proving that this strategy is general and limited only by the photophysical properties of the fluorophore couples, which must be suitable to allow an efficient FRET, we have tried to push the system further towards larger and larger separations between the excitation and emission maxima. Our goal was to obtain high luminescence in the NIR region, but also a very high signal-to-noise ratio, thanks to the reduced interferences from Rayleigh–Tyndall and Raman bands. We have therefore prepared and characterized multi-fluorophoric PluS NPs, each containing up to four different kinds of fluorophores [35], and we could observe unprecedented efficiencies (higher than 94%) in the ET processes in the doping dyes (a sort of cascade ET). This leads to a very large and tuneable pseudo-Stokes shift (up to 435 nm) for each NP which is also characterized by a single colour emission (that of the reddest dye of the four, the final acceptor) with a negligible residual emission intensity from all the donor dyes, and a very high brightness, even exciting the "bluest" donor. These multicomponent PluS NPs can thus be seen as very promising materials for multiplex analysis in medical diagnostic techniques.

We conclude this section by discussing an example of silica multi-dye nanoparticles designed as possible theranostic agents. S. R. Grobmyer and co-workers have proposed mesoporous silica nanoparticles around 110 nm in diameter as theranostic platforms for bioimaging and cancer therapy [75]. They first covalently doped these species with a properly silanized cyanine (IR780) which, after derivatization and incorporation, showed a highly enhanced luminescence quantum yield (>300-fold) and Stokes shift (>110 nm). In a second step they also loaded a Si-naphthalocyanine in the pores of the NPs. After the optimization of both dye concentrations, the resulting multi-dye structures, thanks to a partial energy transfer, allowed the imaging of a murine mammary tumour in vivo after intratumoral injection and then its photothermal ablation via laser irradiation. The non-fluorescent metallo-naphthalocyanine, in fact, acts as a rather efficient photothermal therapeutic agent as first evidenced by bioluminescence analysis and then proved by the histological analysis. The same authors have also proved that these NPs remain within the tumour for several days and are suitable for successive activations, allowing the application of image-guided fractionated photothermal ablation.

A second aspect to be carefully evaluated and addressed is the clearance time, the fate of these imaging or theranostic agents being still mostly unknown, although it is a fundamental aspect that needs to be clarified before considering any possible clinical translation. The multi-dye bright NIR absorbing and emitting nanoparticles described in this chapter can, in our opinion, be a key tool in making significant steps forward in this area, allowing NPs tracking in vivo in real time via fluorescence imaging.

4 Luminescent Chemosensors Based on Silica Nanoparticles

Luminescent chemosensors have already found wide applications in many fields of great economic and social importance, such as environmental monitoring, process control, and food and beverage analysis. In addition, thanks to the high sensitivity and the high spatial and temporal resolution granted by luminescence [76–78], fluorescence-based chemosensors can enable the investigation of the functions and possible misregulation of crucial target analytes in living systems [79, 80], especially at the cellular level. To face the latest challenges of biology and medicine, however, very sophisticated systems are needed, and this is the reason why nanostructured architectures are so widely studied in this context. In particular, silica nanoparticles can offer many advantages compared to their molecular and supramolecular counterparts together with the above-mentioned high brightness and photostability.

The first is related to the possibility of engineering collective energy- and electron-transfer processes which, if designed to modify the luminescence properties of several dyes upon a single complexation event, can lead to large signal amplification effects and thus lower detection limits [76]. Moreover, as discussed in the previous section, collective energy-transfer processes can also lead to an even higher brightness and a very large Stokes shift (up to hundreds of nanometres), thus minimizing the interferences coming from the scattered light, the Raman band, and the auto-fluorescence of the biological sample.

Another advantage is the possibility offered by silica nanoparticles to obtain ratiometric systems easily. Quantitative intensity measurements using fluorescent chemosensors can indeed be affected at a single wavelength by their dependence not only on the concentration of the analyte but also on the concentration of the chemosensors itself and, possibly, on the optical path and on environmental conditions. This problem can generally be overcome by using time-resolved techniques or by properly designing the system to be able to compare the signals coming from two different bands. Using silica nanoparticles, this can be achieved by inserting a fluorescent probe together with an analyte-independent fluorophore in the silica-based nanosensor. The most common methods use a design that aims to prevent all electronic interactions between the two fluorescent units, thus eliminating the possibility of energy transfer processes. Typically, the reference dye is included in the core of the nanoparticles, whereas the probe is inserted in an outer shell or bound to the surface [79, 81].

Moreover, using DDSNPs, the solubility and stability of the system in solution is conferred mainly by the silica structure; in this way, many interesting compounds that, because of their limited solubility, cannot be used in an aqueous environment can instead be conveniently inserted in the DDSNP structure and also work in water [82–84]. In these cases, however, we recommend that the possibility be considered that the photophysical properties of the chemosensor in its free form and/or when interacting with the analyte could be drastically changed when inserted into the silica matrix based on what is observed in aqueous or organic solvents. At the same time, the affinity towards the analyte can also change because of the insertion of the chemosensor into the silica matrix. Because of cooperative effects, a considerable affinity increase can usually be obtained, which can be used to push their sensitivity to even lower levels. Interestingly, we have also proved that even the stoichiometry of complexation of a chemosensor can be changed when inserted inside a DDSNP [83].

In biology the ideal probe should allow precise compartmentalization, without perturbing the equilibrium present in the cell, for example not depleting cellular ion pools when detecting metals [85]. To achieve this, a high signal-to-noise ratio and a suitable affinity and selectivity are, for a chemosensor, mandatory. In addition, the possibility of derivatizing the DDSNPs with suitable biomolecules can help to increase selectivity towards a specific hat can be acquired cellular compartment, improving the information acquirable by the system [85].

To date, the use of fluorescent probes based on silica nanoparticles for cell imaging has been reported for determining the concentration of protons (pH), metal ions (with a possible future extension to their speciation), anions, small molecules (including molecular oxygen), reducing agents (such as glutathione), and enzymes. We have selected here few examples with the aim of showing the best features of silica nanoparticles and their possible future developments.

4.1 pH Probes

Tissue acidosis is a clear indication of inflammatory diseases, and low pH values have been found, for example, in cardiac ischemia and inside and around malignant tumours [86]. For this reason, although pH chemosensors are, at least in principle, among the simplest, their full development is of particular interest. In this regard, it is important to mention the pioneering work of Wiesner and coworkers [87] who, based on the Stöber strategy, designed a core/shell architecture. In particular, the silica core was doped with the reference dye, a conjugate of tetramethylrhodamine, whose fluorescence properties do not depend on pH over a large range (Fig. 7a). The outer silica shell was instead doped with fluorescein, which assumes different pH-dependent equilibria; of particular importance between the monoanionic and dianionic forms ($\Phi = 0.36$ and 0.93, respectively; $pK_a = 6.4$).

The ratiometric pH response of the two signals were used to prepare a calibration curve (Fig. 7b). The nanoparticles were than internalized in RBL-2H3 cells, and it was possible, after correction for taking into account a certain amount of autofluorescence, to measure the pH in different compartments (from 6.5 in early endosome to ca. 5.0 in late endosome/lysosome). Many examples of fluorescent NPs designed for pH sensing have since been reported. A similar couple of dyes (fluorescein as the pH sensitive dye and Rhodamine B as the reference dye), grafted to mesoporous silica NPs, were used by Chen and coworkers [88]. These authors used two differently charged ethoxysilane derivatives to functionalize the NPs surface and observed how the charge influenced the final location of the NPs in HeLa cells. In particular, positively charged NPs were found to prefer higher pH regions (mostly cytosol) whereas negatively charged NPs accumulated in acidic endosomes. It should be noted that, as expected, the internalization of the NPs also depends on their charge.

Doussineau and coworkers [89], for example, prepared NPs with a diameter of 100 nm, again containing rhodamine as a reference and, as a pH probe, a fluorescent naphthalimide derivative grafted to the NPs surface through linkers of different lengths. The two-dye structure showed good pH sensitivity in a physiologically relevant pH range with an ON–OFF response of the naphthalimide units. Wang and coworkers prepared another, smaller (diameter 42 nm) ratiometric system, able to sense pH changes in a range between 4 and 7 in murine macrophages and in HeLa cells during apoptosis [90]. In this case the NPs were doped with the tris(2,2′-bipyridyl)ruthenium(II) complex as reference and with a fluorescein derivative. The changes in lysosomal pH were monitored in real time after exposure to the antimalaria drug chloroquine. Ratiometric fluorescence detection allowed the

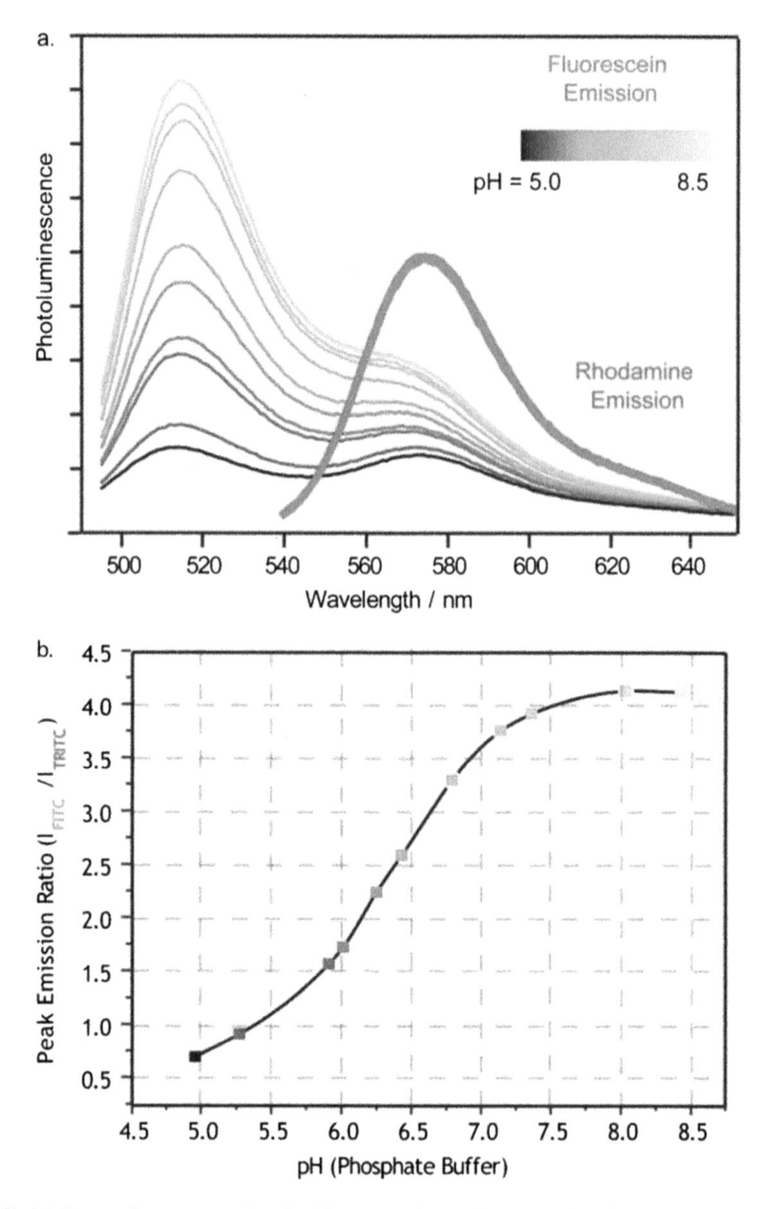

Fig. 7 (**a**) Spectrofluorometry data for 70 nm dual-emission core-shell fluorescent silica nanoparticle pH sensors showing fluorescein (*green*) and TRITC (*red*) in phosphate pH calibration buffers from pH 5–8.5. (**b**) A ratiometric calibration curve based on the peak intensity ratio between fluorescein and TRITC across the pH range under investigation (reproduced, in part, with permission from Burns et al. [87])

authors to conclude that chloroquine stimulates lysosomal pH changes. Upon incubation of HeLa cells with the same nanosensors, it was instead possible to observe, in real time, intracellular acidification in apoptotic cancer cells after treatment with dexametasone, a synthetic glucocorticoid commonly used as an anti-inflammatory agent.

A possible strategy to obtain a higher sensitivity can be developed, based on trying to maximize the signal changes under the same pH change. This approach has been used by Shoufa Han et al. [91], who doped mesoporous silica nanoparticles (diameter 100 nm) with fluorescein (whose fluorescence, as we have already discussed, decreases in acidic media) and the acid activable Rhodamine 6G lactam, whose fluorescence increases with decreasing pH. Because the two signals change in opposite directions, small pH changes can induce large variations in the ratio between the two intensities. It was possible in this way to monitor lysosomal pH in live L929 cells.

4.2 Ion Chemosensors

After pH, the Cu^{2+} probes are the most studied chemosensors based on silica nanoparticles. This research trend can be easily explained by considering the ability of this ion to form stable complexes and to perturb dramatically (typically, quenching) the fluorescence properties of the dye, together with the importance of copper ions in environmental and biological matrices. A very interesting example is represented by the derivatization of mesoporous silica (SBA-15) with a 1,8-naphthalimide-based receptor [92]. In aqueous solution, Cu^{2+} quenched the fluorescence of 1,8-naphthalimide with high efficiency and selectivity (detection limit 0.01 μM), with Hg^{2+} being the most important interfering species. Interestingly, the same receptor, but conjugated with a treholose unit to confer the right water solubility to the system, showed a ratiometric response towards copper ions, without an appreciable interference by mercury ions. In vitro (MCF-7 human breast cancer cells) and in vivo (5-day-old zebrafish) experiments showed in both cases the internalization of the nanostructures and a strong quenching of the fluorescence after incubation with Cu^{2+} ions. In a second example, the recognition was achieved by capping silica nanoparticles having a diameter of 10 nm with a trialkoxysilane derivative of fluorescein bearing two carboxylic functions as coordinating sites [93]. The fluorescence of this nanosensor showed a selective and reversible (upon addition of EDTA) quenching in the presence of copper ions in water at pH 7.4, affording an interesting detection limit of 0.5 μM. Furthermore, this nanoarchitecture was able to permeate into HeLa cells, and its fluorescence decreased, as expected, when the cells were incubated with Cu^{2+}.

Copper(I) is also a crucial species to monitor inside cells, and plays a role, among others, in neurodegenerative diseases. In this framework we have developed a ratiometric chemosensor based on core-shell silica NPs (Fig. 8) [65], synthesized using Pluronic F127 as a template and doped with a coumarin derivative.

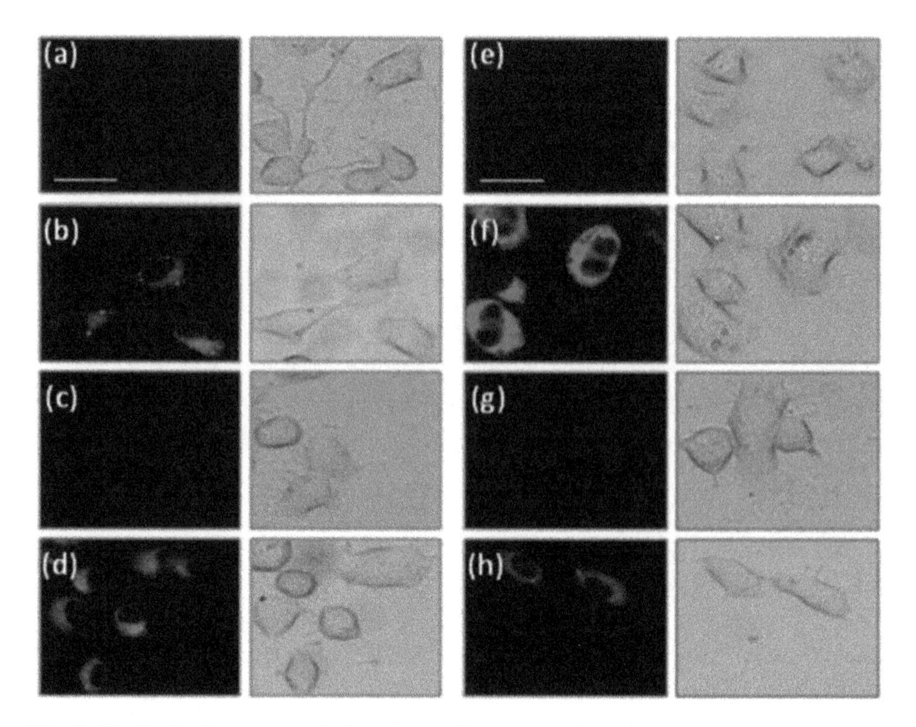

Fig. 8 Confocal microscopy emission ($\lambda_{ex} = 543$ nm; $\lambda_{em} = 572$ nm) and corresponding bright-field images for the SHSY5Y cell line: (**a–d**) cultured in DMEM medium for 24 h, and (**e–h**) supplemented with $CuCl_2$ (100 mm) for 5 h, incubated at 378 °C. The probe treatments (5 min incubation at 25.8 °C, then rinsed with PBS) are as follows: (**a**, **e**) control PBS; (**b**, **f**) molecular sensor (5 μM); (**c**, **g**) nanoparticles (5 μM); (**d**, **h**) sensor inside NPs (5 μM; 1:1 molar ratio); *scale bar*: 30 mm (copyright Rampazzo et al. [65])

The addition of the selective probe designed by Chang et al. [94] to a water dispersion of the NPs leads, because of its lipophilic nature, to its inclusion into the nanostructure, making possible an efficient energy transfer process between the coumarin dye bound in the silica matrix and the bodipy of the probe located in the shell close by. The complexation of Cu^{+} by the probe led to changes in the absorption and luminescence properties of the bodipy which affected both the coumarin signal (because of a change in the efficiency of the energy transfer process) and the bodipy signal (because of the change in its fluorescence quantum yield).

This resulted in a ratiometric system with an amplified response. We were also able to demonstrate that the inclusion of the probe led to a tenfold increase of its affinity towards the target analyte. Experiments conducted in SHSY5Y cell lines also confirmed the validity of this approach for biological systems. For these reasons, we strongly believe we could significantly push further the application of chemosensors in biomedical analysis, considering that it enables the use of otherwise water insoluble species without demanding synthetic efforts.

In the development of chemosensors, the design of efficient systems for monitoring the concentration of anions has been typically quite challenging. In this context, it is relevant that the example reported by Bau et al. [95] involved the preparation of a ratiometric probe for chloride anions based on silica NPs grafted with 6-methoxyquinolinium as the chloride-sensitive component and fluorescein as the reference dye. The measured Stern–Volmer constant was 50 M^{-1} at pH 7.2, sufficient to monitor chloride ion changes around physiological concentrations. In fact, these nanostructures were able to penetrate neuronal cells at submillimolar concentrations and to signal in real time chemically induced chloride currents in hippocampal cells.

4.3 Probing Molecular Processes

Interestingly, silica nanoparticles can also be used with biological processes. In particular, the possibility of following the triggering of apoptosis, programmed cell death, has attracted the attention of several researchers. Malfunctioning apoptosis can, in fact, lead to several pathological conditions, such as neurodegenerative disorders, cardiovascular diseases, and cancer. A popular extracellular method for the detection of apoptosis involves monitoring the distribution of phospholipids on the cell surface; in particular, the appearance of phosphatidylserine (PS), which usually constitutes less than 10% of the total phospholipids in cell membranes, on the outer part of cell membranes is a universal indicator of the initial stage of apoptosis.

Yeo, Hong and coworkers have grafted to the surface of silica nanoparticles doped with $Ru(bpy)_3^{2+}$ the phenoxo-bridged Zn(II)-di-2-picolylamime complex (pbZn(II)-DPA, Fig. 9), which selectively recognizes PS among the different animal-cell-membrane phospholipids [96].

Interestingly, the authors showed through fluorescence microscopy that Jurtkat T cells were not stained by the nanoparticles except when treated with camptothecin,

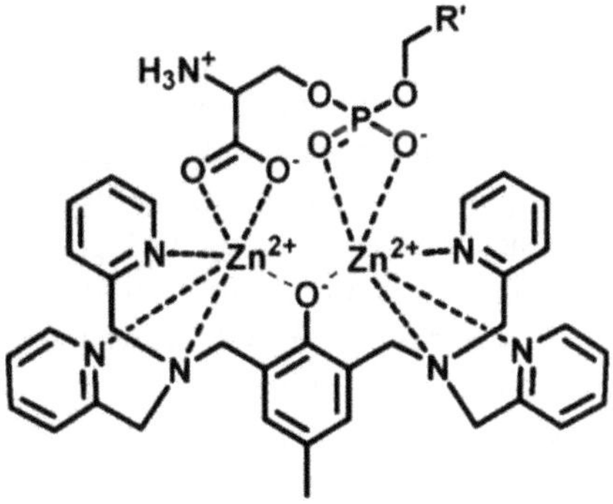

Fig. 9 pbZn(II)-DPA complex

an apoptosis inducer. Of note, the orange luminescence of the nanoparticles was localized on the cell surface, as expected in the case of binding to externalized PS.

A different example of monitoring apoptosis is to follow the activity of caspases, a family of cysteine proteases which play critical roles in this process. Lee, Chen and their colleagues prepared a very interesting nanosensor which made possible the multiplex imaging of intracellular proteolytic cascades [97]. The design of this nanosensor is very elegant (Fig. 10). Mesoporous silica nanoparticles, with a diameter of 60 nm, were doped, after their suitable derivatization with a silane group, with three different kind of commercially available molecules belonging to the series of the so-called black hole quenchers (BHQ) which, having large and intense absorption bands in the visible region of the spectrum, can efficiently quench the fluorescence of many kinds of organic dyes through a FRET mechanism.

The nanoparticle surface was grafted, through a short PEG spacer, with specific substrate (an oligopeptides) for caspase-3 derivatized with a fluorophore (Cy5.5GDEVDGC). The fluorescence of Cy5.5, because of the efficient quenching of BHQs, was strongly reduced (up to ca. 360-fold, depending on the doping degree). Incubation with caspase-3 led to a strong fluorescence recovery, proportional to the concentration of the enzyme (in the range 0.4–7 nM), because of the separation of the dye from the quenching silica nanoparticle. The authors also proved the suitability of this system for multiplex imaging: they conjugated the same NP with three substrates specific for caspase-3, caspase-8 and caspase-9, each derivatized with a different dye, in such a way that each substrate would correspond to an emission in a different spectral region (490, 560 and 675 nm). After internalization of this nanosensor into HCT116 cells, fluorescence microscopy revealed very low intensities from all three channels, and a large increase of fluorescence was observed upon treatment of the same cells with TRAIL, an apoptosis-inducing ligand. As expected, low signals were observed when inhibitors for the three caspases were also added, demonstrating the possibility of designing nanostructured systems that can be used for imaging at the single cell level in the presence of multiple intracellular enzymes.

5 Conclusions

Luminescent silica nanoparticles play an important role in the development of nanomedicine because their intrinsic features well match most essential pre-requisites for their possible use in vivo. On the other hand, their brightness can reach intensities among the highest reported so far in the field of nanomaterials, allowing one to profit fully from the sensitivity and resolution offered by fluorescence spectroscopy. However, what really makes silica nanoparticles unique is, in our opinion, their versatility. For example, their ζ potential can be finely tuned to match the needs of a specific application (such as cell internalization or lymph node mapping), or their surface can be rather easily derivatized with suitable ligands to

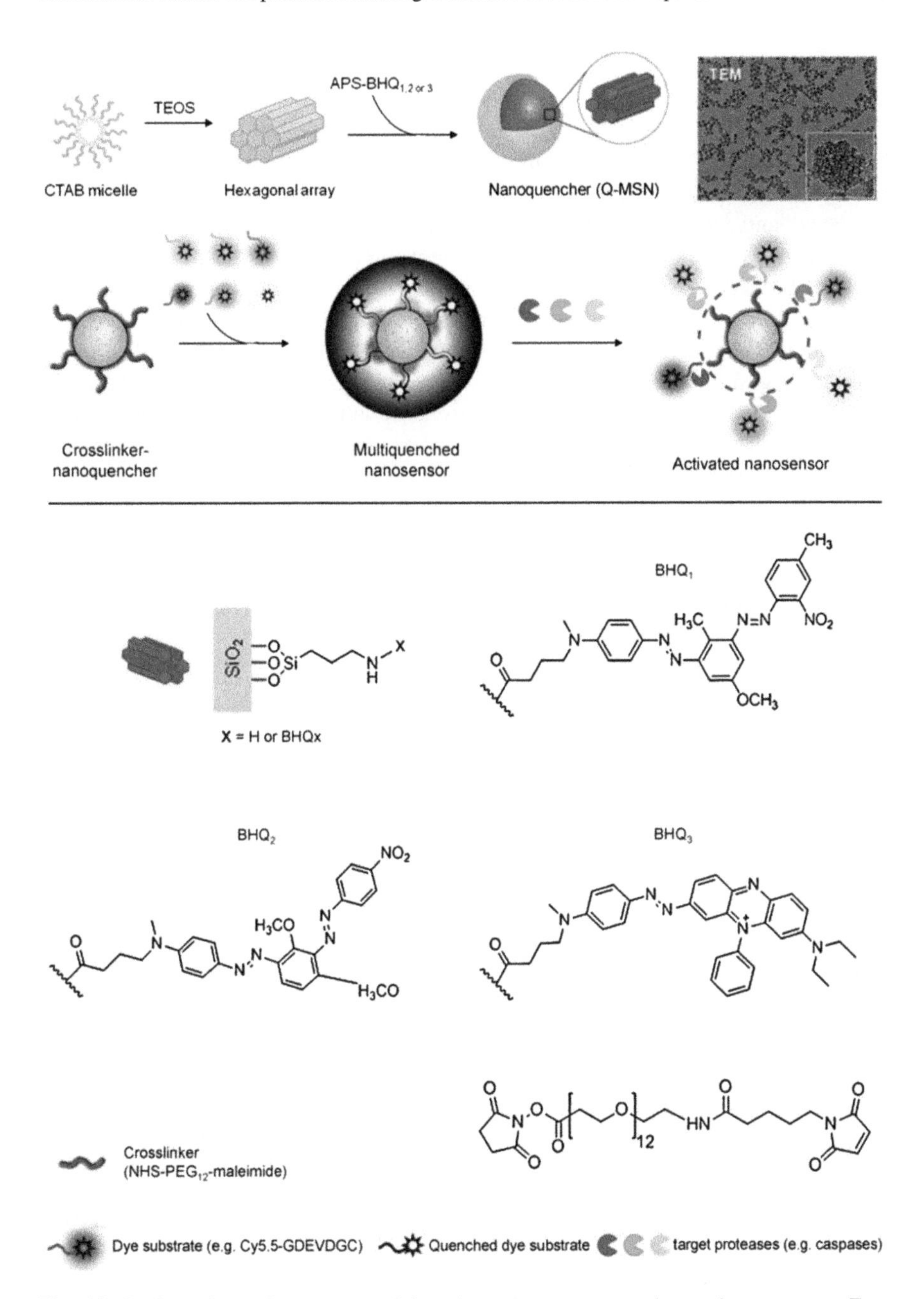

Fig. 10 Design of non-fluorescent and broad-spectrum nanoquencher and nanosensor. *Top*: diagram showing the synthesis of Q-MSNs. *Image*: TEM image of Q-MSNs. *Bottom*: chemical structures of the components of Q-MSNs (copyright Huang et al. [97])

obtain selective targeting. A highly synthetic versatility also offers the rare chance to design multichromophoric structures to obtain collective energy- and electron-transfer processes. These processes, as we have described in this review, are the basis of extremely valuable features. For example, with controlled, incomplete, energy transfer processes, nanoparticles presenting two or more emission signals can be advantageously used for barcoding applications or for carrying out in vivo and ex vivo experiments on the same subject using the same nanostructured probe. On the other hand, an almost complete energy transfer, when obtained with careful design, can be greatly beneficial, yielding a set of nanostructures which, excited at the same wavelength, can give bright luminescence of different colours, favouring multiplexing in microscopy and cytofluorimetry. These nanoparticles also present the additional advantages of minimizing self-quenching processes and increasing the separation between the excitation and the emission wavelengths, highly increasing the signal to noise ratio.

Finally, when designing nanoparticle-based chemosensors, collective energy- and electron-transfer processes can allow signal amplification, extending the possibility of monitoring important biological species and processes.

Notably, although many advantages are becoming evident, as we have discussed, additional potentials remain largely unexplored for these materials. We hope that the data discussed in this review article inspire a growing number of scientists to proceed in this direction, addressing more and more ambitious goals.

Acknowledgments We acknowledge financial support from the University of Bologna (FARB Project Advanced Ultrasensitive Multiplex Diagnostic Systems Based on Luminescence Techniques) and from Regione Lombardia-INSTM (Sinfonia project).

References

1. Hussain T, Nguyen QT (2014) Molecular imaging for cancer diagnosis and surgery. Adv Drug Deliv Rev 66:90–100
2. Tirotta I et al (2015) (19)F magnetic resonance imaging (MRI): from design of materials to clinical applications. Chem Rev 115(2):1106–1129
3. Abramczyk H, Brozek-Pluska B (2013) Raman imaging in biochemical and biomedical applications. Diagnosis and treatment of breast cancer. Chem Rev 113(8):5766–5781
4. Cutler CS et al (2013) Radiometals for combined imaging and therapy. Chem Rev 113 (2):858–883
5. Verwilst P et al (2015) Recent advances in Gd-chelate based bimodal optical/MRI contrast agents. Chem Soc Rev 44(7):1791–1806
6. Nie L, Chen X (2014) Structural and functional photoacoustic molecular tomography aided by emerging contrast agents. Chem Soc Rev 43(20):7132–7170
7. Zhu L, Ploessl K, Kung HF (2014) PET/SPECT imaging agents for neurodegenerative diseases. Chem Soc Rev 43(19):6683–6691
8. Montalti M et al (2014) Dye-doped silica nanoparticles as luminescent organized systems for nanomedicine. Chem Soc Rev, 43(12):4243–4268
9. Merian J et al (2012) Fluorescent nanoprobes dedicated to in vivo imaging: from preclinical validations to clinical translation. Molecules 17(5):5564–5591

10. Buckland J (2015) Experimental arthritis: in vivo noninvasive molecular optical imaging of disease. Nat Rev Rheumatol 11(5):258
11. Broome AM et al (2015) Optical imaging of targeted beta-galactosidase in brain tumors to detect EGFR levels. Bioconjug Chem 26(4):660–668
12. Anderson RR, Parrish JA (1981) The optics of human skin. J Invest Dermatol 77(1):13–19
13. Rocha U et al (2014) Neodymium-doped LaF(3) nanoparticles for fluorescence bioimaging in the second biological window. Small 10(6):1141–1154
14. Bi S et al (2015) A hot-spot-active magnetic graphene oxide substrate for microRNA detection based on cascaded chemiluminescence resonance energy transfer. Nanoscale 7(8):3745–3753
15. Nepomnyashchii AB, Bard AJ (2012) Electrochemistry and electrogenerated chemiluminescence of BODIPY dyes. Acc Chem Res 45(11):1844–1853
16. Wu P et al (2014) Electrochemically generated versus photoexcited luminescence from semiconductor nanomaterials: bridging the valley between two worlds. Chem Rev 114 (21):11027–11059
17. Di Fusco M et al (2015) Organically modified silica nanoparticles doped with new acridine-1,2-dioxetane analogues as thermochemiluminescence reagentless labels for ultrasensitive immunoassays. Anal Bioanal Chem 407(6):1567–1576
18. Gong H et al (2014) Engineering of multifunctional nano-micelles for combined photothermal and photodynamic therapy under the guidance of multimodal imaging. Adv Funct Mater 24 (41):6492–6502
19. Lv RC et al (2015) Multifunctional anticancer platform for multimodal imaging and visible light driven photodynamic/photothermal therapy. Chem Mater 27(5):1751–1763
20. Yi XM et al (2014) Near-infrared fluorescent probes in cancer imaging and therapy: an emerging field. Int J Nanomedicine 9:1347–1365
21. Wang R, Zhang F (2014) NIR luminescent nanomaterials for biomedical imaging. J Mater Chem B 2(17):2422–2443
22. Wu H et al (2014) Self-assembly-induced near-infrared fluorescent nanoprobes for effective tumor molecular imaging. J Mater Chem B 2(32):5302–5308
23. Li K, Liu B (2014) Polymer-encapsulated organic nanoparticles for fluorescence and photoacoustic imaging. Chem Soc Rev 43(18):6570–6597
24. Liberman A et al (2014) Synthesis and surface functionalization of silica nanoparticles for nanomedicine. Surf Sci Rep 69(2–3):132–158
25. Cassette E et al (2013) Design of new quantum dot materials for deep tissue infrared imaging. Adv Drug Deliv Rev 65(5):719–731
26. Jing LH et al (2014) Magnetically engineered semiconductor quantum dots as multimodal imaging probes. Adv Mater 26(37):6367–6386
27. Vivero-Escoto JL, Huxford-Phillips RC, Lin WB (2012) Silica-based nanoprobes for biomedical imaging and theranostic applications. Chem Soc Rev 41(7):2673–2685
28. Bonitatibus PJ et al (2012) Preclinical assessment of a Zwitterionic tantalum oxide nanoparticle X-ray contrast agent. ACS Nano 6(8):6650–6658
29. Caltagirone C et al (2014) Silica-based nanoparticles: a versatile tool for the development of efficient imaging agents. Chem Soc Rev 44(14):4645–4671
30. Toy R et al (2014) Targeted nanotechnology for cancer imaging. Adv Drug Deliv Rev 76:79–97
31. Li CH, Hu JM, Liu SY (2012) Engineering FRET processes within synthetic polymers, polymeric assemblies and nanoparticles via modulating spatial distribution of fluorescent donors and acceptors. Soft Matter 8(27):7096–7102
32. Ray PC et al (2014) Nanoscopic optical rulers beyond the FRET distance limit: fundamentals and applications. Chem Soc Rev 43(17):6370–6404
33. Genovese D et al (2014) Energy transfer processes in dye-doped nanostructures yield cooperative and versatile fluorescent probes. Nanoscale 6(6):3022–3036
34. Genovese D et al (2013) Prevention of self-quenching in fluorescent silica nanoparticles by efficient energy transfer. Angew Chem Int Ed 52(23):5965–5968

35. Biffi S et al (2014) Multiple dye-doped NIR-emitting silica nanoparticles for both flow cytometry and in vivo imaging. Rsc Adv 4(35):18278–18285
36. Meng HM et al (2015) Multiple functional nanoprobe for contrast-enhanced bimodal cellular imaging and targeted therapy. Anal Chem 87(8):4448–4454
37. Helle M et al (2013) Surface chemistry architecture of silica nanoparticles determine the efficiency of in vivo fluorescence lymph node mapping. ACS Nano 7(10):8645–8657
38. Jeon YH et al (2010) In vivo imaging of sentinel nodes using fluorescent silica nanoparticles in living mice. Mol Imaging Biol 12(2):155–162
39. Bonacchi S et al (2011) Luminescent silica nanoparticles: extending the frontiers of brightness. Angew Chem Int Ed 50(18):4056–4066
40. Montalti M et al (2013) Luminescent chemosensors based on silica nanoparticles for the detection of ionic species. New J Chem 37(1):28–34
41. Stöber W, Fink A, Bohn E (1968) Controlled growth of monodisperse silica spheres in the micron size range. J Colloid Interface Sci 26(1):62–69
42. Van Blaaderen A, Vrij A (1992) Synthesis and characterization of colloidal dispersions of fluorescent, monodisperse silica spheres. Langmuir 8(12):2921–2931
43. Wang J et al (2011) Two-phase synthesis of monodisperse silica nanospheres with amines or ammonia catalyst and their controlled self-assembly. ACS Appl Mater Interfaces 3(5):1538–1544
44. Larson DR et al (2008) Silica nanoparticle architecture determines radiative properties of encapsulated fluorophores. Chem Mater 20(8):2677–2684
45. Bagwe RP, Hilliard LR, Tan W (2006) Surface modification of silica nanoparticles to reduce aggregation and nonspecific binding. Langmuir 22(9):4357–4362
46. Zanarini S et al (2009) Iridium doped silica – PEG nanoparticles: enabling electrochemiluminescence of neutral complexes in aqueous media. J Am Chem Soc 131(40):14208–14209
47. Yong KT et al (2009) Multifunctional nanoparticles as biocompatible targeted probes for human cancer diagnosis and therapy. J Mater Chem 19(27):4655–4672
48. Bagwe RP et al (2004) Optimization of dye-doped silica nanoparticles prepared using a reverse microemulsion method. Langmuir 20(19):8336–8342
49. Kim S et al (2007) Organically modified silica nanoparticles co-encapsulating photosensitizing drug and aggregation-enhanced two-photon absorbing fluorescent dye aggregates for two-photon photodynamic therapy. J Am Chem Soc 129(9):2669–2675
50. Kim S et al (2007) Intraparticle energy transfer and fluorescence photoconversion in nanoparticles: an optical highlighter nanoprobe for two-photon bioimaging. Chem Mater 19(23):5650–5656
51. Selvestrel F et al (2013) Targeted delivery of photosensitizers: efficacy and selectivity issues revealed by multifunctional ORMOSIL nanovectors in cellular systems. Nanoscale 5(13):6106–6116
52. Kumar R et al (2008) Covalently dye-linked, surface-controlled, and bioconjugated organically modified silica nanoparticles as targeted probes for optical imaging. ACS Nano 2(3):449–456
53. Kumar R et al (2010) In vivo biodistribution and clearance studies using multimodal organically modified silica nanoparticles. ACS Nano 4(2):699–708
54. Seddon A, Li Ou D (1998) CdSe quantum dot doped amine-functionalized ormosils. J Solgel Sci Technol 13(1–3):623–628
55. Law W-C et al (2008) Optically and magnetically doped organically modified silica nanoparticles as efficient magnetically guided biomarkers for two-photon imaging of live cancer cells. J Phys Chem C 112(21):7972–7977
56. Ma K et al (2013) Controlling growth of ultrasmall sub-10 nm fluorescent mesoporous silica nanoparticles. Chem Mater 25(5):677–691
57. Rampazzo E et al (2010) Energy transfer from silica core – surfactant shell nanoparticles to hosted molecular fluorophores. J Phys Chem B 114(45):14605–14613

58. Chi F et al (2010) Size-tunable and functional core – shell structured silica nanoparticles for drug release. J Phys Chem C 114(6):2519–2523
59. Liu J et al (2009) Tunable assembly of organosilica hollow nanospheres. J Phys Chem C 114 (2):953–961
60. Rampazzo E et al (2012) Multicolor core/shell silica nanoparticles for in vivo and ex vivo imaging. Nanoscale 4(3):824–830
61. Cushing BL, Kolesnichenko VL, O'Connor CJ (2004) Recent advances in the liquid-phase syntheses of inorganic nanoparticles. Chem Rev 104(9):3893–3946
62. Gallagher D, Ring TA (1989) Chimia 43:298
63. Pedone A et al (2013) Understanding the photophysical properties of coumarin-based Pluronic-silica (PluS) nanoparticles by means of time-resolved emission spectroscopy and accurate TDDFT/stochastic calculations. Phys Chem Chem Phys 15(29):12360–12372
64. Valenti G et al (2012) A versatile strategy for tuning the color of electrochemiluminescence using silica nanoparticles. Chem Commun 48(35):4187–4189
65. Rampazzo E et al (2011) A versatile strategy for signal amplification based on core/shell silica nanoparticles. Chem Eur J 17(48):13429–13432
66. Genovese D et al (2011) Reversible photoswitching of dye-doped core-shell nanoparticles. Chem Commun 47(39):10975–10977
67. Soster M et al (2012) Targeted dual-color silica nanoparticles provide univocal identification of micrometastases in preclinical models of colorectal cancer. Int J Nanomedicine 7:4797–4807
68. Rampazzo E et al (2013) Proper design of silica nanoparticles combines high brightness, lack of cytotoxicity and efficient cell endocytosis. Nanoscale 5(17):7897–7905
69. Verhaegh NAM, Blaaderen AV (1994) Dispersions of rhodamine-labeled silica spheres: synthesis, characterization, and fluorescence confocal scanning laser microscopy. Langmuir 10(5):1427–1438
70. Smith JE et al (2007) Aptamer-conjugated nanoparticles for the collection and detection of multiple cancer cells. Anal Chem 79(8):3075–3082
71. Wang L et al (2007) Fluorescent nanoparticles for multiplexed bacteria monitoring. Bioconjug Chem 18(2):297–301
72. Wang L, Yang CY, Tan WH (2005) Dual-luminophore-doped silica nanoparticles for multiplexed signaling. Nano Lett 5(1):37–43
73. Chen XL et al (2009) Using aptamer-conjugated fluorescence resonance energy transfer nanoparticles for multiplexed cancer cell monitoring. Anal Chem 81(16):7009–7014
74. He XX et al (2012) Fluorescence resonance energy transfer mediated large Stokes shifting near-infrared fluorescent silica nanoparticles for in vivo small-animal imaging. Anal Chem 84 (21):9056–9064
75. Gutwein LG et al (2012) Fractionated photothermal antitumor therapy with multidye nanoparticles. Int J Nanomedicine 7:351–357
76. Prodi L (2005) Luminescent chemosensors: from molecules to nanoparticles. New J Chem 29 (1):20–31
77. Bonacchi S et al (2011) Luminescent chemosensors based on silica nanoparticles. Top Curr Chem 300:93–138
78. Lodeiro C et al (2010) Light and colour as analytical detection tools: a journey into the periodic table using polyamines to bio-inspired systems as chemosensors. Chem Soc Rev 39 (8):2948–2976
79. Montalti M et al (2014) Dye-doped silica nanoparticles as luminescent organized systems for nanomedicine. Chem Soc Rev 43(12):4243–4268
80. Que EL, Domaille DW, Chang CJ (2008) Metals in neurobiology: probing their chemistry and biology with molecular imaging (2008, 108:1517). Chem Rev 108(10):4328
81. Doussineau T et al (2010) On the design of fluorescent ratiometric nanosensors. Chem Eur J 16 (34):10290–10299

82. Bazzicalupi C et al (2013) Multimodal use of new coumarin-based fluorescent chemosensors: towards highly selective optical sensors for Hg2+ probing. Chem Eur J 19(43):14639–14653
83. Ambrosi G et al (2015) PluS nanoparticles as a tool to control the metal complex stoichiometry of a new thio-aza macrocyclic chemosensor for Ag(I) and Hg(II) in water. Sens Actuat B Chem 207:1035–1044
84. Arca M et al (2014) A fluorescent ratiometric nanosized system for the determination of Pd-II in water. Chem Commun 50(96):15259–15262
85. Qin Y et al (2013) Direct comparison of a genetically encoded sensor and small molecule indicator: implications for quantification of cytosolic Zn2+. ACS Chem Biol 8(11):2366–2371
86. Tsai Y-T et al (2014) Real-time noninvasive monitoring of in vivo inflammatory responses using a pH ratiometric fluorescence imaging probe. Adv Healthc Mater 3(2):221–229
87. Burns A et al (2006) Core/shell fluorescent silica nanoparticles for chemical sensing: towards single-particle laboratories. Small 2(6):723–726
88. Chen Y-P et al (2012) Surface charge effect in intracellular localization of mesoporous silica nanoparticles as probed by fluorescent ratiometric pH imaging. Rsc Adv 2(3):968–973
89. Doussineau T, Trupp S, Mohr GJ (2009) Ratiometric pH-nanosensors based on rhodamine-doped silica nanoparticles functionalized with a naphthalimide derivative. J Colloid Interface Sci 339(1):266–270
90. Peng J et al (2007) Noninvasive monitoring of intracellular pH change induced by drug stimulation using silica nanoparticle sensors. Anal Bioanal Chem 388(3):645–654
91. Wu S et al (2011) Dual colored mesoporous silica nanoparticles with pH activable rhodamine-lactam for ratiometric sensing of lysosomal acidity. Chem Commun 47(40):11276–11278
92. Meng Q et al (2010) Multifunctional mesoporous silica material used for detection and adsorption of Cu2+ in aqueous solution and biological applications in vitro and in vivo. Adv Funct Mater 20(12):1903–1909
93. Seo S et al (2010) Fluorescein-functionalized silica nanoparticles as a selective fluorogenic chemosensor for Cu2+ in living cells. Eur J Inorg Chem 2010(6):843–847
94. Zeng L et al (2005) A selective turn-on fluorescent sensor for imaging copper in living cells. J Am Chem Soc 128(1):10–11
95. Bau L, Tecilla P, Mancin F (2011) Sensing with fluorescent nanoparticles. Nanoscale 3(1):121–133
96. Bae SW et al (2010) Apoptotic cell imaging using phosphatidylserine-specific receptor-conjugated Ru(bpy)32 + -doped silica nanoparticles. Small 6(14):1499–1503
97. Huang X et al (2012) Multiplex imaging of an intracellular proteolytic cascade by using a broad-spectrum nanoquencher. Angew Chem Int Ed 51(7):1625–1630

Top Curr Chem (2016) 370: 29–60
DOI: 10.1007/978-3-319-22942-3_2
© Springer International Publishing Switzerland 2016

Self-Assembling Nanoparticles of Amphiphilic Polymers for In Vitro and In Vivo FRET Imaging

Jaume Garcia-Amorós, Sicheng Tang, Yang Zhang, Ek Raj Thapaliya, and Françisco M. Raymo

Abstract Self-assembling nanoparticles of amphiphilic polymers are viable delivery vehicles for transporting hydrophobic molecules across hydrophilic media. Noncovalent contacts between the hydrophobic domains of their macromolecular components are responsible for their formation and for providing a nonpolar environment for the encapsulated guests. However, such interactions are reversible and, as a result, these supramolecular hosts can dissociate into their constituents amphiphiles to release the encapsulated cargo. Operating principles to probe the integrity of the nanocarriers and the dynamic exchange of their components are, therefore, essential to monitor the fate of these supramolecular assemblies in biological media. The co-encapsulation of complementary chromophores within their nonpolar interior offers the opportunity to assess their stability on the basis of energy transfer and fluorescence measurements. Indeed, the exchange of excitation energy between the entrapped chromophores can only occur if the nanoparticles retain their integrity to maintain donors and acceptors in close proximity. In fact, energy-transfer schemes are becoming invaluable protocols to elucidate the transport properties of these fascinating supramolecular constructs in a diversity of biological preparations and can facilitate the identification of strategies to deliver

J. Garcia-Amorós (✉)
Department of Chemistry, Laboratory for Molecular Photonics, University of Miami, 1301 Memorial Drive, Coral Gables, FL 33146-0431, USA

Departament de Química Orgànica, Grup de Materials Orgànics, Institut de Nanociència i Nanotecnologia (IN2UB), Universitat de Barcelona, Martí i Franqués 1, 08028 Barcelona, Spain
e-mail: jgarciaamoros@ub.edu

S. Tang, Y. Zhang, E.R. Thapaliya, and F.M. Raymo (✉)
Department of Chemistry, Laboratory for Molecular Photonics, University of Miami, 1301 Memorial Drive, Coral Gables, FL 33146-0431, USA
e-mail: fraymo@miami.edu

contrast agents and/or drugs to target locations in living organisms for potential diagnostic and/or therapeutic applications.

Keywords Amphiphilic polymers • Energy transfer • Fluorescence • Polymer nanoparticles • Self-assembly

Contents

1 Amphiphilic Polymers

1.1 Structural Designs

The word *amphiphile* was coined by Paul Winsor 50 years ago to indicate compounds with polar and nonpolar fragments within the same molecular skeleton [1, 2]. This term derives from the two Greek roots *amphi* (on both sides) and *philos* (friendly). Indeed, amphiphilic molecules incorporate two distinct components with different solvent affinities. The part of the molecule with an affinity for polar solvents, such as water, is said to be *hydrophilic* (water friendly) and, generally, comprises one or more polar functional groups (e.g., alcohols, amines, ammonium cations, carboxylate anions, carboxylic acids, ethers, esters, phosphates, sulfates, sulfonates, or thiols). By contrast, the part of the molecule with an affinity for nonpolar solvents is termed *hydrophobic* (water unfriendly) or *lipophilic* (fat friendly) and, in most instances, consists of one or more nonpolar chains (e.g., alkyl chains). Thus, an amphiphile has both hydrophilic and hydrophobic components within its covalent backbone. For example, sodium dodecyl sulfate (Fig. 1) satisfies this structural requirement and, probably, is the most common amphiphilic ingredient in commercial formulations such as shampoo, shower gel, soap, and toothpaste.

The structural composition of amphiphilic molecules can be replicated at the macromolecular level. Specifically, hydrophobic and hydrophilic segments can be

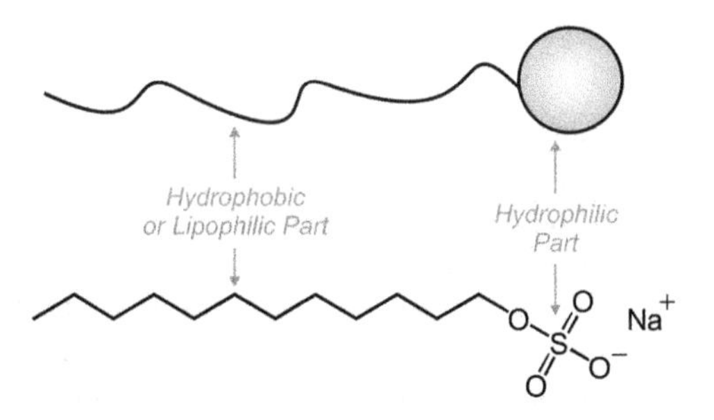

Fig. 1 Schematic representation of an amphiphilic molecule and chemical structure of dodecyl sulfate, a common amphiphilic component of commercial formulations

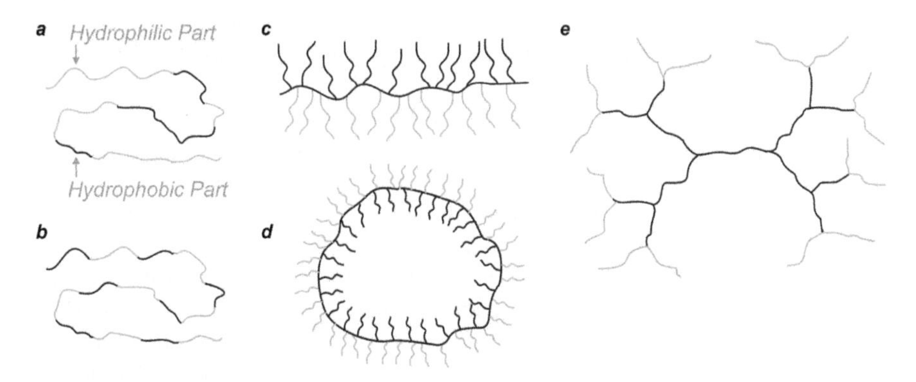

Fig. 2 Schematic representations of linear (**a–c**), cyclic (**d**) and branched (**e**) amphiphilic macromolecules

incorporated within the same macromolecular backbone to produce an amphiphilic polymer [3–9]. The two functional components can be integral parts of a linear polymer backbone either in a random fashion or in defined blocks (Fig. 2a, b). Alternatively, they can be appended to either a linear macromolecular scaffold or a cyclic polymer backbone (Fig. 2c, d). Similarly, the hydrophilic and hydrophobic building blocks can be interconnected through a multitude of branching points to produce a dendritic structure (Fig. 2e). In all instances, the final result is always a covalent assembly with an appropriate balance of polar and nonpolar functional groups to display amphiphilic character. In fact, such macromolecular designs are attracting significant attention because of their potential industrial applications and biomedical implications. Indeed, amphiphilic polymers can be viable emulsifiers/de-emulsifiers [10–14] and foamers/de-foamers [15–17] in a diversity of industrial formulations as well as valuable drug delivery vehicles in therapeutic applications and probe carriers for in vitro and in vivo imaging schemes [18–43].

1.2 Synthetic Strategies

Free-radical polymerizations are routinely employed to assemble amphiphilic polymers [44, 45]. In these processes, a macromolecular backbone forms after the sequential connection of radical building blocks. Such reactive components can be generated through a number of different mechanisms, usually involving the assistance of appropriate initiators. As an illustrative example (Fig. 3), azobisisobutyronitrile (AIBN) can be combined with a mixture of hydrophilic (**1**) and hydrophobic (**2**) methacrylate monomers [46]. Upon warming the mixture, AIBN cleaves homolytically to generate a pair of radicals. The resulting species add to the [C=C] bond of either monomer to produce further radicals. The latter add to pristine monomers to initiate the polymer growth until pairs of radicals eventually recombine to terminate the process. The final result is the random assembly of hydrophilic and hydrophobic side chains along a common poly(methacrylate) backbone in the shape of **3**.

Atom-transfer radical-polymerization (ATRP) is also commonly employed to assemble amphiphilic polymers [47–50]. In this method, a transition-metal catalyst, in combination with an alkyl or aryl halide initiator, facilitates the formation of [C–C] bonds between the reacting monomers. Generally, an electron transfer step oxidizes the metal and activates the "dormant" alkyl or aryl halide. The resulting radicals then react with the monomers to initiate the polymer growth. However, the electron-transfer step is reversible and an equilibrium is rapidly established between the redox partners which maintains a low concentration of free radicals. As a result, most polymer chains in the reaction mixture grow with similar rates to produce macromolecules with a narrow molecular weight distribution. As an example, Fig. 4 shows the sequential polymerization of styrene (**4**) and *N*,*N*-dimethylacrylamide (**7**) through ATRP to produce amphiphilic polymer **8**, which has two well-separated hydrophilic and hydrophobic domains along its poly

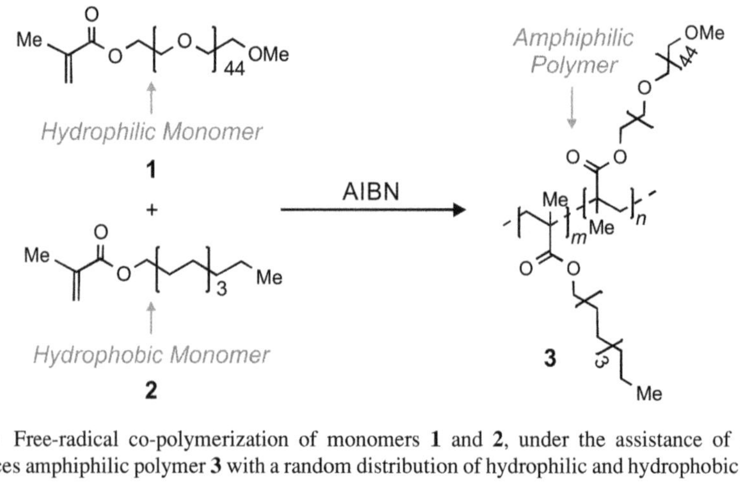

Fig. 3 Free-radical co-polymerization of monomers **1** and **2**, under the assistance of AIBN, produces amphiphilic polymer **3** with a random distribution of hydrophilic and hydrophobic chains

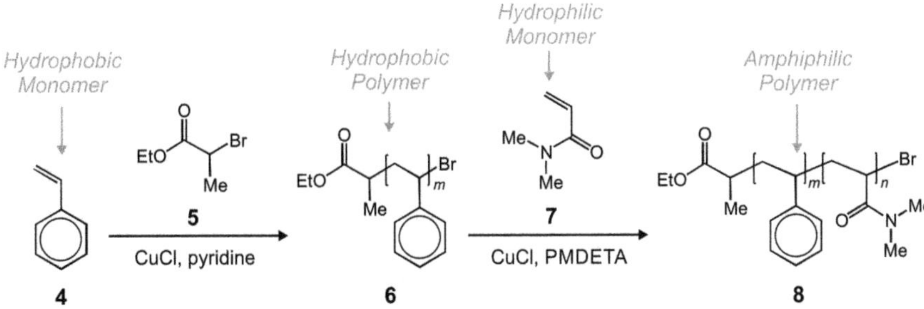

Fig. 4 Sequential ATRP steps of monomers **4** and **7** produce amphiphilic polymer **8** with two well-separated hydrophobic and hydrophilic domains along its macromolecular structure (PMDETA = *N,N,N′,N′,N″*-pentamethyldiethylenetriamine)

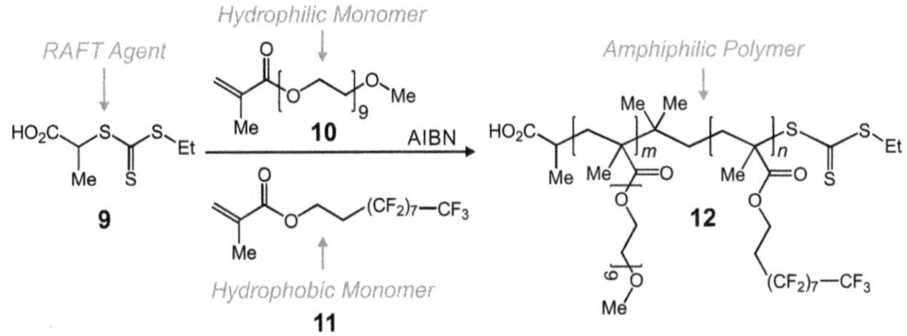

Fig. 5 RAFT polymerization of PEG-based (**10**) and perfluorinated (**11**) monomers to produce amphiphilic polymer **12**

(methacrylate) backbone [51]. Initially, Cu(I) oxidizes to Cu(II) with concomitant activation of the dormant species, ethyl 2-bromopropanoate (**5**). Right after activation, the radical species of **5** can add to the [C=C] bond of styrene and start the chain growth. Finally, a termination step eventually occurs, stopping the polymerization reaction, and the metal is concurrently reduced to its former oxidation state. As a result of this process, hydrophobic polymer **6** is obtained. In a second polymerization step, **6** reacts through the very same procedure with the hydrophilic monomer **7** to give amphiphilic polymer **8**.

Reversible addition-fragmentation chain-transfer (RAFT) polymerizations are particularly useful for the preparation of amphiphilic polymers with alternating hydrophobic and hydrophilic blocks [52–59]. These reactions are free-radical polymerizations performed in the presence of a suitable thiocarbonylthio species, the so-called RAFT agent (**9** in Fig. 5), which is generally a dithiocarbamate, a dithioester, a trithiocarbonate, or a xanthate [60]. Radicals are first generated with conventional initiators (e.g., AIBN), photoinitiators, under the influence of γ irradiation, or even by simple thermal initiation. As an example, Fig. 5 shows the polymerization of PEG-based hydrophilic monomer (**10**) and perfluorinated hydrophobic monomer (**11**) through RAFT to produce amphiphilic macromolecule **12**

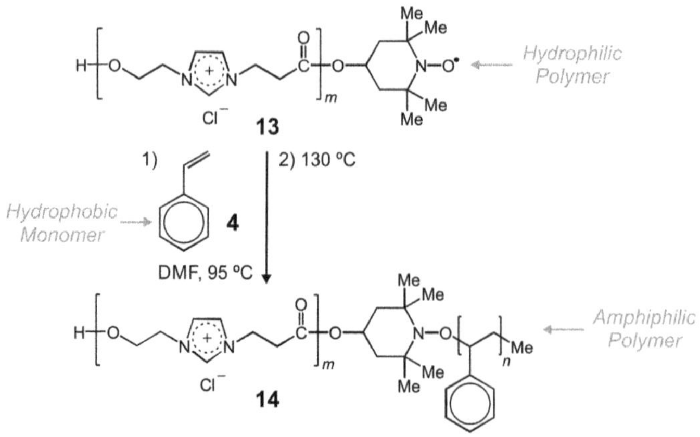

Fig. 6 NMP polymerization of styrene (4) and hydrophilic polymer (13) produces amphiphilic polymer 14 with two well-defined hydrophilic and hydrophobic blocks

[61]. In the first step, the free radicals of the initiator are produced. These species react further with either the monomers or the RAFT agent and generate free radicals of **10** or **11** to start the main-chain growth. In fact, such growing chains keep reacting with the RAFT agent to add an increasing number of monomeric units to the polymer backbone. Finally, a termination step eventually occurs to stop the polymerization reaction and yields amphiphilic polymer **12**.

Nitroxide-mediated radical polymerizations (NMP) rely on alkoxyamine initiators to induce the formation of radicals and generate amphiphilic polymers with controlled stereochemistry and low polydispersity index [62, 63]. As an example (Fig. 6), a nitroxyl capping agent was attached to one end of a poly(ionic) liquid, incorporating multiple imidazolium rings along a macromolecular backbone, to produce hydrophilic polymer **13** [64]. In the presence of styrene (**4**), and at an appropriate temperature, the terminal nitroxyl radical adds onto the [C=C] bond of the hydrophobic monomer to initiate the growth of a hydrophobic segment and eventually produce amphiphilic polymer **14**.

Ring-opening metathesis polymerization (ROMP) is another common method for the preparation of amphiphilic polymers and other industrially important products [65–67]. Specifically, the catalytic cycle of ROMP requires a strained cyclic olefin (e.g., norbornene and cyclopentene). Indeed, the driving force for the process is the relief of ring strain. After the formation of a metal-carbene species, the carbene attacks the double bond in the ring structure, forming a highly strained metallacyclobutane intermediate. The ring then opens, giving a reactive species capable of initiating polymer growth, which is a linear chain double bonded to the metal with a terminal double bond as well. The new carbene reacts with the double bond of the next monomer, propagating the reaction responsible for growing the polymer chain. As an example, Fig. 7 shows the polymerization of norbornene-based hydrophobic monomer (**15**) and PEG-substituted hydrophilic monomer (**18**) through ROMP, using the third generation Grubb's catalyst (**16**) as initiator, to

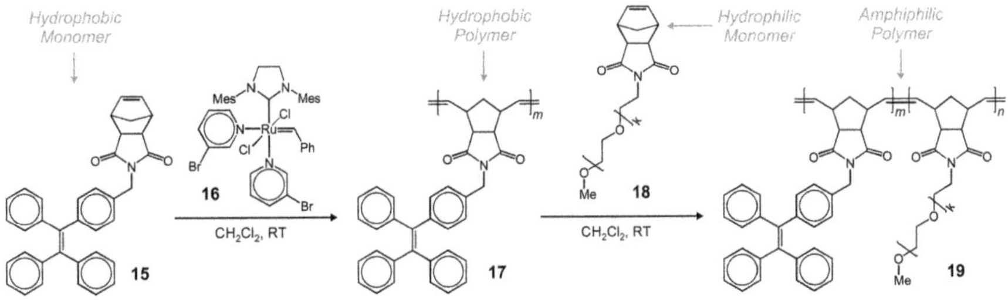

Fig. 7 ROMP polymerization of hydrophobic (**15**) and hydrophilic (**18**) monomers produces amphiphilic polymer **19** with two well-defined hydrophilic and hydrophobic blocks

produce amphiphilic macromolecule **19**, which has two well-defined hydrophilic and hydrophobic blocks [68].

1.3 Critical Concentration

When mixed with water at relatively low concentrations, most amphiphiles migrate to the air/water interface (Fig. 8a) to avoid direct exposure of their hydrophobic components to the aqueous environment [69–71]. In the resulting arrangement, the hydrophilic segment of the molecule remains within the aqueous phase, whereas the hydrophobic counterpart protrudes above the aqueous surface. If the amphiphile concentration is gradually increased, the air/water interface becomes increasingly crowded until additional molecules are forced to reside within the aqueous phase (Fig. 8b). Above a certain concentration threshold, denoted critical concentration (CC), the amphiphiles in the aqueous phase self-assemble into micellar aggregates (Fig. 8c) to prevent the direct exposure of their hydrophobic parts to water. In fact, the hydrophobic portions of the many molecules in each supramolecular assembly point toward the interior of the construct, whereas the hydrophilic parts are on the surface to be directly exposed to the aqueous solvent. Such an arrangement ensures the most appropriate environment around the hydrophobic and hydrophilic fragments of each molecular component of the self-assembling supramolecular constructs. Any further addition of amphiphiles slightly beyond the CC then translates into an increase of the number of micelles dissolved in the aqueous phase. Under these conditions, the many supramolecular assemblies remain isolated from each other with no spatial correlation. At concentrations significantly greater than the CC, however, liquid-crystalline phases with long-range orientational and positional order may appear [72–75].

The morphology of micellar aggregates of amphiphilic components is related to the structural composition and concentration of the individual building blocks and can vary with the ionic strength and temperature of the solution [76–80]. Generally, these supramolecular assemblies are spherical (Fig. 9) and, only occasionally, adopt globular or rod-like shapes. In all three instances, the interior of the aggregate is

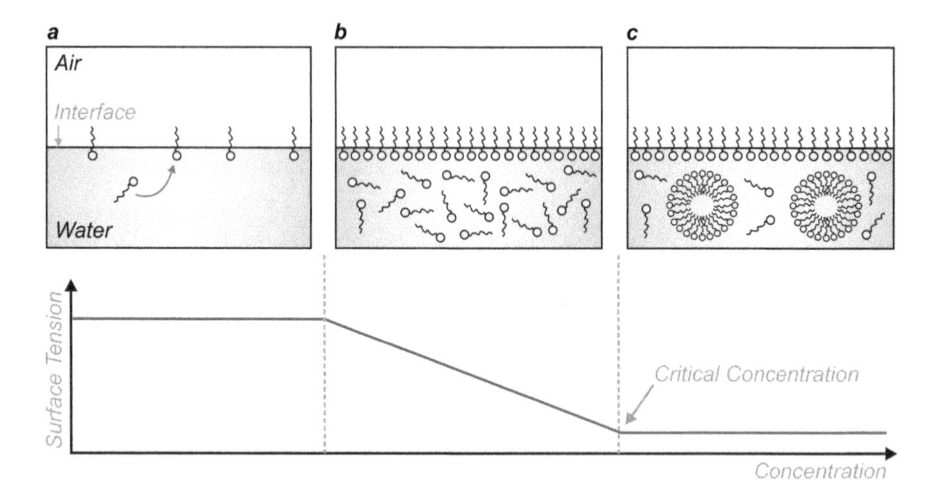

Fig. 8 The addition of increasing amounts of amphiphilic compounds to water encourages the formation of a layer of molecules at the air/water interface (**a, b**) and, eventually, results in the self-assembly of micellar aggregates above a critical concentration (**c**)

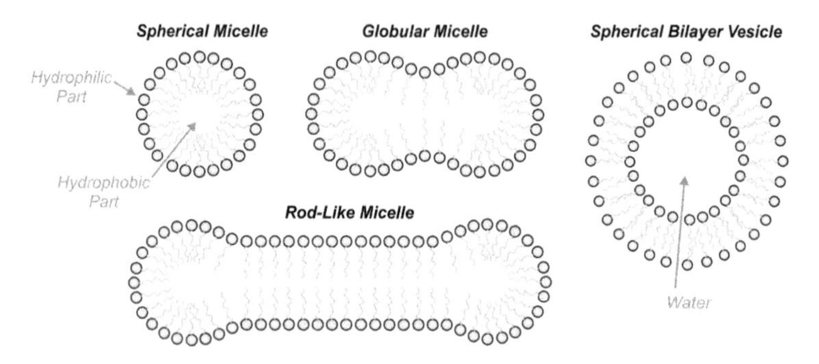

Fig. 9 Common morphologies of supramolecular assemblies of multiple amphiphilic components

hydrophobic and the surface hydrophilic. However, certain amphiphiles tend to form bilayer assemblies capable of curling into spherical vesicles and encircling solvent molecules within their inner hydrophilic face.

The value of CC varies with the structural composition of the self-assembling building blocks and its determination for any amphiphile is essential to identify a range of optimal concentrations to induce the spontaneous formation of micellar aggregates. This crucial parameter can be measured with diverse experimental procedures based on its influence on certain physical variables, including chemical shift, absorbance, fluorescence intensity, ionic conductivity, osmotic pressure, and surface tension [81–90]. Of these many methods, those relying on fluorescence and surface tension measurements are relatively simple and, as a result, appear to be the most common protocols for CC determination in literature reports.

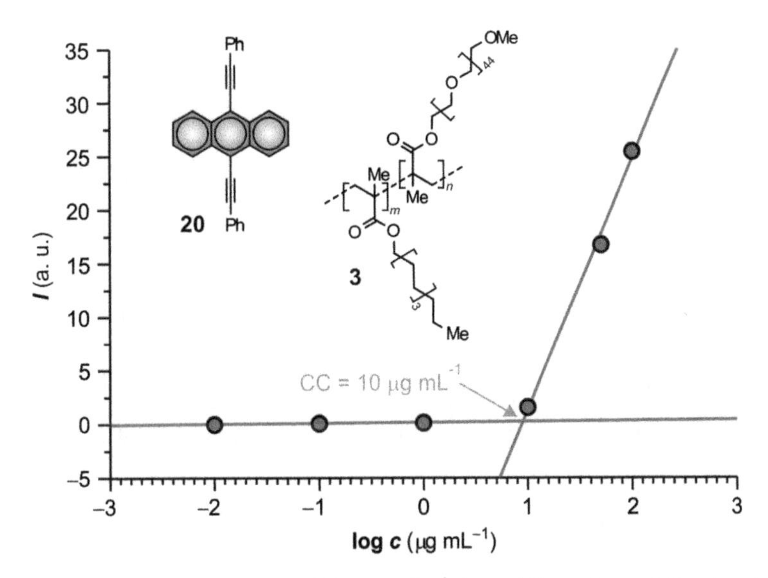

Fig. 10 Plot of the emission intensity of **20** (1 μg mL^{-1}, $\lambda_{Ex} = 440$ nm, $\lambda_{Em} = 476$ nm) against the concentration of **3** in PBS at 25 °C

Fluorescence methods for CC determination rely on the ability of the micellar aggregates to encapsulate emissive and hydrophobic probes in their nonpolar interior [91–94]. These molecular guests must be selected to lack any significant solubility in aqueous environments on their own, but readily transfer into an aqueous phase after entrapment within their supramolecular hosts. Under these conditions, their fluorescence can be detected in water only if a given amphiphile is present at a concentration greater than the corresponding CC. For example, 9,10-bis (diphenylethynyl)anthracene (**20** in Fig. 10) is essentially insoluble in phosphate buffer saline (PBS), but readily dissolves in the presence of significant amounts of an amphiphilic polymer (**3**) [95]. As a result, the treatment of a fixed amount of **20** with identical volumes of PBS solutions containing increasing concentrations of **3** can be exploited to identify the CC value of the polymer. Indeed, no emission can be detected at low polymer concentrations (Fig. 10). Once CC is reached, however, the amphiphiles assemble into micellar aggregates. The resulting supramolecular constructs capture the fluorescent species, transfer them into the aqueous phase, and allow the detection of their fluorescence. Consistently, a sudden increase in emission intensity is observed above this particular concentration threshold (Fig. 10).

In principle, the encapsulation of fluorescent probes in the interior of the micellar aggregates might have an influence on the ability of the amphiphilic components to assemble and, therefore, affect the actual CC value. This potential limitation can be overcome with protocols based on measurements of surface tension, which do not require the addition of any molecular guest [96, 97]. Indeed, this parameter can be determined experimentally with a variety of instrumental setups, including Traube's stalagmometer, modified manometers or computer-controlled tensiometers calibrated with liquids of known surface tension.

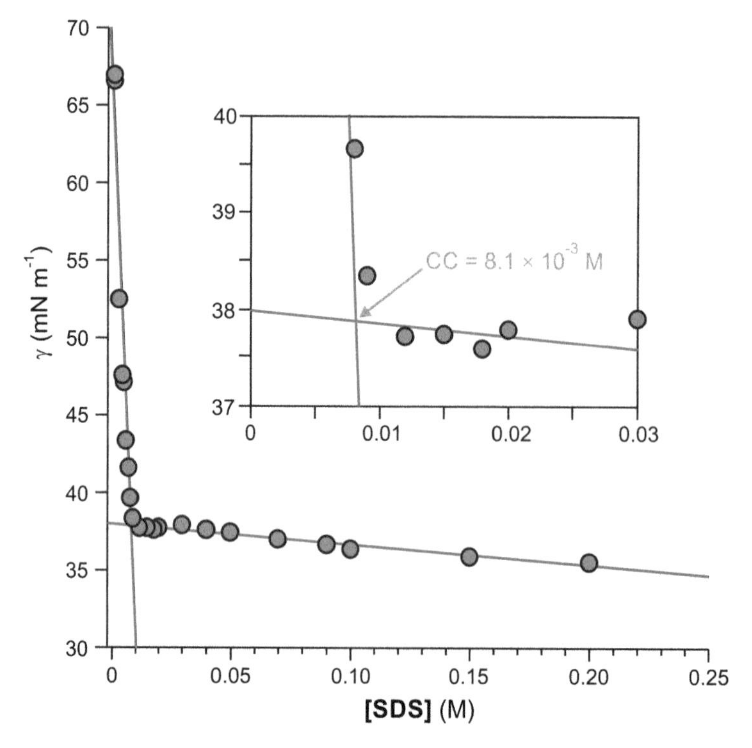

Fig. 11 Plot of the surface tension of a solution of sodium dodecyl sulfate (SDS) in water against the amphiphile concentration at 25 °C. (Reprinted with permission from *J. Chem. Ed.* © 2001, American Chemical Society)

Molecules in a liquid attract each other. These interactions are balanced in all directions for a given molecule within the bulk of the liquid. However, they are unbalanced for an equivalent molecule on the surface of the very same liquid, resulting in forces that tend to hold the liquid together. The surface tension (γ) is a measurement of such a cohesive energy which exists at the air/liquid interface. The presence of amphiphilic molecules at the interface disrupts these cohesive interactions with a concomitant decrease in surface tension. In fact, amphiphiles are often called "surface active" compounds or "surfactants" because of this effect. Specifically, the surface tension depends on the concentration of a given amphiphile, as shown schematically in Fig. 8. It remains approximately constant at low concentrations, but drops rapidly with the crowding of the air/water interface and then shows a negligible change once the micellar aggregates start assembling. As a result, CC can be estimated from the intersection of the linear correlations associated with the last two segments of the plot (Fig. 8). For example, the dependence of the surface tension on the concentration of sodium dodecyl sulfate in water and the corresponding CC are illustrated in Fig. 11 [98].

1.4 Hydrodynamic Diameter

Dynamic light scattering measurements can provide a quantitative assessment of the size distribution of spherical nanoparticles dispersed in a solvent from the analysis of their random thermal motion (Brownian motion) [99–102]. Indeed, the translational diffusion coefficient (D_t) of the moving nanoparticles is inversely related to their hydrodynamic diameter (D_h) according to (1), where k_B, T, and η are Boltzmann's constant, temperature, and viscosity, respectively. As a result of this correlation, the displacement of small diffusing nanoparticles over time (Fig. 12a) is more pronounced than that of large counterparts (Fig. 12b). In fact, the sequential tracking of the nanoparticle positions with very short probing intervals (<100 μs) permits the determination of their physical dimensions, even when the moving objects are micellar aggregates of amphiphilic building blocks.

$$D_h = \frac{k_B T}{3\pi\eta D_t} \tag{1}$$

Experimentally, this particular technique determines the aggregate size by measuring the random changes in the intensity of light scattered from the corresponding solution. Generally, a laser source illuminates the sample maintained in a transparent cell and the scattered light (Fig. 13) is collected on a photomultiplier tube. As the diffusing aggregates are constantly in motion, the obtained optical signal shows random changes because of the constructive and destructive phase addition of the scattered light with time. Their size can then be obtained from this signal after an appropriate mathematical treatment. Specifically, the detected signal can be interpreted in terms of an autocorrelation function [$G(\tau)$] of the delay time (τ),

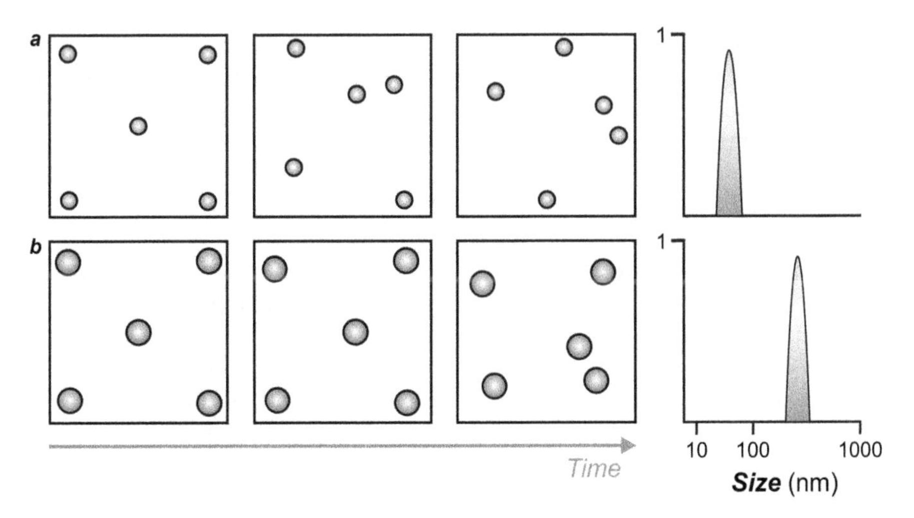

Fig. 12 Illustration of the Brownian motions of small (**a**) and large (**b**) particles

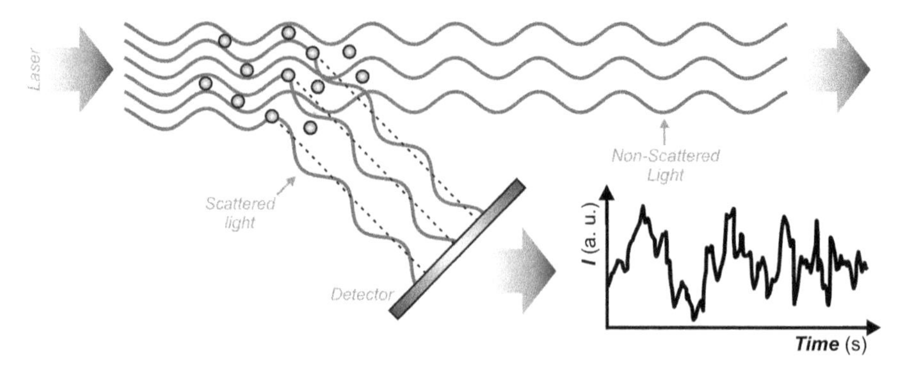

Fig. 13 Schematic representation of the optical setup used for dynamic light scattering measurements of nanoparticle sizing. (Adapted from [103])

i.e., the time that elapses between two consecutive data collections. If all the particles are identical in size, $[G(\tau)]$ decays monoexponentially with τ, according to (2), and fitting of the experimental data provides an estimate of the decay constant (Γ). In turn, this parameter is directly related to D_t, according to (3), where the scattering vector (q) is a function of the refractive index (n) of the liquid, the wavelength (λ) of the laser, and the scattering angle (θ), according to (4). Thus, a measurement of $[G(\tau)]$ against τ ultimately provides a value of D_t and, in combination with (1), also of D_h.

$$G(\tau) = \exp(-2\Gamma\tau) \tag{2}$$

$$\Gamma = D_t q^2 \tag{3}$$

$$q = (4\pi n/\lambda)\sin(\theta/2) \tag{4}$$

Nonetheless, most samples have a distribution of nanoparticle sizes and, therefore, the correlation of $[G(\tau)]$ and τ becomes a power series, according to (5). Once again, a decay constant ($\overline{\Gamma}$), which is the sum of all the individual exponential decays contained in the correlation function and is proportional to D_t, is extracted to obtain a weighted average of D_h, termed z-average size.

$$G(\tau) = \exp\left(-2\overline{\Gamma}\tau + \mu_2\tau^2 - \cdots\right) \tag{5}$$

Besides fitting a single exponential decay to the correlation function (5) to obtain the z-average size, a multiexponential function can be used instead to obtain the distribution of particle sizes. After applying such a mathematical algorithm, one gets the size distribution as a plot of the relative intensity of light scattered by the particles in various size classes. As an example, Fig. 14) shows the intensity distributions for amphiphilic polymers **3** and **21**. Specifically, hydrodynamic diameters of ca. 10 and 130 nm are registered for these two particular amphiphilic macromolecules, respectively.

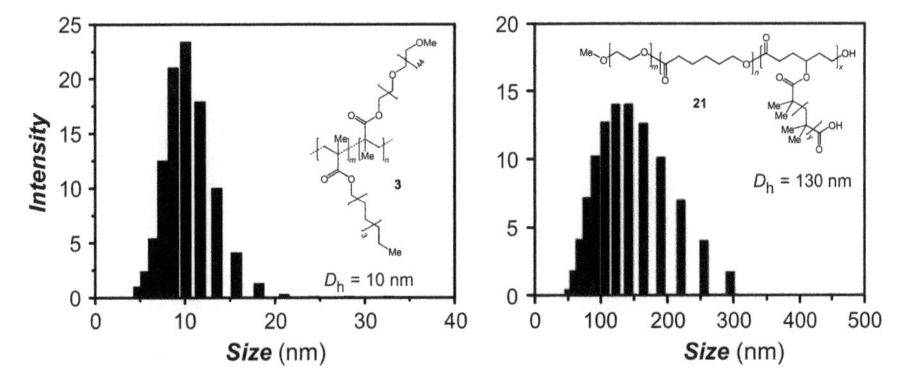

Fig. 14 Chemical structure and hydrodynamic diameters, D_h, for amphiphilic polymers **3** and **21**. (Reprinted with permission from *Polym Chem* © 2013, Royal Society of Chemistry)

2 Energy Transfer Within Self-Assembling Nanoparticles of Amphiphilic Polymers

2.1 Förster Resonance Energy Transfer

Organic molecules can absorb radiation across the ultraviolet and visible regions of the electromagnetic spectrum with concomitant electronic transitions from their ground state (S_0) to their first singlet excited state (S_1) (Fig. 15a). The excited species can then relax from S_1 to S_0 either radiatively, by emitting light in the form of fluorescence, or nonradiatively, by releasing heat into the surrounding environment. Alternatively, the excitation energy can be transferred to a complementary chromophore, if present at an appropriate distance (Fig. 15b, c). As a result of this process, termed Förster resonance energy transfer (FRET) [104–109], the energy donor returns to S_0 and the energy acceptor is excited to S_1. The latter species can then relax radiatively or nonradiatively back to S_0. Under these conditions, the excitation of the donor results in either acceptor emission (Fig. 15b) or fluorescence quenching (Fig. 15c), respectively.

The resonant transfer of energy from a donor to an acceptor can only occur if appropriate spectral and geometrical requirements are satisfied [110–112]. Specifically, the bands associated with the $S_1 \rightarrow S_0$ transition of the donor and the $S_0 \rightarrow S_1$ transition of the acceptor in the corresponding emission and absorption spectra, respectively, must overlap significantly. In addition, the transition moments for the donor emission and acceptor absorption cannot be orthogonal to each other. When both conditions are satisfied, energy is transferred from one species to the other with a rate constant (k_{FRET}) inversely related to the sixth power of their distance (R), according to (6):

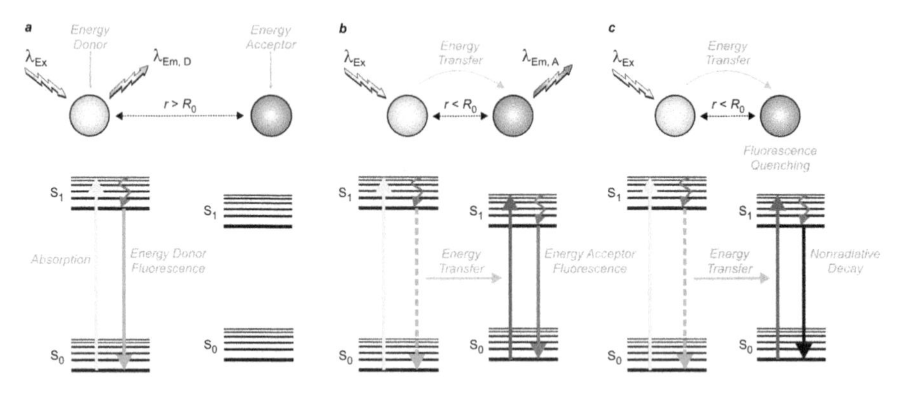

Fig. 15 Schematic representation of FRET and corresponding Jablonski's diagrams. Generally, organic fluorophores absorb radiation with concomitant electronic transitions from their ground state (S_0) to their first singlet excited state (S_1) and then relax back from S_1 to S_0 by emitting light in the form of fluorescence (**a**). In addition, the fluorophore can also transfer its excitation energy to a complementary chromophore, if present at an appropriate distance (**b, c**). As a result of this process, the energy donor returns to S_0 and the energy acceptor is excited to S_1. Under these conditions, the excitation of the donor results in either acceptor emission (**b**) or fluorescence quenching (**c**) respectively

$$k_{\text{FRET}} = \frac{1}{\tau_{\text{D}}} \left(\frac{R_0}{R} \right)^6 \tag{6}$$

Additionally, k_{FRET} is also related to the fluorescence lifetime (τ_{D}) of the donor in the absence of the acceptor and Förster's distance (R_0). The latter parameter is the donor–acceptor separation at which the FRET efficiency (E) is equal to 50%, according to (7), and can be estimated from the emission intensity measured for the donor without (F_{D}) and with (F_{DA}) the acceptor:

$$E \equiv 1 - \frac{F_{\text{D}}}{F_{\text{DA}}} = \frac{R_0^6}{R_0^6 + R^6} \tag{7}$$

The value of R_0 is related to the fluorescence quantum yield (Q_0) of the donor in the absence of the acceptor, the refractive index (n) of the solvent, Avogadro's number (N_{A}), the orientation factor (k^2), and the overlap integral (J):

$${R_0}^6 = \frac{9000\, Q_0 (\ln 10)\, k^2 J}{128 \pi^5 n^4 N_{\text{A}}} \tag{8}$$

The value of J quantifies the spectral overlap between the bands associated with the $S_1 \rightarrow S_0$ transition of the donor and the $S_0 \rightarrow S_1$ transition of the acceptor. It can be estimated from the normalized emission of the donor (f_{D}) and the molar absorption coefficient (ε_{A}) of the acceptor measured over the wavelength (λ) range covered by the two bands with (9):

$$J = \int f_{\mathrm{D}}(\lambda)\, \varepsilon_{\mathrm{A}}(\lambda)\, \lambda^4\, d\lambda \tag{9}$$

The value of k^2 quantifies the relative orientation of the transition moments of donor and acceptor and can vary from 0 and 4. If the two species exchanging energy are isotropically oriented, k^2 averages to 2/3.

2.2 *Nanoparticle Stability*

Self-assembling nanoparticles of amphiphilic building blocks are promising vehicles for the delivery of drugs [19, 26–32, 42, 43]. The nonpolar environment in their interior can promote the encapsulation of hydrophobic molecules, and the polar groups on their surface ensure optimal solvation and permit the transport of the entrapped cargo across hydrophilic media. These supramolecular constructs, however, are held together solely by noncovalent forces. Indeed, the hydrophobic domains of the amphiphilic components come into contact to minimize direct exposure to the surrounding aqueous environments. Such interactions are relatively weak and external perturbations can easily reverse them to encourage the disassembly of the nanostructured constructs and the release of their cargo. For example, the introduction of the loaded nanocarriers into a biological sample might result in dilution below the corresponding CC and/or nonspecific adsorption on certain biomolecules with their concomitant disassembly and leakage of their cargo. Therefore, the identification of strategies to monitor the stability of these supramolecular assemblies within biological specimens is of crucial importance in view of possible therapeutic applications. In this context, imaging schemes based on FRET are particularly convenient.

A particularly promising strategy to transport and operate synthetic chromophores in biological media involves their encapsulation within lipid nanoparticles, called Lipidots [113–116]. These nanostructured carriers have an oily core (essentially soybean oil and wax), surrounded by a shell of phospholipids (e.g., lecithin) and an outer poly(ethylene glycol) (PEG)-based coating (Fig. 16). The diameters of the resulting supramolecular assemblies generally range from 30 to 120 nm and such relatively large dimensions enable the encapsulation of multiple hydrophobic guests in the lipophilic interior. As a result, the brightness of these fluorescent constructs is comparable to that of semiconductor quantum dots. Nonetheless, the lack of heavy atoms in the chemical composition of the former translates in negligible cytotoxicity ($IC_{50} > 75$ nM), making them optimal probes for in vitro, but especially in vivo, fluorescence imaging applications.

The fate of fluorescent Lipidots in vitro can be monitored conveniently by fluorescence spectroscopy on the basis of FRET schemes [117]. For example, donor **22** and acceptor **23** (Fig. 17) can be co-encapsulated in the interior of the same Lipidot to promote the transfer of energy from the former chromophore to the

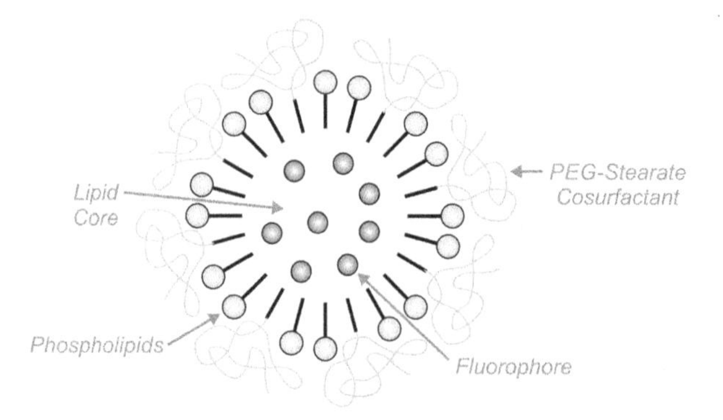

Fig. 16 Schematic representation of a fluorescent Lipidot

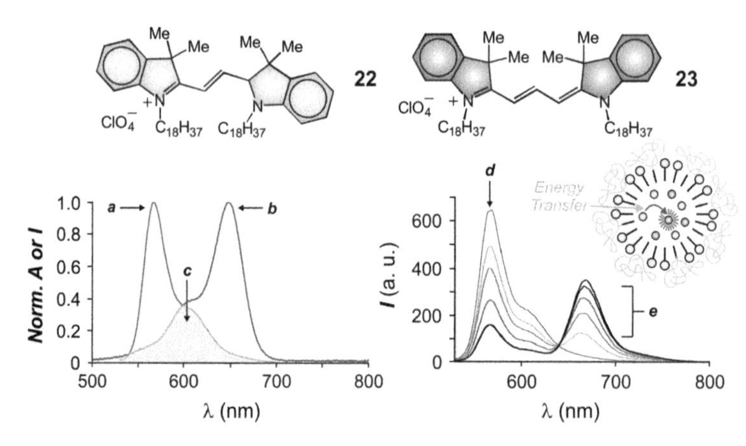

Fig. 17 Normalized emission and absorption spectra of Lipidots loaded with **22** (***a***) and **23** (***b***) respectively together with the corresponding spectral overlap (***c***). Emission spectra ($\lambda_{Ex} = 520$ nm) of Lipidot containing **22** without (***d***) and with (***e***) increasing amounts of **23**

latter. The emission spectrum of **22** (***a*** in Fig. 17) reveals pronounced fluorescence between 550 and 700 nm. Similarly, the absorption spectrum of **23** (***b*** in Fig. 17) shows significant absorbance within the very same range of wavelengths. This particular spectral output ensures the degree of overlap (***c*** in Fig. 17) required to promote FRET upon excitation and corresponds to a R_0 of 52 Å. Consistently, the emission of **22** at 567 nm (***d*** in Fig. 17) decreases with the co-encapsulation of increasing amounts of **23** in the same Lipidot (***e*** in Fig. 17). In addition, a second emission band, corresponding to the acceptor fluorescence, grows at 667 nm on increasing the concentration of **23**.

As an alternative to **23**, **24** can be co-encapsulated together with **22** in the lipophilic core of the same Lipidot [118]. This particular chromophore also absorbs in the same range of wavelengths where **22** emits (***a*** and ***b*** in Fig. 18). In this instance, the pronounced spectral overlap corresponds to a value of 62 Å for R_0 and,

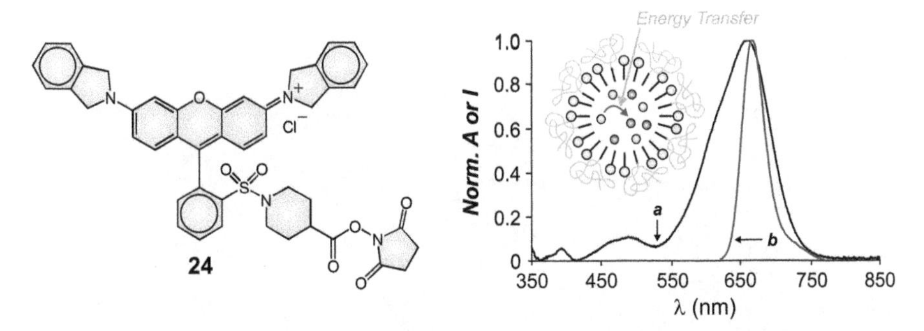

Fig. 18 Normalized emission and absorption spectra of Lipidots loaded with **22** (***a***) and **24** (***b***) respectively

once again, ensures the efficient transfer of energy from donor to acceptor within the Lipidot interior. However, the excitation of **24** is followed exclusively by its nonradiative deactivation from S_1 to S_0. As a result, the co-entrapment of increasing amounts of **24** with a fixed quantity of **22** translates eventually into complete fluorescence suppression.

In both donor–acceptor pairs, FRET occurs efficiently because the complementary components are maintained in close proximity within the core of their supramolecular container [113–118]. The disassembly of the nanostructured construct would result in the physical separation of the components and the suppression of FRET. As a consequence, the structural integrity of these supramolecular assemblies can be monitored in a given biological specimen simply by probing the ability of donors and acceptors to exchange energy. For example, human embryonic kidney (HEK) cells, expressing $\alpha_V\beta_1$ (HEKβ1) or $\alpha_V\beta_3$ (HEKβ3) integrins, were incubated with two sets of Lipidots, co-encapsulating **22** and **23** [118]. One set was tagged with cRGD peptides, which are known to bind integrins, and the other was not. Fluorescence images of HEKβ3 cells, treated with Lipidots lacking the targeting peptides, show emission exclusively in the extracellular space (Fig. 19a, d). These particular images were recorded exciting the donor at 543 nm and collecting fluorescence between 640 and 704 nm, where the acceptor emits. They demonstrate that the lack of the targeting peptides prevents cellular uptake and that the nanostructured assemblies retain their integrity outside the cells to permit energy transfer. Under otherwise identical experimental conditions, the cRGD-labeled Lipidots are instead taken up by both $\alpha_V\beta_1$- and $\alpha_V\beta_3$- expressing cells. In the former case, the acceptor fluorescence was mostly distributed across the whole cytoplasm (Fig. 19b, e). However, the energy transfer was gradually suppressed with the recovery of the donor emission, indicating that the nanocarriers disassemble with relatively slow kinetics to separate the complementary chromophores. In the other instance, significant acceptor fluorescence is observed on the cell periphery (Fig. 19c, f) instead. Furthermore, the acceptor emission changes in intensity over time, indicating that the nanocarriers retain their integrity and maintain the original FRET efficiency. A similar experiment was performed with HEKβ3 cells and

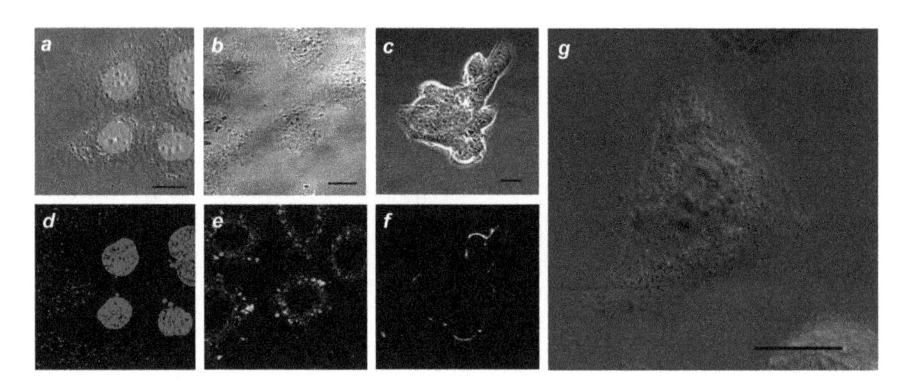

Fig. 19 Normal phase (**a–c**) and merged (**d–g**) confocal images (*scale bar* = 20 μm) of HEKβ3 cells recorded after incubation with either Lipidots, containing **22** and **23** (**a, d**), or cRGD-labeled Lipidots, containing **22** and **23** (**b, e**), for 1 h, HEKβ1 cells recorded after incubation with cRGD-labeled Lipidots, containing **22** and **23** (**c, f**), for 1 h and HEKβ3 cells recorded after incubation of Lipidots, containing **22** and **24** (**g**)

Lipidots, containing **22** and **24**, without peptide labels. This particular combination of chromophores results in complete fluorescence suppression if the nanocarriers retain their integrity. Nonetheless, the corresponding image (Fig. 19g) clearly reveals the intense donor fluorescence in the cytoplasm. These observations demonstrate that the uptaken nanocarriers disassemble in the intracellular space with the concomitant separation of the donors from the acceptor and the suppression of energy transfer.

Operating principles, based on FRET, to assess the stability of self-assembling nanoparticles can be extended to supramolecular constructs of amphiphilic polymers [119–135]. For example, PEG and poly(γ-propargyl-L-glutamate) (PPLG) blocks can be integrated within the same macromolecular backbone together with cyanine chromophores to generate **25** and **26** (Fig. 20) [136]. The absorption and emission bands of **25** appear at 685 and 725 nm, respectively, in the corresponding spectra (***a*** and ***c*** in Fig. 20), whereas those of **26** are at 770 and 810 nm (***b*** and ***d*** in Fig. 20). The optimal spectral overlap between the emission of **25** and the absorption of **26** suggests that energy can be transferred efficiently from one to the other if the two components are brought into close proximity.

Mixing appropriate amounts of **25** and **26** in aqueous solution results in the spontaneous assembly of nanoparticles with D_h of 130 nm. The resulting absorption spectrum (***e*** in Fig. 20) shows the characteristic bands of both chromophores. However, the corresponding emission spectrum (***f*** in Fig. 20) reveals predominantly the fluorescence of **26** at 800 nm. These observations demonstrate that the two independent polymers co-assembled within the same supramolecular construct enable the transfer of energy from **25** to **26**. After administration in animal models, however, the donor fluorescence is partially recovered. Specifically, images of a mouse, recorded after injection of an aqueous solution of the nanoparticles with a concentration three times greater than the corresponding CC, show a gradual

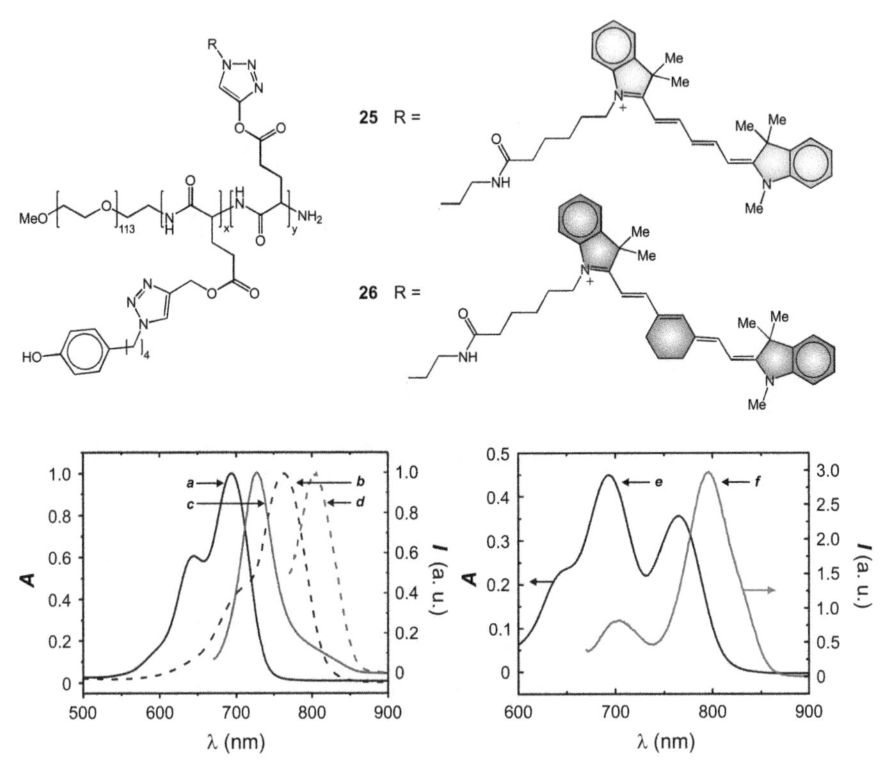

Fig. 20 Normalized absorption (***a*** and ***b***) and emission (***c*** and ***d***) spectra of amphiphilic polymers **25** (***a*** and ***c***) and **26** (***b*** and ***d***)

increase in donor emission (Fig. 21a), relative to the acceptor fluorescence (Fig. 21b). Thus, the supramolecular nanoparticles partially disassemble when in circulation within the animal, and approximately 44 % of the donor emission is recovered after 24 h. Biodistribution analysis, however, revealed different fluorescence recoveries in different organs. In particular, 33%, 11%, and 4% of the donor emission was recovered in the liver, spleen, and kidneys, demonstrating that the nanoparticle fate varies significantly with the nature of the organ.

2.3 *Nanoparticle Dynamics*

The reversibility of the noncovalent bonds holding multiple amphiphilic components together in a single nanostructure imposes a dynamic character on these supramolecular assemblies [137–149]. In fact, amphiphilic building blocks can dissociate from one nanoparticle and associate with another in solution. Similarly, the cargo of the nanocarriers is maintained within their hydrophobic interior, also on the basis of such reversible noncovalent interactions. As a result, the molecular

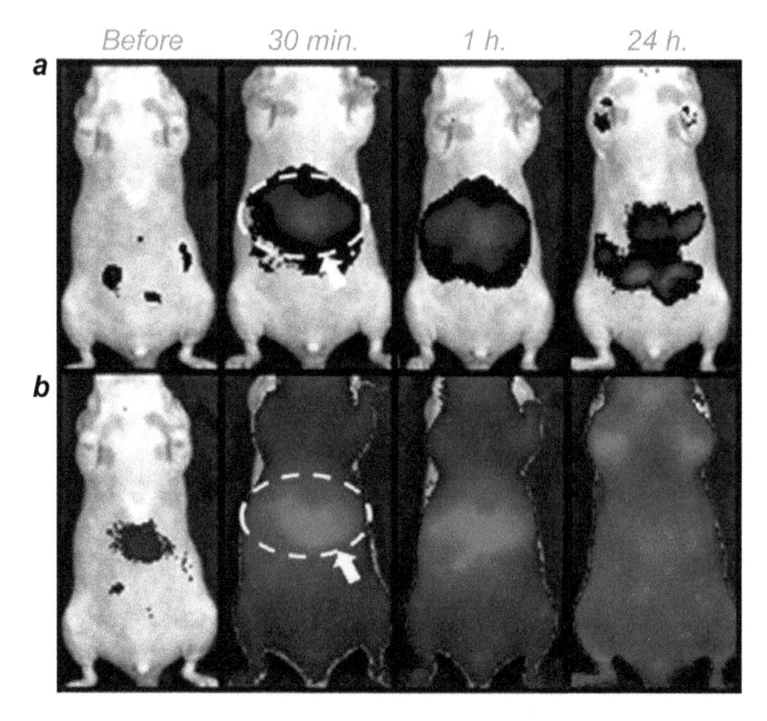

Fig. 21 Fluorescence images of a mouse injected with nanoparticles of **25** and **26** recorded after excitation of the donor at 640 nm and collection of the emission for either the donor at 700 nm (**a**) or the acceptor at 800 nm (**b**)

guests entrapped in one supramolecular host can escape into the bulk solution to be recaptured by another. Thus, these supramolecular containers can exchange their constituent amphiphilic components as well as their cargo on relatively short timescales under appropriate experimental conditions. Such dynamic processes can be monitored conveniently by relying, once again, on FRET between appropriate chromophoric units [119–135].

The pronounced hydrophobicity of **20** and **27** (Fig. 22) translates into negligible aqueous solubility [95]. In the presence of amphiphilic polymer **3**, however, both molecules dissolve readily in PBS. In fact, this particular macromolecule forms nanoparticles with a D_h of ca. 15 nm which can capture these hydrophobic guests and transfer them into the aqueous medium. The characteristic emission of the boron dipyrromethene (BODIPY) chromophore of **20** extends from 450 to 600 nm. The absorption bands associated with the $S_0 \rightarrow S_1$ transitions of the anthracene chromophore of **27** are positioned between 450 and 550 nm. The pronounced spectral overlap corresponds to a J of 9.1×10^{-14} M^{-1} cm^3 and a R_0 of 47 Å. These values ensure the efficient transfer of energy from **20** to **27**, when both species are co-encapsulated within the interior of nanoparticles of **3**. Consistently, the emission spectrum shows predominantly the fluorescence of the acceptor

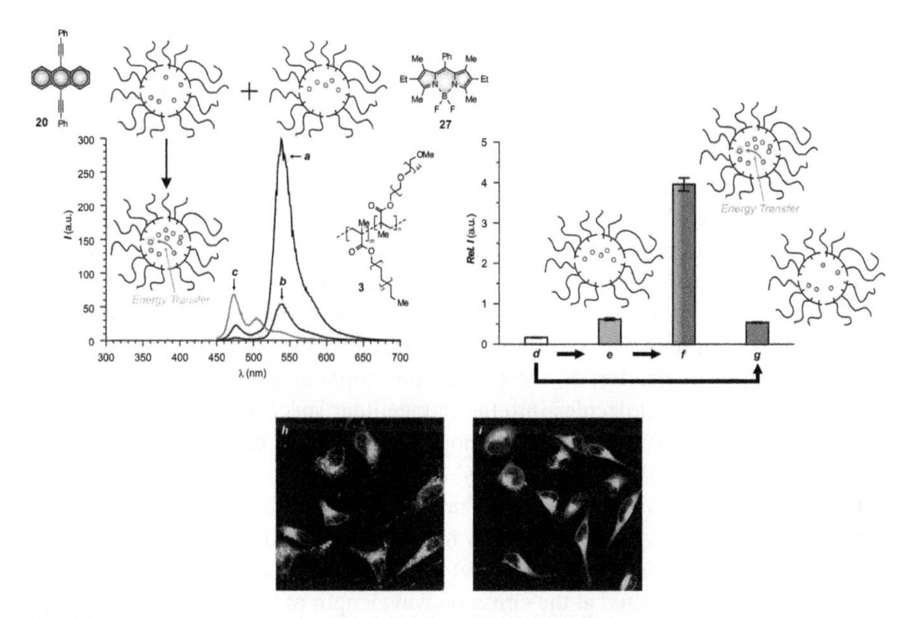

Fig. 22 Emission spectra ($\lambda_{Ex} = 430$ nm) recorded after mixing identical volumes of two PBS dispersions of nanoparticles of **3** (500 μg mL^{-1}), loaded with **20** (5 μg mL^{-1}) or **27** (5 μg mL^{-1}), respectively, before (***a***) and after tenfold dilution with either PBS (***b***) or THF (***c***). Emission intensities ($\lambda_{Ex} = 430$ nm, $\lambda_{Em} = 540$ nm), reported relative to that of an indocyanine green standard (50 μM, $\lambda_{Ex} = 730$ nm, $\lambda_{Em} = 780$ nm) added 30 min prior to termination of incubation, recorded with a plate reader before (***d***) and after incubation of HeLa cells with a PBS dispersion of nanoparticles of **3** (125 μg mL^{-1}), containing **27** (1.25 μg mL^{-1}), for 3 h and washing (***e***) and subsequent incubation with a PBS dispersion of nanocarriers of **3** (125 μg mL^{-1}), containing **20** (1.25 μg mL^{-1}), for a further 3 h and washing (***f***) or after incubation with the same dispersion of nanoparticles, containing **1**, for 3 h and washing (***g***). Fluorescence images ($\lambda_{Ex} = 458$ nm, $\lambda_{Em} = 540$–640 nm) of HeLa cells recorded after incubation with a PBS dispersion of nanoparticles of **3** (125 μg mL^{-1}), containing **27** (1.25 μg mL^{-1}), for 3 h, washing and subsequent incubation with a PBS dispersion of nanocarriers of **3** (125 μg mL^{-1}), containing **20** (1.25 μg mL^{-1}), and washing (**h**) or after the same treatment but inverting the order of addition of the two components (**i**)

between 540 and 640 nm, and its comparison to those of the separate components indicates that energy is transferred with an E of 0.95.

In these FRET experiments, donor, acceptor and amphiphilic polymer are mixed in chloroform and, after the evaporation of the organic solvent, the residue is dispersed in aqueous medium [95]. Under these conditions, the two complementary chromophoric guests are captured within the same supramolecular host and the transfer of energy from one to the other occurs efficiently upon excitation. The very same result, however, is obtained if an aqueous solution of nanoparticles containing the donors exclusively is combined with an aqueous solution of nanocarriers entrapping only the acceptors. Upon mixing, the two sets of supramolecular assemblies exchange their components with relatively fast kinetics to co-localize donors and acceptors within the same containers. Indeed, the emission spectrum (***a*** in

Fig. 22), recorded immediately after mixing the two solutions, shows predominantly the acceptor fluorescence and scales linearly with a tenfold dilution of the sample with PBS (***b*** in Fig. 22). These observations demonstrate that FRET occurs after mixing and that the efficiency of the process does not change with dilution. In turn, the latter result indicates that the increase in physical separation between the nanoparticles, occurring with dilution, has no depressive effect on FRET and, therefore, confirms that donors and acceptors are co-entrapped within the same containers. By contrast, a tenfold dilution of the sample with THF disrupts the hydrophobic interactions responsible for the integrity of the nanoparticles to separate donors from acceptors and prevent FRET. As a result, the corresponding emission spectrum (***c*** in Fig. 22) shows predominantly the donor fluorescence.

The intriguing behavior of these dynamic supramolecular systems can be exploited to transport molecules into the intracellular space and, only there, allow their mutual interaction [95]. Indeed, nanoparticles of **3** can cross the membrane of cervical cancer (HeLa) cells and carry either **20** or **27** along in the process. Fluorescence measurements performed before (***d*** in Fig. 22) and after (***e*** and ***g*** in Fig. 22) incubation, however, reveal only negligible intracellular emission. In these experiments, the sample is illuminated at the excitation wavelength of the donor and fluorescence is detected at the emission wavelength of the acceptor. Therefore, the presence of only the donor or only the acceptor cannot produce any significant fluorescence, under these excitation and detection conditions. If the cells are incubated sequentially with the two sets of nanoparticles, then intense emission is detected instead under the very same conditions (***f*** in Fig. 22). These observations demonstrate that the internalized nanoparticles exchange their components with fast kinetics within the intracellular environment to co-localize donors and acceptors in close proximity and enable FRET. Consistently, fluorescence images (Fig. 22h, i) recorded after sequential incubation with nanoparticles containing the donor and then nanocarriers containing the acceptor, or vice versa, clearly show intracellular fluorescence.

2.4 Cell Apoptosis

As cancer treatment progresses, tumors may develop drug resistance, decreasing the effectiveness of chemotherapy [150–153]. Therefore, the development of strategies to assess the ability of drugs to kill tumors is as essential as the identification of protocols to deliver them to their targets in living organisms. In particular, the efficacy of a cancer drug can be probed by monitoring apoptosis (i.e., programmed cell death). This process is associated with the production of several enzymes. One of them, caspase-3, is able to cleave the Asp-Glu-Val-Asp (DEVD) peptide sequence. In turn, the cleavage of DEVD can be designed to control the efficiency of FRET and, therefore, enable the monitoring of cell death with fluorescence measurements.

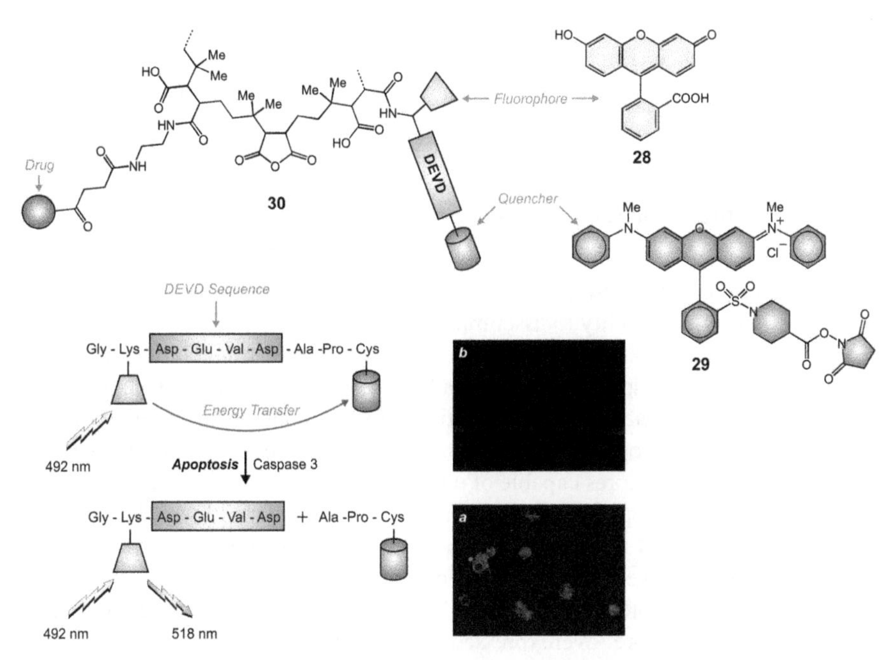

Fig. 23 Polymer **30** incorporates a drug (paclitaxel), a donor (fluorescein, **28**) and an acceptor (QSY 7, **29**) within its macromolecular backbone and does not produce any significant fluorescence because of the efficient transfer of energy from the donor to the nonemissive acceptor. When exposed to caspase-3, the peptide sequence (DEVD) connecting donor and acceptor cleaves to prevent energy transfer and activates the fluorescence of the donor. Fluorescence images of responsive (**a**) and resistant (**b**) cell lines after treatment with **30**

Fluorescein and QSY 7 (**28** and **29** in Fig. 23, respectively) have optimal spectral overlap to implement FRET schemes [154]. These complementary chromophores can be connected through a DEVD bridge and integrated covalently within amphiphilic polymer **30** (Fig. 23). This particular macromolecule also incorporates a chemotherapeutic agent, in the shape of paclitaxel, at one of its ends. In the resulting assembly, the fluorescein donor (**28**) and the QSY 7 acceptor (**29**) are sufficiently close to each other to ensure efficient energy transfer from one to the other upon excitation. The excited acceptor, however, relaxes back from S_1 to S_0 nonradiatively. As a result, the emission of this particular macromolecular construct is completely suppressed. After internalization in cancer cells, the paclitaxel appendage initiates apoptosis with the concomitant production of caspase-3. This enzyme then cleaves the DEVD sequence embedded in the macromolecular backbone, separates the donor from the acceptor, and prevents FRET. The overall result of this sequence of events is the appearance of the characteristic fluorescence of the donor. Thus, fluorescence activation can signal cell death on the basis of these operating principles, providing the opportunity to assess the efficacy of the chemotherapeutic agent. Indeed, comparison of paclitaxel-responsive (Lewis lung carcinoma, LLC) and taxane-resistant (TR) cells clearly confirms the viability of this strategy. In the former instances, the internalization of **30** induces apoptosis and activates fluorescence (Fig. 23a). In the latter, the nanostructured construct has no influence on the cells and no fluorescence can be detected (Fig. 23b).

2.5 Photoinduced Drug Release

The introduction of stimuli-responsive groups within amphiphilic polymers permits the regulation of their ability to assemble into nanoparticles under external control [155–158]. These operating principles can be particularly valuable to induce the capture and release of species and can lead to the realization of controllable drug-delivery systems for therapeutic applications. In this context, the spatiotemporal control possible with optical stimulations is especially convenient. In fact, the attachment of chromophores capable of undergoing photochemical transformations to self-assembling nanoparticles, together with ligands able to direct the resulting constructs to biological targets, appears to be a promising strategy for the realization of light-responsive nanocarriers. For example, spiropyrans are known to undergo reversible photoisomerizations [159, 160] and folic acid binds the many folate receptors that are overexpressed on the surface of cancer cells [161–166]. Both groups can be incorporated within the macromolecular backbone of an amphiphilic polymer, in the shape of **31** (Fig. 24), together with rhodamine B. When combined with hollow mesoporous silica nanoparticles pre-coated with

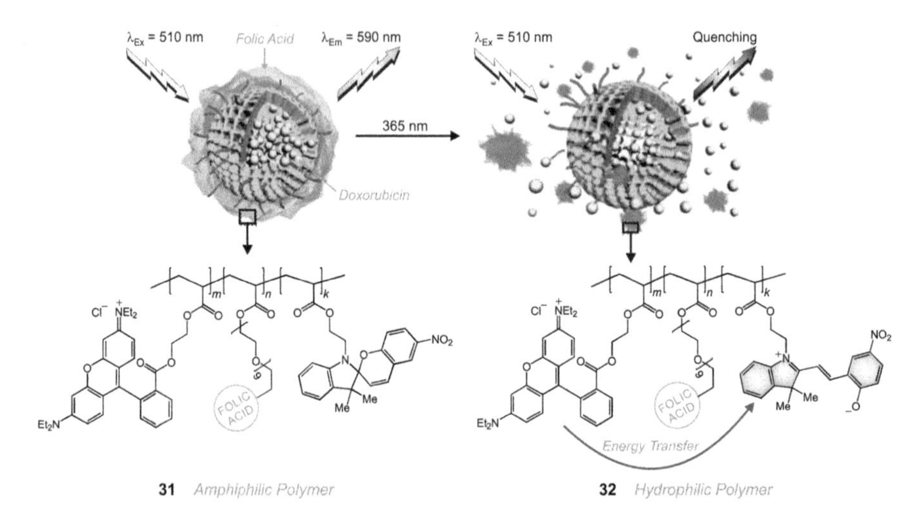

Fig. 24 Photoresponsive polymer **31** incorporates a tumor-site binder (folic acid), a donor (rhodamine B), and a photoactivatable acceptor (spiropyran) within its macromolecular backbone and produces significant fluorescence because of the lack of transfer of energy from the donor to the spiropyran form. When exposed to UV light (365 nm), the spiropyran photochrome switches to the colored merocyanine form (**32**), enabling FRET between donor and acceptor and quenching the fluorescence of the former

a hydrophobic layer, multiple copies of **31** assemble on the surface of the inorganic scaffold to generate a composite construct. In addition, the porous silica matrix can be pre-loaded with hydrophobic drugs (doxorubicin), before coating the inorganic core with the polymer envelope, and traps them in place after self-assembly of the amphiphilic building blocks. Upon ultraviolet illumination, the many spiropyran components switch to the corresponding merocyanine isomers (**32** in Fig. 24). These photochemical products have an extended π-system which absorbs significantly (***b*** in Fig. 25) in the same region of the electromagnetic spectrum where the rhodamine components emit (***e*** in Fig. 25). As a result of the pronounced spectral overlap, the photoinduced spiropyran → merocyanine transformation activates FRET. Specifically, the excitation of the rhodamine donors is followed by the transfer of energy to the merocyanine acceptors, which then relax nonradiatively (***c*** in Fig. 25). Under visible illumination, the merocyanine acceptors revert back to the original spiropyran components and the FRET pathway is suppressed resulting in significant fluorescence (***d*** in Fig. 25). It follows that the reversible interconversion of the photoswitchable component can be exploited to suppress and reactivate the donor fluorescence.

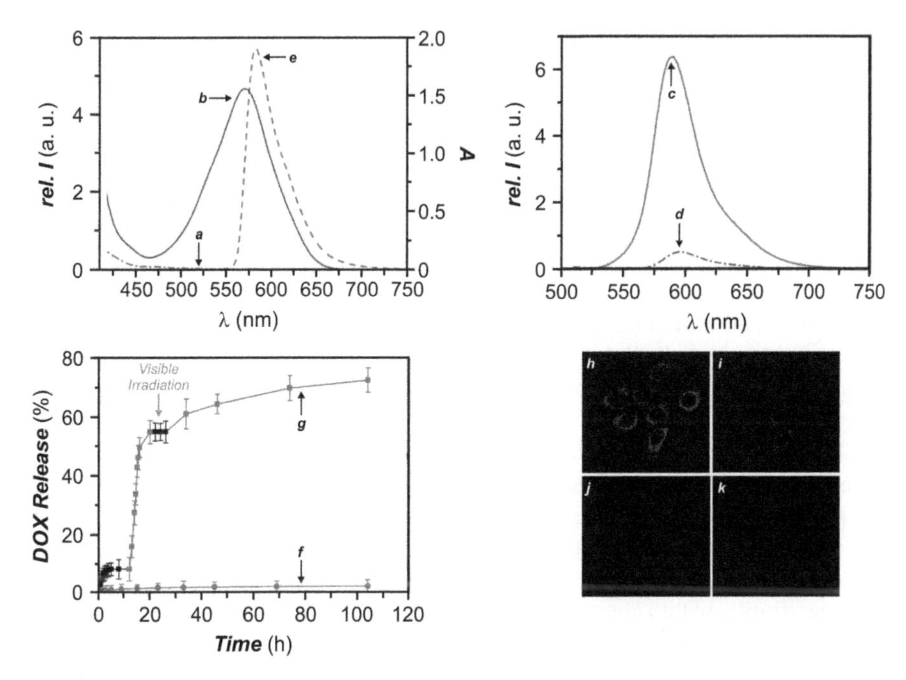

Fig. 25 Absorption (***a*** and ***b***) and emission (***c*** and ***d***) spectra of **31** before (***a*** and ***c***) and after (***b*** and ***d***) ultraviolet irradiation together with the emission spectrum (***e***) of a model rhodamine fluorophore. Release of DOX in vitro without (***f***) and with (***g***) irradiation. DOX-loaded nanocarriers (500 mg mL^{-1}) with KB cells for 2 h in the dark at first. Then the cells were exposed to UV light ($\lambda = 365$ nm, 0.2 W cm^{-2}) for time periods ranging from 0 to 20 min and the cells were examined by confocal laser scanning microscopy (***h–k***)

The amphiphilic polymers assembled on the surface of the inorganic core block the portals of the porous silica scaffold and prevent the leakage of doxorubicin in the bulk aqueous solution [167]. Consistently, no doxorubicin can be detected in solution (***f*** in Fig. 25) if the sample is maintained in the dark. Under visible illumination for up to 8 h, less than 10 wt% of doxorubicin was released in solution (***g*** in Fig. 25). However, a sudden increase in doxorubicin concentration occurs after only a few minutes of ultraviolet illumination and up to 70 wt% of drug is released into the surrounding aqueous environment after 100 h. In all instances, the amount of released DOX was determined spectroscopically by measuring the characteristic absorbance of this compound in a withdrawn portion of the nanoparticle suspension. These observations demonstrate that the photoinduced transformation of the relatively hydrophobic spiropyrans into the zwitterionic merocyanines alters the morphology of the polymer coating around the porous silica core and allows the escape of the entrapped drugs. In addition, FRET imaging was exploited to monitor the light-controlled release of DOX from the nanocarriers. Fluorescence images for KB cells reveal significant fluorescence in the dark right after incubation (Fig. 25h). Exposure to ultraviolet radiation for 5, 10, and 20 min results in a gradual fluorescence suppression (Fig. 25i–k), demonstrating that the structural modification of the photoresponsive copolymer in the nanocarriers not only enables the FRET process but also triggers the drug release.

3 Conclusions

Free-radical polymerizations can be exploited to incorporate hydrophilic and hydrophobic components within the main chain or the side chains of a common macromolecular backbone to generate polymers with amphiphilic character. Above a given concentration threshold in aqueous solutions, these amphiphilic polymers self-assemble into micellar aggregates with nanoscaled dimensions. Noncovalent contacts bring the hydrophobic domains of distinct amphiphilic components together to avoid their direct exposure to water. At the same time, solvation of the hydrophilic domains maintains the resulting nanoparticles in solution and prevents their further aggregation into macroscopic clusters. These nanostructured assemblies can accommodate hydrophobic molecules in their nonpolar interior and transport them across hydrophilic media. Furthermore, these supramolecular hosts can co-encapsulate pairs as distinct molecular guests and constrain them in close proximity within their hydrophobic core. If the emission spectrum of one entrapped species overlaps the absorption spectrum of the other, then efficient energy transfer from the former species to the latter can occur upon excitation. Such a process, however, demands the overall construct to retain its integrity and keep donors and acceptors at distances compatible with energy transfer. As a result, the occurrence of energy transfer can be exploited to signal nanoparticle stability. Similarly, the encapsulation of complementary donors and acceptors in distinct nanoparticles offers the opportunity to assess the ability of these supramolecular constructs to exchange their components. These operating principles are especially convenient to probe nanoparticle stability and nanoparticle dynamics in biological samples with the aid of convenient fluorescence measurements. Furthermore, they can be adapted to monitor biochemical transformations which eventually lead to cell apoptosis. Thus, such energy transfer schemes can contribute to the fundamental understating of the fate of these promising delivery vehicles in biological preparations and can guide the design of strategies for the transport of contrast agents and/or drugs in view of potential diagnostic and/or therapeutic applications.

Acknowledgements J. G.-A. is grateful for a *Beatriu de Pinós* postdoctoral grant from the *Generalitat de Catalunya* (Spain, 2011 BP-A-00270 and 2011 BP-A2-00016). F. M. R. acknowledges the National Science Foundation (CAREER Award CHE-0237578, CHE-0749840 and CHE-1049860) for financial support.

References

1. Winsor PA (1948) Trans Faraday Soc 44:376–382
2. Winsor PA (1954) Solvent properties of amphiphilic compounds. Butterworths, London
3. Nagarajan R (2011) Amphiphilic surfactants and amphiphilic polymers: principles of molecular assembly. In: Amphiphiles: molecular assembly and applications, ACS symposium series. American Chemical Society, Washington, DC, pp 1–22, Ch. 1
4. Alexandridis P, Lindman B (2000) Amphiphilic block copolymers: self-assembly and applications. Elsevier, Amsterdam
5. McCormick CL (2000) Stimuli-responsive water soluble and amphiphilic polymers, ACS symposium series. American Chemical Society, Washington, DC
6. Müller AHE, Borisov O (2011) Self-organized nanostructures of amphiphilic block copolymers I. Springer, Berlin
7. Müller AHE, Borisov O (2011) Self-organized nanostructures of amphiphilic block copolymers II. Springer, New York
8. Hamley IW (2000) Introduction to soft matter: polymers, colloids, amphiphiles, and liquid crystals. Wiley, New York
9. Holmberg K, Jonsson B, Kronberg B, Lindman B (2002) Surfactants and polymers in aqueous solution. Wiley, Chichester
10. Yamashita H, Oosawa M (1995) Patent 7309714 A2 951128, Kanebo Ltd., Japan
11. O'Lenick AJ Jr (1995) US Patent 5475125 A 951212, Siltech Inc.
12. Dupuis C (1995) Patent 684041 A1 951129, Oreal S.A., Europ
13. Sakuta K (1995) Patent 7304627 A2 951121, Shinetsu Chem. Ind. Co., Japan
14. Mohammed RA, Bailey AI, Luckham PF, Taylor SE (1994) Colloids Surf A Physicochem Eng Asp 83:261–271
15. Amaravathi M, Pandey BP (1991) Res Ind 36:169
16. Blease TG, Evans JG, Hughes L, Loll P (1993) In: Garrett P (ed) Defoaming, vol 45, Surfactant science series. Marcel Dekker, New York, p 299
17. Schmidt DL (1996) In: Prudhomme RK, Khan SA (eds) Foams, vol 57, Surfactant science series. Marcel Dekker, New York, p 287
18. Bader H, Ringsdorf H, Schmidt B (1984) Angew Makromol Chem 123:457–485
19. Fox ME, Szoka FC, Fréchet JMJ (2009) Acc Chem Res 42:1141–1151
20. Kataoka K, Kwon GS, Yokoyama M, Okano T, Sakurai YJ (1993) Control Release 24:119–132
21. Torchilin VP (2007) AAPS J 9:E-128–E-147

22. Jones M-C, Leroux J-C (1999) Eur J Pharm Biopharm 48:101–111
23. Adams ML, Lavasanifar A, Kwon GS (2003) J Pharm Sci 92:1343–1355
24. Husseini AG, Pitt WG (2008) Adv Drug Deliv Rev 60:1137–1152
25. Mondon K, Gurny R, Möller M (2008) Chimia 62:832–840
26. Park JH, Lee S, Kim JH, Park K, Kim K, Kwon IC (2008) Prog Polym Sci 33:113–137
27. Kim S, Shi Y, Kim JY, Park K, Cheng J-X (2010) Expert Opin Drug Deliv 7:49–62
28. Chacko RT, Ventura J, Zhuang J, Thayumanavan S (2012) Adv Drug Deliv Rev 64:836–851
29. Wang Y, Grayson SM (2012) Adv Drug Deliv Rev 64:852–865
30. Jin Q, Maji S, Agarwal S (2012) Polym Chem 3:2785–2793
31. Lalatsa A, Schatzlein AG, Mazza M, Thi BHL, Uchegbu IFJ (2012) Control Release 161:523–536
32. Nicolas J, Mura S, Brambilla D, Mackiewicz N, Couvreur P (2013) Chem Soc Rev 42:1147–1235
33. Lu Y, Park K (2013) Int J Pharm 452:198–214
34. Tyler JY, Xu X-M, Cheng J-X (2013) Nanoscale 5:8821–8836
35. Wang DR, Wang XG (2013) Prog Polym Sci 38:271–301
36. Gu L, Faig A, Abdelhamid D, Uhrich K (2014) Acc Chem Res 47:2867–2877
37. Yang Y, Pan D, Luo K, Li L, Gu Z (2013) Biomaterials 34:8430–8443
38. Alakhova DY, Kabanov AV (2014) Mol Pharm 11:2566–2578
39. Wei T, Chen C, Liu J, Liu C, Posocco P, Liu X, Chengk Q, Huo S, Liang Z, Fermeglia M, Pricl S, Liang X-J, Rocchi P, Peng L (2015) Proc Natl Acad Sci 112:2978–2983
40. Makino A (2014) Polym J 46:783–791
41. Torchilin VP (2001) J Control Release 73:137–172
42. Zhuang J, Gordon MR, Ventura J, Li L, Thayumanavan S (2013) Chem Soc Rev 42:7421–7435
43. Mura S, Nicolas J, Couvreur P (2013) Nat Mater 12:991–1003
44. Lv M-Q, Shi Y, Yang W-T, Fu Z-F (2013) J Appl Polym Sci 128:332–339
45. Smeets NMB (2013) Eur Polym J 49:2528–2544
46. Yildiz I, Impellizzeri S, Deniz E, McCaughan B, Callan JF, Raymo FM (2011) J Am Chem Soc 133:871–879
47. Wang Y, Al AM, He J, Grayson SM (2014) Polym Chem 5:622–629
48. Muhlebach A, Gaynor SG, Matyjaszewski K (1998) Macromolecules 31:6046–6052
49. Rikkou MD, Kolokasi M, Matyjaszewski K, Patrickios CS (2010) J Polym Sci A 48:1878–1886
50. Rikkou-Kalourkoti M, Loizou E, Porcar L, Matyjaszewski K, Patrickios CS (2012) Polym Chem 3:105–116
51. Hua M, Kaneko T, Liu X-Y, Chen M-Q, Akashi M (2005) Polym J 37:59–64
52. York AW, Kirkland SE, McCormick CL (2008) Adv Drug Deliv Rev 60:1018–1036
53. Huang Z, Zhang X, Zhang X, Fu C, Wang K, Yuan J, Tao L, Wei Y (2015) Polym Chem 6:607–612
54. Barner-Kowollik C (2008) Handbook of RAFT polymerization. Wiley, Weinheim
55. Shi X, Zhou W, Qiu Q, An Z (2012) Chem Commun 48:7389–7391
56. Rikkou-Kalourkoti M, Elladiou M, Patrickios CS (2015) J Polym Sci 53:1310–1319
57. Fu J, Cheng Z, Zhou N, Zhu J, Zhang W, Zhu X (2009) e-Polymers 18:1–11
58. Garnier S, Laschewsky A (2006) Colloid Polym Sci 284:1243–1254
59. Smith AE, Xu X, McCormick CL (2010) Prog Polym Sci 35:45–93
60. Lowe AB, McCormick CL (2007) Prog Polym Sci 32:283–351
61. Koda Y, Terashima T, Sawamotoa M, Maynard HD (2015) Polym Chem 6:240–247
62. Hawker CJ, Bosman AW, Harth E (2001) Chem Rev 101:3661–3688
63. Grubbs RB (2011) Polym Rev 51:104–137
64. Li X, Ni X, Liang Z, Shen Z (2012) J Polym Sci A Polym Chem 50:2037–2044
65. Sutthasupa S, Shiotsuki M, Sanda F (2010) Polym J 42:905–915
66. Nomura K, Abdellatif MM (2010) Polymer 51:1861–1881

67. Nuyken O, Pask SD (2013) Polymers 5:361–403
68. Zhao Y, Wu Y, Yan G, Zhang K (2014) RSC Adv 4:51194–51200
69. Hartley GS (1941) Trans Faraday Soc 37:130–133
70. Dukhin SS, Kretzchmar G, Miller B (1995) Dynamics of adsorption at liquid interfaces. Elsevier, Amsterdam
71. van Oss CJ (2006) Interfacial forces in aqueous media, 2nd edn. CRC/Taylor & Francis, Boca Raton
72. Gray GW (1998) Handbook of liquid crystals. Wiley, Weinheim
73. Figueiredo Neto AM, Salinas SRA (2005) The physics of lyotropic liquid crystals: phase transitions and structural properties. Oxford University Press, Oxford
74. Garti N, Somasundaran P, Mezzenga R (2012) Self-assembled supramolecular architectures: lyotropic liquid crystals. Wiley, Weinheim
75. Gelbart WM, Ben-Shaul A, Roux D (1994) Micelles, membranes, microemulsions, and monolayers. Springer, New York
76. Hamley IW (2007) Introduction to soft matter: synthetic and biological self-assembling materials. Wiley, Chichester
77. Förster S, Konrad M (2003) J Mater Chem 13:2671–2688
78. Bucknall DG, Anderson HL (2003) Science 302:1904–1905
79. Zvelindovsky AV (2007) Nanostructured soft matter: experiment, theory, simulation and perspectives. Springer, Dordrecht
80. Torchilin V, Amiji MM (2010) Handbook of materials for nanomedicine. Pan Stanford, Singapore
81. Zhao J, Fung BM (1993) Langmuir 9:1228–1231
82. Al-Soufi W, Piñeiro L, Novo M (2012) J Colloid Interface Sci 370:102–110
83. Chakraborty T, Chakraborty I, Ghosh S (2011) Arab J Chem 4:265–270
84. Topel Ö, Çakır BA, Budama L, Hoda N (2013) J Mol Liq 177:40–43
85. Prazeres TJV, Beija M, Fernandes FV, Marcelino PGA, Farinha JPS, Martinho JMG (2012) Inorg Chim Acta 381:181–187
86. Fendler JH (1982) Membrane mimetic chemistry. Wiley, New York, pp 6–47
87. Berthod A, Garcia-Alvarez-Coque C (eds) (2000) Micellar liquid chromatography, vol 83, Chromatography series. Dekker, New York, pp 503–525
88. Van Os NM, Haak JR, Rupert LAM (1993) Physico-chemical properties of selected anionic, cationic and nonionic surfactants. Elsevier, Amsterdam
89. Hinze WL, Armstrong DW (eds) (1987) Ordered media in chemical separations, vol 342, ACS symposium series. American Chemical Society, Washington, DC, pp 2–82
90. Armstrong DW (1985) Sep Purif Methods 14:213–304
91. Kalyanasundaram K, Thomas JK (1977) J Am Chem Soc 99:2039–2044
92. Ananthapadmanabhan KP, Goddard ED, Turro NJ, Kuo PL (1985) Langmuir 1:352–355
93. Aguilar J, Carpena P, Molina-Bolívar JA, Carnero Ruiz C (2003) J Colloid Surf Sci 258:116–122
94. Bhaisare ML, Pandey S, Khan MS, Talib A, Wu H-F (2015) Talanta 132:572–578
95. Swaminathan S, Fowley C, McCaughan B, Cusido J, Callan JF, Raymo FM (2014) J Am Chem Soc 136:7907–7913
96. Mukherjee I, Moulik SP, Rakshit AK (2013) J Colloid Interface Sci 394:329–336
97. Mittal KL (1972) J Pharm Sci 61:1334–1335
98. Castro MJL, Ritacco H, Kovensky J, Fernández-Cirelli A (2001) J Chem Educ 78:347–348
99. Berne BJ, Pecora R (2000) Dynamic light scattering: with applications to chemistry, biology, and physics. Courier Dover, Mineola
100. Pecora R (1985) Dynamic light scattering: applications of photon correlation spectroscopy. Plenum, New York
101. Meurant G (1990) Introduction to dynamic light scattering by macromolecules. Academic, London

102. Jonasz M, Fournier GR (2007) Light scattering by particles in water: theoretical and experimental foundations. Elsevier, San Diego
103. Zetasizer Nano Series User Manual (2004) Man0317, Issue 1.1, © Malvern Instruments Ltd., Malvern, Feb 2004
104. Förster T (1946) Naturwissenschaften 6:166–175
105. Förster T (1948) Ann Phys 2:55–75
106. Förster T (1949) Z Naturforsch A 4:321–327
107. Förster T (1951) Fluoreszenz Organischer Verbindungen. Vandenhoeck & Ruprecht, Gottingen
108. Förster T (1959) Discuss Faraday Soc 27:7–17
109. Förster T (1960) Radiat Res Suppl 2:326–339
110. Van Der Meer BV, Coker G III, Simon Chen S-Y (1994) Resonance energy transfer: theory and data. Wiley, New York
111. Braslavsky SE, Fron E, Rodríguez HB, San Roman E, Scholes GD, Schweitzer G, Valeur B, Wirz J (2008) Photochem Photobiol Sci 7:1444–1448
112. Broussard JA, Rappaz B, Webb DJ, Brown CM (2013) Nat Protoc 8:265–281
113. Gravier J, Navarro FP, Delmas T, Mittler F, Couffin AC, Vinet F, Texier I (2011) J Biomed Opt 16:096013
114. Navarro FP, Mittler F, Berger M, Josserand V, Gravier J, Vinet F, Texier I (2012) J Biomed Nanotechnol 8:594–604
115. Navarro FP, Berger M, Guillermet S, Josserand V, Guyon L, Neumann E, Vinet F, Texier I (2012) J Biomed Nanotechnol 8:730–774
116. Navarro FP, Creusat G, Frochot C, Moussaron A, Verhille M, Vanderesse R, Thomann J-S, Boisseau P, Texier I, Couffin A-C, Barberi-Heyob M (2014) J Photochem Photobiol B Biol 130:161–169
117. Mérian J, Gravier J, Navarro F, Texier I (2012) Molecules 17:5564–5591
118. Gravier J, Sancey L, Coll JL, Hirsjärvi S, Benoit JP, Vinet F, Texier I (2011) Proc SPIE 7910:79100W-1–79100W-12
119. Cao T, Munk P, Ramireddy C, Tuzar Z, Webber SE (1991) Macromolecules 24:6300–6305
120. Stepanek M, Krijtova K, Prochazka K, Teng Y, Webber SE, Munk P (1998) Acta Polym 49:96–102
121. Hu Y, Kramer MC, Boudreaux CJ, McCormick CL (1995) Macromolecules 28:7100–7106
122. Chen H, Kim S, He W, Wang H, Low PS, Park K, Cheng JX (2008) Langmuir 24:5213–5217
123. Chen HT, Kim SW, Li L, Wang SY, Park K, Cheng JX (2008) Proc Natl Acad Sci U S A 105:6596–6601
124. Njikang GN, Gauthier M, Li JM (2008) Polymer 49:5474–5481
125. Jiwpanich S, Ryu JH, Bickerton S, Thayumanavan S (2010) J Am Chem Soc 132:10683–10685
126. Ryu JH, Chacko RT, Jiwpanich S, Bickerton S, Babu RP, Thayumanavan S (2010) J Am Chem Soc 132:17227–17235
127. Bickerton S, Jiwpanich S, Thayumanavan S (2012) Mol Pharm 9:3569–3578
128. Chen KJ, Chiu YL, Chen YM, Ho YC, Sung HW (2011) Biomaterials 32:2586–2592
129. Lu J, Owen SC, Shoichet MS (2011) Macromolecules 44:6002–6008
130. Hua P, Tirelli N (2011) React Funct Polym 71:303–314
131. McDonald TO, Martin P, Patterson JP, Smith D, Giardiello M, Marcello M, See V, O'Reilly RK, Owen A, Rannard S (2012) Adv Funct Mater 22:2469–2478
132. Li YP, Budamagunta MS, Luo JT, Xiao WW, Voss JC, Lam KS (2012) ACS Nano 6:9485–9495
133. Li YP, Xiao WW, Xiao K, Berti L, Luo JT, Tseng HP, Fung G, Lam KS (2012) Angew Chem Int Ed 51:2864–2869
134. Javali NM, Raj A, Saraf P, Li X, Jasti B (2012) Pharm Res 29:3347–3361
135. Klymchenko AS, Roger E, Anton N, Anton H, Shulov I, Vermot J, Mely Y, Vandamme TF (2012) RSC Adv 2:11876–11886

136. Morton SW, Zhao X, Quadir MA, Hammond PT (2014) Biomaterials 35:3489–3496
137. Lehn J-M (1999) Chem Eur J 5:2455–2463
138. Lehn J-M (2000) Chem Eur J 12:2097–2102
139. Lehn J-M (2007) Chem Soc Rev 36:151–160
140. Lehn J-M (2012) Top Curr Chem 322:1–32
141. Lehn J-M (2013) Angew Chem Int Ed 52:2836–2850
142. Rowan SJ, Cantrill SJ, Cousins GRL, Sanders JKM, Stoddart JF (2002) Angew Chem Int Ed 41:899–958
143. Belowich M, Stoddart JF (2012) Chem Soc Rev 41:2003–2024
144. Stoddart JF (2012) Angew Chem Int Ed 51:12902–12903
145. Cheeseman JD, Corbett AD, Gleason JL, Kazlauskas RJ (2005) Chem Eur J 11:1708–1716
146. Corbett PT, Leclaire J, Vial L, West KR, Wietor J-L, Sanders JKM, Otto S (2006) Chem Rev 106:3652–3711
147. Cougnon FBL, Sanders JKM (2012) Acc Chem Res 45:2211–2221
148. Ladame S (2008) Org Biomol Chem 6:219–226
149. Nicolai T, Colombani O, Chassenieux C (2010) Soft Matter 6:3111–3118
150. Selby P (1984) Br Med J 288:1251–1253
151. Luqmani YA (2005) Med Princ Pract 14:35–48
152. Baguley BC, Kerr DJ (2002) Anticancer drug development. Academic, San Diego
153. Housman G, Byler S, Heerboth S, Lapinska K, Longacre M, Snyder N, Sarkar S (2014) Cancers 6:1769–1792
154. Miller JN (2005) Analyst 130:265–270
155. Son S, Shin E, Kim B-S (2014) Biomacromolecules 15:628–634
156. Wang B, Chen K, Yang R, Yang F, Liu J (2014) Carbohydr Polym 103:510–519
157. Shen H, Zhou M, Zhang Q, Keller A, Shen Y (2015) Colloid Polym Sci. doi:10.1007/s00396-015-3550-7
158. Zhao Y, Ikeda T (2009) Smart light-responsive materials: azobenzene-containing polymers and liquid crystals. Wiley, New Jersey
159. Crano JC, Guglielmetti R (1999) Organic photochromic and thermochromic compounds. Plenum, New York
160. Horspool WM, Lenci F (2004) Handbook of organic photochemistry and photobiology. CRC, Boca Raton
161. Leamon CP, Low PS (1991) Proc Natl Acad Sci U S A 88:5572–5576
162. Leamon CP (2008) Curr Opin Investig Drugs 9:1277–1286
163. Low PS, Kularatne SA (2009) Curr Opin Chem Biol 13:256–262
164. Kelemen LE (2006) Int J Cancer 119:243–250
165. Zwicke GL, Mansoori GA, Jeffery CJ (2012) Nano Rev 3:18496-1–18469-11
166. Crooke ST (2001) Antisense drug technology: principles, strategies, and applications. Marcel Dekker, New York
167. Xing Q, Li N, Chen D, Sha W, Jiao Y, Qi X, Xu Q, Lu J (2014) J Mater Chem B 2:1182–1189

Top Curr Chem (2016) 370: 61–112
DOI: 10.1007/978-3-319-22942-3_3
© Springer International Publishing Switzerland 2016

Polymeric Nanoparticles for Cancer Photodynamic Therapy

Claudia Conte, Sara Maiolino, Diogo Silva Pellosi, Agnese Miro, Francesca Ungaro, and Fabiana Quaglia

Abstract In chemotherapy a fine balance between therapeutic and toxic effects needs to be found for each patient, adapting standard combination protocols each time. Nanotherapeutics has been introduced into clinical practice for treating tumors with the aim of improving the therapeutic outcome of conventional therapies and of alleviating their toxicity and overcoming multidrug resistance.

Photodynamic therapy (PDT) is a clinically approved, minimally invasive procedure emerging in cancer treatment. It involves the administration of a photosensitizer (PS) which, under light irradiation and in the presence of molecular oxygen, produces cytotoxic species. Unfortunately, most PSs lack specificity for tumor cells and are poorly soluble in aqueous media, where they can form aggregates with low photoactivity. Nanotechnological approaches in PDT (nanoPDT) can offer a valid option to deliver PSs in the body and to solve at least some of these issues. Currently, polymeric nanoparticles (NPs) are emerging as nanoPDT system because their features (size, surface properties, and release rate) can be readily manipulated by selecting appropriate materials in a vast range of possible candidates commercially available and by synthesizing novel tailor-made materials. Delivery of PSs through NPs offers a great opportunity to overcome PDT drawbacks based on the concept that a nanocarrier can drive therapeutic concentrations of PS to the tumor cells without generating any harmful effect in non-target tissues. Furthermore, carriers for nanoPDT can surmount solubility issues and the tendency of PS to aggregate, which can severely affect photophysical, chemical, and biological properties. Finally, multimodal NPs carrying different drugs/bioactive species

C. Conte, S. Maiolino, A. Miro, F. Ungaro, and F. Quaglia (✉)
Department of Pharmacy, University of Napoli Federico II, Via D. Montesano 49, 80131 Naples, Italy
e-mail: quaglia@unina.it

D.S. Pellosi
Department of Chemistry, State University of Maringá, Av. Colombo 5.790, 87020-900 Maringá, PR, Brazil

with complementary mechanisms of cancer cell killing and incorporating an imaging agent can be developed.

In the following, we describe the principles of PDT use in cancer and the pillars of rational design of nanoPDT carriers dictated by tumor and PS features. Then we illustrate the main nanoPDT systems demonstrating potential in preclinical models together with emerging concepts for their advanced design.

Keywords Cancer • Chemotherapy • Drug delivery • Photodynamic therapy • Photosensitizers • Polymeric nanoparticles

Contents

Abbreviations

$^1O_2$	Singlet oxygen
ABC	Amphiphilic block copolymers
AFPAA	Amine-functionalized PAA
AFPMMA	Amine-functionalized polyacrylamide
ALG	Alginate
AuNR	Gold nanorods
c(RGDfK)	Tumor targeting peptide
Ce6	Chlorin E6
CHA2HB	Cyclohexane-1,2-diamino hypocrellin B

CpG-ODN	5′-Purine–purine/T-CpG–pyrimidine–pyrimidine–3′–oligodeoxynucleotide
CS	Chitosan
DOX	Doxorubicin
DR5	Antibody targeting death receptor 5
DTX	Docetaxel
EPR	Enhanced permeability and retention
GA	Glutaraldehyde
GC	Glycol chitosan
HA	Hyaluronic acid/hyaluronan
HB	Hypocrellin B
HMME	Hematoporphyrin
HpD	Hematoporphyrin derivative
HPPH	2-(1-Hexyloxyethyl)-2-devinyl pyropheophorbide A
HSA	Human serum albumin
ICG	Indocyanine green
MB	Methylene blue
MDR	Multidrug resistance
MRI	Magnetic resonance imaging
NIR	Near infrared
NPs	Nanoparticles
PAA	Poly(acrylic acid)
PAAm	Poly(acrylamide)
Pc4	Silicon phthalocyanine
PCL	Poly(ε-caprolactone)
PDEAEMA	Poly(diethylaminoethyl methacrylate)
PDLLA	Poly(D,L-lactic acid)
PDT	Photodynamic therapy
PEG	Polyethylene glycol
PEG-GEL	Poly(ethylene glycol)-modified gelatin
PEI	Poly(ethylenimine)
PheoA	Pheophorbide A
PLA	Poly(lactic acid)
PLGA	Poly(lactic-*co*-glycolic acid)
PLL	Poly(L-lysine)
PMA	Poly(methacrylic acid)
pNIPAM	Poly(*N*-isopropylacrylamide)
PpIX	Protoporphyrin IX
PS	Photosensitizer
PTT	Photothermal therapy
ROS	Reactive oxygen species
TCPP	*meso*-tetra(Carboxyphenyl) porphyrin
THPP	tetra(Hydroxyphenyl)porphyrin
TMP	*meso*-tetra(*N*-Methyl-4-pyridyl) porphine tetratosylate

$TPPS_4$	tetrasodium-*meso*-tetra(4-sulfonatophenyl) porphyrine
UDCA	Ursodeoxycholic acid
ZnPc	Zinc phthalocyanine

1 Introduction

Cancer treatment is currently based on a combination of surgery, radiotherapy, chemotherapy, and, more recently, immunotherapy. Each treatment modality bears advantages and drawbacks and needs to be established depending on tumor location, stage of tumor growth, and presence of metastasis. Chemotherapy is one of the principal modes of treatment for cancer. The main Achilles' heel in a chemotherapeutic regimen lies in poor selectivity of the treatment that generates severe side effects, contributing to decreased patient compliance and quality of life. Indeed, most chemotherapeutics are administered by the intravenous route, distribute in the whole body according to the physical chemical features (which drive interactions with plasma proteins), and reach healthy organs as well as diseased tissue. A fine balance between therapeutic and toxic effects needs to be found for each patient, adapting standard combination protocols each time. Nevertheless, the effectiveness of chemotherapy is limited by intrinsic or acquired drug resistance [1, 2].

In the past 20 years, nanotherapeutics has been introduced in the clinical practice for treating tumors with the aim of improving the therapeutic outcome of conventional pharmacological therapies and alleviating their toxicity, as well as overcoming multidrug resistance (MDR) [3–9]. By providing a protective housing for the drug, nanoscale delivery systems can in theory offer the advantages of drug protection from degradation and efficient control of pharmacokinetics and accumulation in tumor tissue, thus limiting drug interaction with healthy cells and, as a consequence, side effects. The delivery of chemotherapeutics through nanocarriers has been mainly focused on the intravenous route to reach remote sites in the body through the blood system [10]. By exploiting the presence of the dysfunctional endothelium of the tumor capillary wall and the absence of effective lymphatic drainage in solid tumors, nanocarriers can extravasate from the blood circulation and can reach the solid tumor interstitium [11–13]. This mechanism, referred as the Enhanced Permeability and Retention (EPR) effect, is the main determinant in passive targeting [14, 15]. Nanocarrier decoration with ligands that specifically recognize peculiar elements of tumors (receptors on endothelial cells of blood vessels, extracellular matrix, cancer cells) or with magnetically sensitive materials can ameliorate drug specificity, allowing its effective accumulation in a solid tumor, an approach known as active targeting [16, 17]. Nanocarriers can also be designed with exquisite responsiveness to the tumor environment (pH, temperature, redox potential) or external stimuli (light, magnetic field, ultrasound, temperature), which can, in theory, trigger drug release only at tumor level [18–25].

In the attempt to find alternative treatment modalities for cancer, photodynamic therapy (PDT) has emerged as an adjuvant therapy to target neoplastic lesions selectively [26, 27]. PDT consists in the administration (local or systemic) of a photosensitizer (PS) which accumulates in different tissue/cells and, under application of light with a specific wavelength and in the presence of molecular oxygen, produces highly reactive oxygen species (ROS), mainly singlet oxygen ($^1O_2$), finally inducing cell death and tumor regression. Selectivity is achieved partly by the accumulation of the PS in the malignant cells/tissue and partly by restricting the application of the incident light to the tumor area. PS is minimally toxic in the non-irradiated zones, although a such phototoxicity and photosensitivity can occur when PS shows a tropism for organs exposed to daylight (skin, eye) or is topically administered, as in the case of skin cancer. PDT has been approved as a primary treatment option for certain neoplastic conditions including inoperable esophageal tumors, head and neck cancers, and microinvasive endo-bronchial non-small cell lung carcinoma [28]. PDT is also being investigated in preclinical and clinical studies for other cancer types including colon, breast, prostate, and ovarian [28–30].

There are several technical difficulties in the application of PDT in cancer, partly shared by most clinically relevant chemotherapeutics. First is the difficulty in preparing pharmaceutical formulations that enable PS parenteral administration because most existing PSs are hydrophobic, aggregate easily under physiological condition, and somewhat lose their photophysical properties. Second is the selective accumulation in diseased tissues, which is often not high enough for clinical use. A third aspect is related to light-activation of PS that generally occurs at a wavelength where radiation is poorly penetrating and unable to reach deep tissues.

Nanotechnological approaches in PDT (nanoPDT) can offer a valid option to deliver a PS and to solve at least part of these issues. Currently, several nanosized carriers made of different materials, such as lipids, polymers, metals, and inorganic materials, have been proposed in nanoPDT, each type of system highlighting pros and cons [31–34]. This review focuses on polymer-based nanoparticles (NPs) specifically designed for cancer PDT. The main advantages of polymeric NPs lie in the ability to manipulate carrier properties readily by selecting polymer type and mode of carrier assembly [35, 36]. In fact, advances in polymer chemistry make it possible to produce an almost infinite number of sophisticated structures and to engineer these structures in light of a strictly defined biological rationale. As a consequence, not only are those features which affect the distribution of drug doses in the body and interaction with target cells controlled, but also spatio-temporal release of the delivered drug is predetermined.

This review covers current trends and novel concepts in the design of passively, actively, and physically targeted NPs proposed in cancer PDT, focusing on those tested in preclinical studies.

2 Photodynamic Therapy in Cancer Treatment

2.1 Principles of a PDT Treatment

PDT is based on photochemical processes between light and an exogenous PS localized at disease level. These components, well tolerated singly by the cells, generate oxygen-based molecular species exerting a number of effects at cell and tissue level. Mechanistically, a photodynamic reaction consists in exciting PS molecules with light of appropriate wavelength, usually visible (VIS) or near-infrared (NIR), preferentially at PS maximum absorption. PS passes from the ground state to the excited state and can at this stage decay to the ground state with concomitant emission of light in the form of fluorescence. The excited PS may also undergo intersystem crossing to form a relatively more stable and long-lived excited triplet state which can either decay to the ground state or transfer electrons/energy to the surroundings through (1) electron transfer to organic molecules and molecular oxygen in cell microenvironment to form radicals finally giving hydrogen peroxide (H_2O_2) and hydroxyl/oxygen radicals (Type I process) or (2) transfer of energy to molecular oxygen leading to the formation of $^1O_2$ which initiates oxidation of susceptible substrates (Type II process).

Both Type I and Type II reactions can occur simultaneously and competitively, and the ratio between these processes depends on the type of PS used, and on the concentrations of substrate and oxygen. Type II reaction, however, appears to play a central role in cytotoxicity, because of the highly efficient interaction of the $^1O_2$ species with various biomolecules [27]. $^1O_2$ has a lifetime of less than 3.5 μs in an aqueous environment and can diffuse only 0.01–0.02 μm during this period. Nevertheless, it should be taken into account that $^1O_2$ senses the inherent heterogeneity of cell environment and its lifetime can consequently be affected [37]. A natural consequence is that the initial extent of the damage is limited to the site of concentration of the PS [38]. This is usually the mitochondria, plasma membrane, Golgi apparatus, lysosomes, endosomes, and endoplasmic reticulum. The nucleus and nuclear membrane are usually spared and DNA damage is rare. Net ionic charge (from −4 to +4), hydrophobicity, and the degree of asymmetry of PS are reported to play a role in cell uptake and intracellular localization.

2.1.1 Generalities on PS

Although an enormous number of chemical structures have been found to act as PSs, only a handful have proceeded to clinical trial and even fewer are commercially available [39]. PSs are generally classified as porphyrinoids and non-porphyrinoids [40]. Within porphyrinoid-based PSs, first, second, and third generation PSs are reported.

The first generation agent hematoporphyrin derivative (HpD) represents the foundation and the reference for novel PSs. Purified HpD is commercialized as porfimer sodium (Photofrin®), a lyophilized concentrated form of monomeric and oligomeric hematoporphyrin derivatives. Photofrin® is characterized by an absorption band at 630 nm (corresponding to a penetration of about 5–10 mm), a low molar extinction coefficient which in turn demands large amounts of Photofrin® and light to obtain adequate tumor eradication, and a long half-life of 452 h, leading to long-lasting photosensitivity. The time delay between drug administration and the time needed to maximize the tumor to normal cell uptake within the target tissue determines the correct delay for light application. Photofrin®-mediated PDT involves intravenous administration of PS followed by irradiation (100–200 J/cm^2 of red light) 24–48 h later. During this period, Photofrin® is cleared from a number of tissues and remains concentrated at the target site [41].

Second generation PSs have been developed with the aim of alleviating certain problems associated with first-generation molecules such as prolonged skin photosensitization and suboptimal tissue penetration. 5-Aminolevulinic acid (5-ALA) is a prodrug enzymatically converted to the active PS protoporphyrin IX (PpIX) during the biosynthesis of heme. 5-ALA (Levulan®) is now approved for the topical treatment of actinic keratosis (AK) and is in clinical trials for other types of cancer [42]. Because of its poor ability to cross the skin, lipophilic derivatives have been proposed such as 5-methyl-aminolevulinate (Metvix®) and hexyl ester of 5-ALA (Hexvix®) [43].

From the porphyrin family, *meta*-tetra(hydroxyphenyl)porphyrin (*m*-THPP) and 5,10,15,20-tetrakis(4-sulfanatophenyl)-21*H*,23*H*-porphyrin ($TPPS_4$) are the main second generation PDT sensitizers. *m*-THPP, although being 25–30 times more potent than HpD in tumor photonecrosis when irradiated at 648 nm, causes severe skin phototoxicity [40, 44].

Various chemical modifications of the tetrapyrrolic ring of the porphyrins characterize the different groups of the second-generation PSs [45, 46]. They have high absorption coefficients/ $^1O_2$ quantum yields and absorption peaks in the IR (660–700 nm) or NIR (700–850 nm) regions. The serum half-life of these compounds is short and tissue accumulation is improved and occurs quickly (within 1–6 h after injection). Thus, the treatment can be carried out on the same day as the administration of the drug. In addition, the risk of burns by accidental sun exposure is low because clearance from normal tissues is rapid. Finally, toxicity to skin and internal organs in the absence of light (so-called 'dark' toxicity) is absent or minimal.

The chlorin family includes benzoporphyrin derivative monoacid ring A (BPD-MA, Verteporfin, Visudyne®), *meta*-tetra(hydroxyphenyl)chlorin (*m*-THPC, Foscan®), tin ethyl etiopurpurin (SnET2, Rostaporfin, Purlytin™), and *N*-aspartyl chlorin e6 (NPe6, Talaporfin, Ls11) which is derived from chlorophyll a. When compared to porphyrins, the structure of chlorins differs by two extra hydrogens in one pyrrole ring. This structural change leads to a bathochromic shift in the absorption band (640–700 nm) and gives $\varepsilon_{max} \sim 40{,}000\ M^{-1}\ cm^{-1}$. Pheophorbides also have two extra hydrogens in one pyrrole unit and can be derived from chlorophyll. 2-(1-Hexyloxyethyl)-2-devinyl pyropheophorbide A (HPPH, Photochlor®) absorbs at 665 nm with $\varepsilon_{max} \sim 47{,}000\ M^{-1}\ cm^{-1}$.

The joining of four benzene or naphthalene rings to the β-pyrrolic positions of porphyrins and the substitution of the methylene-bridge carbons with nitrogen produce phthalocyanine and naphthalocyanines, respectively. The presence of Al (III), Zn(II), Si(IV), Ru(II), and other diamagnetic metal ions with axial ligands gives hexacoordination and guarantees a satisfactory yield of $^1O_2$ generation, thus decreasing the tendency to form PS self-aggregates and inducing high photodynamic efficiency and reduced phototoxic side effects [46].

Non-porphyrin derivatives, including hypericin, hypocrellins, methylene blue (MB), toluidine Blue, and merocyanine 540, are other potential PSs for cancer PDT [40].

Currently, research efforts are focusing on the development of third generation PSs, characterized by a higher specificity to target cells and minimal accumulation in healthy tissues. The basic approach consists in the conjugation of a PS with a targeting component, such as an antibody directed against the tumor antigens, to promote the localization and the accumulation of the drug at the diseased site [47, 48]. As discussed in the following, the most advanced strategy to ameliorate PS therapeutic outcomes relies in their delivery through engineered nanosystems.

2.1.2 Light Sources

Besides the type of PS, the selection of a light source plays a central role in achieving effective PS excitation in the bioenvironment. PS maximum absorption range, disease location, size of the area to treat, and cost are the main determinants to identify an appropriate illuminating system. Furthermore, the clinical efficacy of PDT is dependent on dosimetry: total light dose, light exposure time, light delivery mode (single vs fractionated or even metronomic), and fluence rate (intensity of light delivery) [49].

The effective excitation light magnitude is determined by the combination of optical absorption and scattering properties of the tissue. Absorption is largely because of endogenous tissue chromophores such as hemoglobin, myoglobin, and cytochromes. On the other hand, the optical scattering of a tissue depends on wavelength. For the spectral range of 450–1750 nm, tissue scattering is, in general, more prevalent than absorption, although, for the range of 450–600 nm, melanin and hemoglobin provide significant absorption, and water plays a similar role for wavelengths >1350 nm. Therefore, the optimal optical window for PDT, and for optical imaging, is in the NIR spectral region (600–1300 nm), where the scattering and absorption by tissue are minimized and, therefore, the longest penetration depth can be achieved. Within this optical window, the longer the wavelength, the deeper the penetration depth. However, light up to only approximately 800 nm can extensively generate $^1O_2$, because longer wavelengths have insufficient energy to initiate a photodynamic reaction [50]. In fact, optical penetration depth of 780-nm light was found to be 3.62 mm in a mammary carcinoma and 2.82 mm in a lung carcinoma [51]. Currently approved PSs absorb in the visible spectral regions below 700 nm, where light penetration into the skin is only a few millimeters, clinically limiting PDT to treat topical lesions. Thus, a PDT treatment is also generally carried out with red light at higher penetration for PSs, which have maximum absorption in the blue region of the absorption spectrum.

Laser systems widely used for treating dermatological conditions allow the selection of a wavelength with a maximal effective tissue penetration of approximately 10 mm, and have been used in combination with all types of PSs [52]. The laser beams can be launched into an optical fiber applicator, enabling light to be delivered directly into internal tumors. These techniques are relatively expensive, require specialized supporting staff, and are space-consuming. It is likely that such systems will eventually be replaced by laser diode arrays, which are very convenient because they can be easily handled, require only a single phase supply, and are also relatively inexpensive. Because they are monochromatic, the choice of laser wavelength becomes crucial as it must be matched with the often narrow absorption band of the PS, with the result that one laser can only be used in combination with one (or a limited number of) PS(s). Lasers at present are the only possible light source to treat malignancies located in sites that can only be reached with optical fibers.

Several PDT treatments use filtered output high power lamps such as Tungsten Filament Quartz Halogen Lamps, Xenon Arc Lamps, and Metal Halide Lamps, especially in clinical settings. In fact, lamps can provide a broad range of wavelengths at reduced fluence rates to avoid thermal effects, not necessarily producing a dramatic increase in the time required for the treatment. A combination of narrowband, longpass, and shortpass filters is often required to select the irradiation wavelength within 10 nm to cut high-power UV radiation and IR emission (causing an undesired increase in the temperature). Because of their broad emission, lamps can be used in combination with several PSs with different absorption maxima within the emission spectrum of the lamp. Moreover, lamps normally also excite the region where photoproducts absorb, thus being responsible for some additional PDT effects. Because of their characteristics, lamps are well suited for treatment of accessible lesions, especially for larger skin lesions (with or without the use of liquid light guides). Moreover, compared to lasers, such sources offer the advantage of being less expensive and easier to handle [49].

Naturally, most of the light sources for PDT application have been developed to optimize the output near the absorption wavelengths of the main PSs. Thus, the tendency of regulatory agencies, such as the US Food and Drug Administration, has been to approve the PS and the light source to be used for its optical excitation.

2.2 Mechanisms of Cancer Cell Death

A PDT treatment is a two-stage process where a PS is administered in the body locally or by intravenous injection. After a certain period, PS accumulates in cancer cells and is activated by application of light at the level of diseased area where biological effects occur (Fig. 1a). Accumulation in solid tumors is especially critical after intravenous administration and largely related to PS physical-chemical

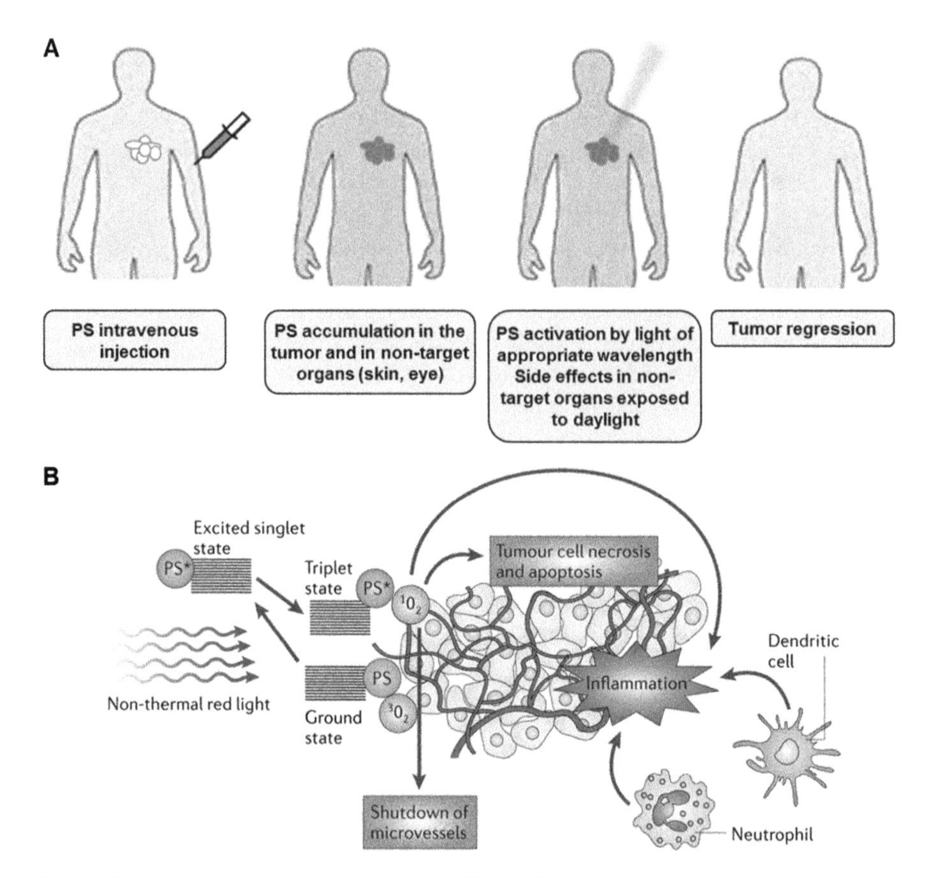

Fig. 1 Treatment of a solid tumor by PDT. (**A**) Steps of an intravenous photodynamic treatment. (**B**) In tumor PDT, PS absorbs light and an electron moves to the first short-lived excited singlet state. This is followed by intersystem crossing, in which the excited electron changes its spin and produces a longer-lived triplet state. The PS triplet transfers energy to ground state triplet oxygen, which produces reactive $^1O_2$ that can (1) directly kill tumor cells by induction of necrosis and/or apoptosis, (2) cause destruction of tumor vasculature, and (3) produce an acute inflammatory response attracting leukocytes. Adapted from [59]

features. PSs probably interact with tumors via low-density lipoprotein (LDL) receptors. Hydrophobic compounds and their aggregates bind to LDL whereas hydrophilic species bind to albumin and globulins [53]. Because cancer cells have elevated levels of LDL receptors, endocytosis of LDL–PS complex is preferred by malignant cells [54, 55]. PS solubility is the main determinant affecting its distribution and location inside tumor cells. Accumulation of PS in the cell organelles also depends on the charge of the sensitizer. Cationic compounds collect in mitochondria, whereas anionic species are found in lysosomes [53]. Dye sensitizers with one or two anionic charges localize in the perinuclear region, vesicles of the cell, and lysosomes, providing multiple sites of PS accumulation [56, 57]. It should be noted that subcellular localization may change with incubation time because PS may relocate to other organelles after illumination.

Although PDT can induce many cellular and molecular signaling pathway events, the final effect is the induction of cell death through the activation of three main cell death pathways: apoptosis, necrosis, and autophagy [27, 38, 58]. The mode and the extent of cell death is related to different elements, including the concentration, the physiochemical properties and subcellular location of the PS, the concentration of oxygen, the wavelength and intensity of the light, and the cell type. For instance, it is recognized that lower doses of PDT lead to more apoptotic cells, whereas higher doses lead to proportionately more necrotic cells [59]. After PDT, cancer cells usually develop a cytoprotective mechanism to limit cytotoxic effects and detoxify from ROS, such as the production of antioxidant molecules (e.g., some amino acids, glutathione, vitamin E) and of enzymes.

Other distinct mechanisms contribute to the reduction or disappearance of tumors after PDT treatment (Fig. 1b). In fact, PDT is also able to damage the tumor-associated vasculature, leading to tumor death via lack of oxygen and nutrients [60]. The higher sensitivity of endothelial cells compared to the other proliferating tumor cells is produced by a greater PS accumulation in the endothelial cells, where biological response occurs at sub-lethal doses of PDT [61, 62]. Although microvascular damage after PDT contributes to greater tumor response, reduction in oxygen during treatment can limit tumor control by inducing the production of proangiogenic markers, creating a favorable environment for tumor recurrence [63, 64].

PDT frequently provokes a strong inflammatory reaction observed as localized edema at the target site caused by oxidative stress. PDT-induced inflammation is orchestrated by innate immune system. The acute inflammation and release of cytokines and stress response proteins induced in the tumor can lead to an invasion of leukocytes contributing both to tumor destruction and to stimulation of the immune system to recognize and destroy tumor cells [59]. Nevertheless, numerous studies have linked PDT to the adaptive immune response. The precise mechanism leading to potentiation vs suppression of adaptive immunity exerted by PDT is unclear as yet; nevertheless, it seems as though the effect of PDT on the immune system is dependent on the PS type, the treatment regimen, and the area treated. Furthermore, recent findings suggest that clinical antitumor PDT can increase antitumor immunity [65, 66].

The relative importance of each mechanism for the overall tumor response is yet to be defined and requires further research. It is clear, however, that the combination of all these components in PDT is required for optimum long-term tumor regression, especially in tumors that may have metastasized.

Finally, two general approaches may increase the antitumor effectiveness of PDT; (1) sensitization of tumor cells to PDT and (2) interference with cytoprotective molecular responses triggered by PDT in surviving tumor or stromal cells [27].

2.3 Clinical PDT for Cancer

PDT has been utilized for preneoplastic and neoplastic diseases in a wide variety of organ systems, including skin, genitourinary, esophagus, prostate, bile duct, pancreas, head and neck, and brain [28]. Several medicines have been approved or are currently in clinical trials (Table 1). At present, 68 open clinical trials on cancer PDT are ongoing.

Successful results for PDT of non-hyperkeratotic actinic keratosis have been achieved with systemically administered porfimer sodium as well as topically applied ALA and methyl-ALA (MAL). Fifty-one randomized clinical trials (RCTs) that reported the use of PDT in the treatment of actinic keratosis have been identified ([67] and clinicaltrial.gov) and aggregated data indicate better rates of complete response and better cosmetic results with PDT than with the other treatments.

Several RCTs on superficial and nodular basal cell carcinoma have been reported, comparing ALA-PDT with surgical excision, cryotherapy, or placebo [68]. In particular, for superficial basal cell carcinoma, the outcome after PDT appears similar to surgery or cryotherapy, whereas for nodular (deep) basal cell

Table 1 Clinically approved PS for cancer PDT

Trade name	Photosensitizer	Structure/ excitation λ	Administration site	Indication
Levulan/ Ameluz	5-Aminolevulinic acid (ALA)	Porphyrin precursor/ 635 nm	Skin	Actinic keratosis (Canada, USA, Europe)
Metvix, Metvixia	Methylester of 5-ALA	Porphyrin precursor/ 635 nm	Skin	Actinic keratosis (Canada, USA)
Photofrin	Porfimer sodium; also called hematoporphyrin derivative (HpD)	Porphyrin/ 630 nm	Intravenous injection	Esophageal, endobronchial, high-grade dysplasia in Barrett's esophagus (USA, Canada)
Foscan	*Meta*-tetrahydroxyphenylchlorin (temoporfin) (*m*-THPC)	Chlorin/ 652 nm	Intravenous injection	Cervical cancer (Japan), esophagus cancer and dysplasia (Canada, EU, USA, Japan), gastric cancer (Japan), advanced head and neck cancer (EU)
Laserphyrin	Mono-(L)-aspartylchlorin-e6 (MACE, NPe6, LS11), (Talaporfin)	Chlorin/ 664 nm	Intravenous injection	Lung cancer (Japan), phase III trials in USA

Adapted from [34]

carcinoma, PDT is less effective than surgery for lesion clearance. Finally, PDT can substantially reduce the size of large squamous-cell carcinoma tumors, reducing morbidity and increasing overall curative response [69].

In the field of head and neck cancer, thousands of patients have been treated with PDT [28, 70] by systemic delivery; in particular, Foscan® was approved in Europe in 2001 for the palliative treatment of patients with advanced head and neck cancer who have exhausted other treatment options. Furthermore, various formulations of porfimer sodium, ALA, and temoporfin are currently undergoing intensive clinical investigation as an adjunctive treatment for brain tumors, such as glioblastoma multiforme, anaplastic astrocytoma, malignant ependymomas or meningiomas, melanoma, lung cancer, brain metastasis, and recurrent pituitary adenomas [71].

PDT is increasingly being used to treat cancers of the airways and other tumors in the thoracic cavity, especially non-small cell lung carcinoma [72, 73]. Different RCTs based on talaporfin or porfimer sodium-mediated PDT showed good results and complete response rate in patients with early stage lung cancer or for whom surgery is not feasible.

In gastroenterology, endoscopically accessible premalignant or malignant lesions located within the esophagus, the stomach, the bile duct, or the colorectum with a high surgical risk have become suitable targets of endoscopic PDT [74]. Photofrin®-PDT has been approved for obstructing esophageal cancer, early-stage esophageal cancer, and Barrett's esophagus in several countries, as an alternative to esophagectomy because these are superficial and large mucosal areas that are easily accessible for light. Recent pilot studies have demonstrated that endoscopic Photofrin®-PDT is also effective in the palliative treatment of cholangiocarcinoma [34, 75], for early duodenal and ampullary cancers, and for advanced adenomas.

Because of advances in light applicators, the interstitial PDT is now becoming a practical option for solid lesions, including those in parenchymal organs such as the liver and pancreas [76, 77]. Talaporfin-mediated PDT may have efficacy in treating hepatocellular carcinoma, whereas Foscan® looked promising in the treatment of pancreatic cancer [78]. In the case of prostate cancer, Foscan® represented a viable minimally-invasive alternative to surgery or radiotherapy, reducing the risk of the post-surgical side effects of incontinence and impotence [79]. Bladder cancer tends to be a superficial condition, and for this reason it is proposed that a superficial treatment with ALA or its ester derivatives by intravesical instillation may be a preferable mean for local therapy [80].

The last PDT application refers to the treatment of gynecological cancers [81]. For cervical intraepithelial neoplasia, PDT based on chlorine e6 (Photolon®) or hexyl-ALA offers a nonscarring alternative to cone biopsy. For vulvar intraepithelial neoplasia, use of Foscan or ALA may ameliorate the need for radical mutilating surgery. Similarly, penile intraepithelial neoplasia and anal intraepithelial neoplasia have been treated with ALA-based PDT, sometimes with complete clearance. Extramammary Paget's disease responds to PDT with porfimer sodium or ALA.

Currently, PSs are being evaluated as intraoperative diagnostic tools both by means of photodetection (PD) and fluorescence guided resection (FGR) during PDT [82]. The most recently published trials that employed PD, FGR, and PDT provided additional encouraging results, but the initial delay in tumor progression did not translate to extended overall survival.

2.4 Combining PDT to Chemotherapy

In a clinical setting, patients treated with anticancer drugs were found to fail the experiences of single agent chemotherapy because it is limited to act on specific cancer survival pathways and showed low response rates and relapse of tumor. To improve the therapeutic potential of cancer chemotherapy, it is essential to establish alternative approaches which could provide a solution to the problems involved in single drug chemotherapy. To this end, much attention has been given to combination approaches for a better long-term prognosis and to decrease side effects associated with high doses of monotherapy. One of the prime benefits of combination therapies is the potential for providing synergistic effects. The overall therapeutic response to drug combinations is generally greater than the sum of the effects of the drugs individually [83]. The best drug combination with maximal antitumor efficacy can be calculated by multiple drug effect/combination index isobologram analysis, an effective way to demonstrate that drugs are working synergistically. The prime mechanism of synergistic effect following combinational drug treatment could act on the same or different signaling pathways to achieve more-favorable outcomes at a lower dose with equal or increased efficacy [84]. Unlike monotherapy, combination therapy can modulate different signaling pathways, maximizing the therapeutic effect while overcoming toxicity and, moreover, can decrease the likelihood that resistant cancer cells develop.

As a complementary therapeutic modality, PDT can be combined with chemotherapy to enhance therapeutic outcome. In PDT, any activity of PDT-sensitizing agents is confined to the illuminated area, thus inducing non-systemic potentiated toxicity of the combinations. This should be of special importance in elderly or debilitated patients who tolerate poorly very intensive therapeutic regimens. Moreover, considering its unique $^1O_2$-dependent cytotoxic effects, PDT can be safely combined with other antitumor treatments without the risk of inducing cross-resistance [85]. Despite this potential, few studies on combinations of PDT with standard antitumor regimens have been published to date [83].

Photochemical internalization (PCI), a specific branch of PDT, is a novel strategy utilized for the site-specific triggered drug/gene release [86–88]. PCI was initially developed at the Norwegian Radium Hospital as a method for light-enhanced cytosolic release of membrane-impermeable molecular therapeutics entrapped in endocytic vesicles. Briefly, the drug or gene of interest colocalizes with a PS in endocytic vesicles. Light-activation of the PS results in ROS-mediated damage of the membranes of these vesicles with subsequent release of the drug or gene to cytosol. This strategy is especially useful for proteins and nucleic acids that are unable to cross biological membranes, and even with a specific delivery system these molecules are taken up by endocytosis and are sequestered in endolysosomal compartments where they are subjected to enzymatic degradation, resulting in lack of biological effect. Furthermore, PCI can facilitate endolysosomal release of anticancer drugs and promote their subcellular redistribution after NP uptake [89]. PCI has been demonstrated to be a feasible drug delivery technology in numerous cancer cell lines and different animal models.

2.5 Drawbacks in Cancer PDT

The efficacy of a PDT treatment depends on multiple factors related to PS photochemical and physicochemical properties ($^1O_2$ production efficiency, tissue penetration of excitation light), PS biodistribution in the body, localization in a specific compartment and dose at target tissue, as well as light parameters (light dose, fluence rate, interval between administration and light exposure). Obviously, cancer tissue characteristics (vascularization, oxygenation level) play an important role in determining the therapeutic outcome of PDT.

Each of the commercially available PSs has specific characteristics, but none of them is an ideal agent. Selectivity remains a key issue in PDT. A PDT treatment can be considered to be selective in that the toxicity to tumor tissue is induced by the local activation of the PS, whereas normal tissues not exposed to light are spared. Second generation PSs show improved selectivity and clearance rate from the body so increased therapeutic efficiency and mostly alleviated toxicity caused by post-PDT photosensitization are experienced. However, Foscan® has failed FDA approval for the treatment of head and neck cancer because of poor tumor selectivity resulting in serious skin burns arising from photosensitivity [90]. Furthermore, most second generation PSs exhibit poor solubility in aqueous media, complicating intravenous delivery into the bloodstream. The low extinction coefficients of PSs often require the administration of relatively large amounts of drug to obtain a satisfactory therapeutic response, thus demanding specific vehicles (Chremophor®, propylene glycol), which can lead to unpredictable biodistribution profiles, allergy, hypersensitivity, and toxicity [91].

Several hydrophobic PSs tend to aggregate in physiological conditions via the strong attractive interactions between π-systems of the polyaromatic macrocycles and, as a consequence, to produce singlet oxygen with very low yields [92]. Aggregation is one of the determining factors which can cause a loss of PS efficacy in vivo by decreasing its bioavailability and limiting its capacity to absorb light [93]. The interactions are affected mainly by the solvent, sample concentration, temperature, and specific interactions with biological structures. Furthermore, the absorption maximum of PSs falls at relatively short wavelengths, leading to poor tissue penetration of light. This has prompted development of alternative strategies to improve quantum yields of $^1O_2$ such as two-photon induced excitation

[82, 94]. This strategy combines the energy of two photons (in the range 780–950 nm) where tissues have maximum transparency to light but where the energy of one photon is not high enough to produce $^1O_2$.

For systemic administration, PS location and extent of PS accumulation in the target tissue depend on post-injection time [41]. At times shorter than PS half-life, the drug predominantly stays in the vascular compartment of the tumor, whereas at longer time, PS can accumulate in extravascular sites because of interstitial diffusion. Therefore, drug-light interval may play a crucial role for the therapeutic outcome. For topical administration there is a need to promote transport through the skin and to accumulate PS in the skin target. In this case there is no need to delay light application except for drugs that need metabolic pathways to become active (such as 5-ALA).

3 Injectable Nanoparticles for Photodynamic Therapy

3.1 *Nanotechnology in Cancer PDT*

Nanotechnology offers a great opportunity in advancing PDT based on the concept that a PS packaged in a nanoscale-carrier can result in optimized pharmacokinetics, enhancing the treatment ability to target and kill cancer cells of diseased tissue/organ while affecting as few healthy cells as possible [95, 96].

Besides improving specificity, nanoPDT is also emerging to surmount solubility issues and the aggregation tendency of PSs, which can severely affect photophysical, chemical, and biological properties. In fact, a carrier specifically engineered for nanoPDT should provide an environment where the PS can be administered in a monomeric form and can also maintain its photochemical properties in an in vivo setting without loss or alteration of photoactivity. Furthermore, a nanocarrier engineered for the therapy of solid tumors is expected to deliver therapeutic concentrations of PS in the diseased tissue and at specific subcellular locations.

NPs can be designed to transport more than one drug/bioactive species with different mechanisms of cancer cell killing. The idea to combine two drugs with different mechanisms of action and pharmacokinetics in a nanocarrier with well-tailored properties can allow control over anticancer drug/PS biological fate and promote co-localization in the same area of the body [97]. This approach is rather recent and demonstrated that cytotoxic drugs can act in concert with PS for tumor killing providing an anticancer synergistic effect, inducing antitumor immunity and sometime reverting MDR.

Another potential application of nanotechnology that has been a research hotspot in the forefront of materials science is the combination of non-invasive PDT and photothermal therapy (PTT) [98]. PTT consists in a NIR irradiation of a photo-absorbing agent which converts electromagnetic energy to local heat producing hyperthermia and subsequently cell death. Generally, noble metal NPs such as gold nanorods coupled with a PS on their surface are used to promote the tumor accumulation and synergistic PDT/PTT. Although high therapeutic outcomes, in this strategy two different wavelength lasers are usually required to allow PDT and PTT because of the absorption mismatch of PS and photothermal agents. Thus, developing a simple and effective strategy for simultaneous PDT and PTT treatment is highly desirable. Recently, researchers have been demonstrated the efficiency of NP loading with a single PS such as chlorins or some phthalocyanines that present a strong NIR absorbance and are capable of both PDT and PTT to kill cancer cells under single wavelength irradiation [99].

Nanocarriers also serve as a multimodal platform to bind/include a great variety of molecules, such as tumor-specific surface ligands for targeted nanoPDT and/or imaging agents integrating in a single platform the unique opportunity for concurrent diagnostic and treatment of cancer tumors, so-called theranostics. Recently, several multifunctional theranostic systems have been developed for real-time imaging-guided PDT of cancer [100, 101].

The general design of a carrier for cancer nanoPDT should be planned on a rational basis in the light of specific needs dictated by (1) tumor features (location, stage, metastatization), (2) selectivity for tumor tissue, which means to accumulating the largest fraction of administered dose at tumor level (cancer cells/tumor interstitium) with little or no uptake by non-target tissue/organs, and (3) stability in the body compartments, withstanding premature disassembly of nanocarrier and release of PS before the target is reached. Rational design is perhaps the most critical step in developing a nanoPDT carrier where a multidisciplinary approach at the interface between chemistry, pharmaceutical technology, biology, and medicine should be planned. In this respect, nanocarrier interactions with the biological environment (protein interaction, blood circulation time, elimination rate, transport through mucus or epithelia, cell internalization just to cite some aspects) can be properly regulated by nanocarrier overall physical-chemical properties (size, surface charge/hydrophilicity, drug loading capacity/release rate).

3.2 Fate of Intravenously Injected Nanocarriers

In analogy to several anticancer drugs, intravenous injection remains the preferred route of PS administration to reach different body compartments. In fact, the unique properties of tumor vasculature and microenvironment result in a natural tendency of a nanocarrier bearing a drug cargo to accumulate in solid tumors referred to as passive targeting [14, 15]. Architectural defectiveness and high degree of vascular density generate abnormal "leaky" tumor vessels, aberrant branching and blind loops of twisted shape. Blood flow behavior, such as direction of blood flow, is also irregular or inconsistent in these vessels. The pore size of tumor vessels varies from 100 nm to almost 1 mm in diameter, depending on the anatomic location of the tumors and the stage of tumor growth. Moreover, solid tumors are characterized by impaired lymphatic drainage which decreases the clearance of locally resident macromolecules. The enhanced permeability and retention (EPR) effect enables nanocarriers to extravasate through these gaps into extravascular spaces and to accumulate inside tumor tissues. Nevertheless, exploiting the EPR effect is complicated by the presence of physiological elimination processes, including both renal clearance and mononuclear phagocyte system (MPS) uptake [14, 15, 102].

Filtration of particles through the glomerular capillary wall (filtration-size threshold) depends on molecular weight and allows molecules with a diameter larger than 15 nm to remain in the circulation [4]. On the other hand, nanocarriers need to escape MPS, which mediates their fast disappearance from blood circulation and accumulation in the MPS organs (liver, spleen, bone marrow). Opsonin adsorption on nanocarrier surface mediates MPS recognition and is considered a key factor in controlling nanocarrier biodistribution in the body [103, 104]. Accumulation in the liver can be of benefit for the chemotherapeutic treatment of MPS localized tumors (e.g., hepatocarcinoma or hepatic metastasis arising from digestive tract or gynecological cancers, bronchopulmonary tumors) but undesirable when trying to target other body compartments. Ideally, an injectable nanocarrier has to be small enough to avoid internalization by the MPS but large enough to avoid renal clearance (100–200 nm). Recent findings highlight that variation of nanocarrier dimension in the scale length >100 nm can heavily affect blood circulation time, whereas the role of geometry in driving in vivo biodistribution has not yet been clarified [105–107].

Although extracellular matrix itself seems not to represent an evident obstacle to NP passage, tissue neighboring tumor cells are surrounded by coagulation-derived matrix gel (fibrin gel or stromal tissues) representing a further barrier to drug transport. Nevertheless, penetration in the remote area of a solid tumor (hypoxic zones) is strictly related to size for drugs, i.e., small drugs penetrate better than a high molecular weight antibody [108].

3.3 Cancer NanoPDT: Biologically-Driven Design Rules

Pharmacokinetics and cell uptake of PSs can be modified by engineering nanocarrier properties (size, surface, shape) to target tumors more specifically, which results in major clinical implications [11, 12, 109]. The general structure of multifunctional NPs for PDT of solid tumors is represented in Fig. 2a. Their rational design relies on appropriate assembling of each building element as dictated by biological requirements.

In order to overcome opsonization, a number of strategies have been investigated to make a nanocarrier "stealth" that is able to evade MPS and long-circulating. Coating with a hydrophilic shell can form a cloud on nanocarrier surface which repels opsonins giving decreased levels of uptake by the MPS and longevity in the blood, finally promoting nanocarrier accumulation in solid tumors through EPR mechanism [110, 111]. To overcome opsonization and rapid

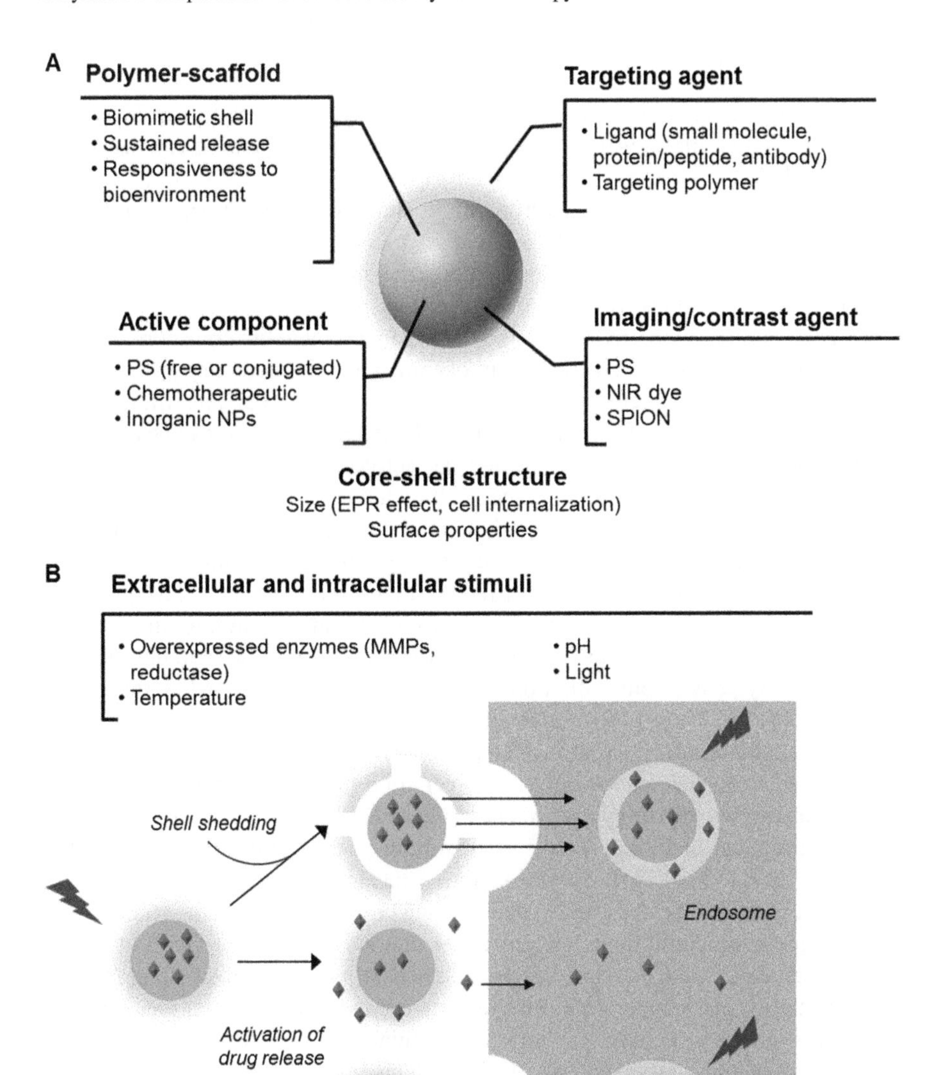

Fig. 2 General structure of multifunctional NPs for PDT treatment of solid tumors. (**A**) Main components of multifunctional NPs. (**B**) Stimuli-sensitive NPs. Extracellular stimuli are suitable for shell shedding (to unmask moieties promoting intracellular transport) and to deliver drug cargo in proximity of cancer cells. Intracellular stimuli can be useful to activate PS intracellularly (dequenching) at specific subcellular locations or to release drug cargo in one pulse

elimination from the bloodstream produced by MPS recognition, coating with a biomimetic shell of polyethylene glycol (PEG) is the most explored strategy to obtain biomimetic long-circulating NPs [112], although other alternative polymers are under investigation [111]. Thus, nanocarriers with hydrophobic surfaces are preferentially taken into MPS organs although long-circulating nanocarriers fulfilling size requirements (less than 100 nm) can accumulate at tumor level. By exploiting passive mechanisms, only a limited nanocarrier fraction can reach the tumor site [13].

Although the presence of a hydrophilic coating allows NP escape from MPS recognition, it decreases the rate and extent of NP uptake inside cancer cells [112, 113]. A strategy to encourage nanocarrier internalization in solid tumors lies in surface decoration with ligands recognizing typical or overexpressed receptors in the tumor microenvironment, which can promote its transport through receptor-mediated endocytosis. Different chemical motifs interacting with specific receptors of endothelial cells in defective tumor vasculature (i.e., integrin receptor) and cancer cells (folate, CD44, transferrin, EGF, and some others) can be exploited for this purpose. This approach, known as active targeting, can aid selective nanocarrier accumulation inside cancer cells while avoiding healthy cells, thus decreasing treatment toxicity. Nevertheless, it has recently been demonstrated that targeted NPs can paradoxically lose targeting ability in a biological environment because of interaction with different high-affinity proteins [103] or can confine their activity to perivascular regions of a tumor (binding site barrier) [114]. Cell cycling also plays a role in NP uptake rate and amount of NPs internalized by cells because of splitting between daughter cells when the parent cell divides [115]. Thus, proper understanding of NP properties at the biointerface is a critical issue needing future investigational efforts [116].

An added sophistication to selective delivery of drug cargo in cancer cells can be brought about by utilizing certain cues inherently characteristic of the tumor microenvironment or by applying certain stimuli to this region from outside the body (Fig. 2b) [18–20, 24]. Stimuli-sensitive nanocarriers based on tailor-made materials can indeed be designed to deliver drug payload sharply and "on demand" by undergoing structural modifications under internal or external stimuli of chemical, biochemical, and physical origin. Internal stimuli typical of solid tumors include mainly pH, temperature, and reductive conditions. In fact, compared to normal/host tissues, pH value in tumor interstitium is lower with an average value of 6.84 because of up-regulated glycolysis producing lactates and protons [22]. Furthermore, once NPs are internalized through endocytic pathways involving lysosomes, pH progressively decreases from early endosomes (pH 5–6) to more acidic late endosomes (pH 4–5) [117], which can strongly alter nanocarrier stability and release features. pH sensitiveness has been widely employed to trigger NP disassembly and drug release [118]. Certain tumor microenvironments are also characterized by mild hyperthermia (1–2 °C above healthy tissues) and some treatment modalities imply rising temperature which, together with pH sensitiveness, can be of help for triggering drug release [25]. In addition, extracellular space is considered oxidative in comparison with intracellular compartment ($\approx$100–1000 folds), mainly in the hypoxic area of tumors, caused by different concentration levels of glutathione [118]. Finally, an array of tumor-associated enzymes, either extracellular or intracellular, can be used as biochemical trigger of drug release to attain a fine control on spatial distribution of the delivered cargo [118].

3.4 Building Carriers for NanoPDT

Polymer-based NPs are matrix-type submicron-sized particles prepared from biodegradable or non-biodegradable materials. In some cases they can be nanocapsules (NCs), where an oily or aqueous core is surrounded by a polymeric shell. The main advantage of polymeric NPs is that their features (size, surface properties, and release rate of drug cargo) can easily be tuned by selecting appropriate materials among a vast number of commercially available candidates as well as by synthesizing novel tailor-made materials [20, 36].

Depending on base material, some NPs are biodegradable/bioeliminable and can be administered by the parenteral route although others are not, and thus are useful only for local applications, assuming that no systemic NP absorption occurs. When dealing with NP entering the systemic circulation, degradability of the polymer is of utmost importance because polymer accumulation in the body above a certain molecular weight can occur. Drug biopharmaceutical properties (solubility, stability, charge, molecular weight, etc.) also guide nanocarrier design and suggest the requirements of specific release features (triggered, sustained) as well as preferential location inside or outside cancer cells.

Numerous materials of synthetic and natural origin have been used to develop PS-loaded NPs. Commonly, the well-defined structure of synthetic polymers results in well-defined and finely-tunable properties of the corresponding NPs. In comparison, natural polymers offer some advantages over synthetic polymers as they are metabolized by enzymes into innocuous side-products. They also take advantage of more than a few drug loading mechanisms including electrostatic attractions, hydrophobic interactions, and covalent bonding. Moreover, NPs from natural polymers offer various possibilities for surface modification caused by the presence of functional groups on the surface of the corresponding NPs, thus enabling conjugation with targeting moieties.

Synthetic polymers commonly employed to produce NPs for drug delivery are reported in Table 2. Polymers that are hydrophobic and insoluble in water form the core of NPs, which can be further modified by depositing one or more layers of hydrophilic polymers, surfactants, or phospholipids to give shells with tailored properties. Instead, hydrophilic polymers can form NP core or shell either by cross-linking or by electrostatic interactions with ions or hydrophilic polymers of opposite charge (nanogels). A further coating of the hydrophilic core with another hydrophilic polymer or a surfactant is possible.

Amphiphilic block copolymers (ABCs) are a class of synthetic materials obtained by the polymerization of more than one type of monomer, typically one hydrophobic and one hydrophilic, so that the resulting molecule is composed of regions with opposite affinities for an aqueous solvent [119, 120]. An advantage of ABCs is their ability to self-assemble in an aqueous environment giving nanostructures spontaneously. Either hydrophobic polymers covalently modified with hydrophilic chains or hydrophilic polymers modified with a hydrophobic moiety (which can also be the active drug) have been synthesized so far. The

Table 2 Main synthetic polymers employed to prepare polymeric NPs for cancer therapy

Polymer type	Location in NPs	Key features	Ref.
Polyesters (PLGA, PLA, PCL)	Core	Hydrophobic and soluble in common organic solvents	[159]
		Biodegraded in the body	
		Encapsulation of hydrophilic/ hydrophobic drugs and macromolecules	
		Protection of drug cargo	
		Sustained release as a function of polymer properties	
PEGylated polyesters (PLGA-PEG, PLA-PEG, PCL-PEG)	Core-shell	Amphiphilic non-ionic copolymers with different segment lengths and architectures forming NPs with a biomimetic/ stabilizing shell	[112, 160, 167]
		Shielding ability depends on molecular weight, architecture and surface density	
		Biodegraded in the body	
		Hydrophilic/hydrophobic ratio and fabrication method control the mode of aggregation (micelles, polymersomes, NPs)	
		Molecular weight affects NP size	
		Molecular weight and hydrophobicity of lipophilic segments affect drug loading and stability inside NPs	
		Conjugation with ligands able to provide NP targeting	
		Low molecular weight copolymers (<500 Da) can revert MDR	
Pluronics	Core-shell	Amphiphilic non-ionic copolymers with different block length and soluble in water	[198, 200]
		Able to form small micelles entrapping hydrophobic drugs above critical micelle concentration	
		Mixed micelles of different Pluronic types can be formed	
		Unimers act on P-gp and allow to overcome MDR	
PAA	Core	Obtained by cross-linking different monomer types to form nanogels	[118, 228]
PAAm		Molecular weight and cross-linker chemistry affect their elimination from the body	
PMA		Some derivatives are protonating/ deprotonating polymers with charge shift from either anionic to neutral or from neutral to cationic	

(continued)

Table 2 (continued)

Polymer type	Location in NPs	Key features	Ref.
PDEAEMA		Acrylic derivatives with hydrazine, hydrazide and acetal linkages swells or collapses for electrostatic reasons and can be employed to get pH-sensitive systems	
PEI/PLL	Shell	Decoration of negatively-charged NPs via electrostatic interactions	[229–231]
		Need of a further polymer coating to shield positive charge of the NP shell	
		For PEI, enhanced tumoricidal capacity of tumor associated macrophages through Toll-like receptor signaling	
pNIPAM and derivatives	Core	Temperature-controlled self-assembly	[25]
		Collapse in the hyperthermic tumor environment	
		Release depending on the MW and nature of the polymer hydrophobic block	
		Encapsulation of both hydrophobic/ hydrophilic drugs	
Polymers sensitive to pH or enzymes - various	Core/ shell	Contain group(s) susceptible to pH variations (hydrazone, hydrazide and acetal) or enzyme degradation (ester or carbamates for proteases, disulfide for reductase)	[22, 118, 228]
		Enzyme- or pH-sensitive sheddable coatings can be designed	

literature abounds with studies encompassing different functional blocks that produce, beside spontaneously formed micelles (spherical, worm-like, crew-cut), an astounding range of other nanoassemblies depending on amphiphile properties [120–125]. The versatility of these materials allows proper design of nanocarriers with specific features depending on the desired application.

Stimuli-responsive polymers, referred to as "environmentally-sensitive," "smart," or "intelligent" polymers, incorporate a chemical motif sharply responding to small changes in physical or chemical conditions with relatively large phase or property changes of the nanocarrier. Over the past 25 years, a huge number of chemical structures and functionalizations have been proposed for numerous biomedical uses [126]. Thus, pH-, redox potential-, and thermo-responsive materials have been applied in the cancer field to build nanocarriers with triggered drug release [24, 118].

PSs can be loaded in NPs through encapsulation, covalent linkage, or post-loading. PSs have also been exposed on NP surface in some NP types. Encapsulation relies on physical entrapment of a PS in NP core or shell based on hydrophobic or electrostatic interactions and hydrogen bond formation. PS loading in the core of

NPs can contribute to achieving a sustained release rate in the biological environment and timing of drug release can be finely tuned by allocating different drugs in the core or the shell, which is especially important in combination therapies. In the post-loading method, PS is added to preformed NPs by equilibrium in solution. The latter method is simple to perform although NPs can suffer premature PS leaching, which can be a drawback for in vivo application. Covalent binding of a PS to NPs is difficult to attain and requires either attachment of a PS to monomers that are then polymerized or self-assembled in NPs or post-modification of preformed NPs. Advantages of this strategy consist in preventing PS leaching from NPs and avoiding PS aggregation in biological environments. Independent of the loading strategy, aggregation of PS inside the matrix needs to be controlled to circumvent loss of PDT efficiency.

From a therapeutic standpoint, timing of drug release is important not only to drive the administration scheme (number of administrations, frequency) but also to optimize the therapeutic outcome. For example, sustained extracellular release can be expected to amplify cell response to some chemotherapeutics and to extend activity to hypoxic zones of certain tumors, resembling a metronomic therapy (subactive doses for longer time frames) [127], whereas responsiveness to external or internal stimuli can be useful to trigger drug release at specific subcellular levels. Nevertheless, timing of drug release can be finely tuned by allocating different drugs in the core or the shell, which is of utmost importance in drug-nucleic acid combination therapies [97, 128]. In all cases, drug amount released from NPs should be reasonably low in the circulation and regulated at tumor level to obtain the optimal therapeutic response.

It should be noted that release of PS from NPs is not considered determinant to achieve a therapeutic effect because molecular oxygen can penetrate polymer matrix and generated $^1O_2$ can diffuse out of NPs to induce photodynamic reactions. In such cases, PDT efficiency depends on NP type (size and oxygen permeability of the matrix) [129].

In general, NPs can be prepared by top-down and bottom-up approaches, each method being useful for a specific material and its combination with others. Bottom-up approaches primarily consist in NP production from monomers or preformed polymers by techniques such as emulsification/solvent evaporation, interfacial deposition after solvent displacement, or salting-out [130–133]. By taking advantage of the unique properties of polymers, such as low melting temperature and the ability to self-aggregate in water, novel preparation methods of NPs based on melting/sonication can be set-up [134]. New approaches, including supercritical technology, electrospraying, premix membrane emulsification, and aerosol flow reactor methods are also under investigation [135]. In the top-down approach, originating from microfabrication tools, monodispersed nanostructures in a range of shapes can be obtained. Among them, particle replication in non-wetting templates (PRINT) technology, involving the use of a nanoscale molds to shape particles, has opened a new avenue to NP production in cancer therapy on an industrial scale [105]. Nevertheless, general principles of applicability of this method to several polymer types and the possibility to engineer surface properties finely are necessary in the near future. It is worth of note that surface properties of NPs are strictly dictated by the production method, which is especially critical when specific targeting elements have to be exposed [136].

The unique nanoscale structure of NPs provides significant increases in surface area to volume ratio which results in notably different behavior compared to larger particles. The stability of colloids, which can be at risk during manufacturing, storage, and shipping, remains a very challenging issue during pharmaceutical product development. To obtain stable NPs, the freeze-drying process is the most useful method for avoiding undesirable changes upon storage. The removal of water from drug-loaded NPs by freeze-drying may be fundamental to avoid the hydrolytic degradation of biodegradable matrix in aqueous suspension and to prevent drug leaching [137]. However, freeze-drying can promote NP aggregation and alter their properties after redispersion in pharmaceutical vehicles. Some sugars such as trehalose, glucose, sucrose, fructose, and sorbitol may be used as cryoprotectants to minimize NP instability upon freeze-drying, preventing their aggregation and protecting them from the mechanical stress of ice crystals. Physical and chemical stability of drug-loaded NPs, including their mechanisms and corresponding characterization techniques, as well as a few common strategies to overcome stability issues, have been reviewed recently [138].

In the following sections we describe different types of nanoPDT polymeric systems, highlighting novel trends in design and specific features achieved.

4 NPs Developed for NanoPDT

4.1 Polysaccharide NPs

Polysaccharides extracted from natural sources or prepared by microorganisms represent the most diffused example of natural polymers employed in the biomedical field. Their use as biomaterials has become much more common as new biological functions are identified. The array of materials that can be investigated has also increased because of new synthetic routes that have been developed for modifying polysaccharides. Their biodegradability, processability, and bioactivity also make polysaccharides very promising natural biomaterials in nanoPDT (Table 3).

Chitosan (CS) is considered one of the most widely used biopolymers for NP preparation because of its unique structural features. CS is a cationic polysaccharide composed of randomly located units of D-glucosamine and *N*-acetylglucosamine. CS is insoluble in water at neutral and basic pH conditions because it contains free amino groups. In contrast, in acidic pH conditions, CS is soluble because the amino groups can be protonated. CS can be cross-linked with various cross-linking agents, such as glutaraldehyde, sodium tripolyphosphate, and geneipin, to provide a hydrated network where drug molecules can be entangled. Its properties make possible the combination with other anionic polymers to provide polyionic

Table 3 NPs made of natural polymers

Polymer	PS/2nd drug/ imaging agent	Intended use	Stage of development	Main finding	Ref.
CS/ALG	TMP (core)	Therapy	In vitro	DR5 antibody-conjugated NPs demonstrated improved TMP uptake and phototoxicity in HCT116 colorectal carcinoma cells	[142]
5β-Cholanic acids-GC or GC-C6	Ce6 (core)	Therapy	In vivo (intravenous injection)	Ce6 was physically entrapped or conjugated	[145]
				Faster release of C6 from HGC NPs	
				In athymic nude mice bearing a subcutaneous xenograft of HT-29 human colorectal adenocarcinoma cells, GC-Ce6 show a prolonged circulation time and efficient accumulation in the tumor resulting in excellent therapeutic efficacy	
CS-UDCA	Ce6 (shell)	Therapy	In vitro	NPs entrapping quenched Ce6 significantly enhance PS uptake and give higher phototoxicity compared with free Ce6 in HuCC-T1 cholangiocarcinoma cells	[144]
GC-SS-PheoA	PheoA (core)	Therapy	In vivo (intravenous injection)	PheoA fluorescence is quenched in NPs	[146]
				Photoactivity is restored in HT-29 human colorectal adenocarcinoma cells because of the dissociation of the self-assembled nanostructure	
				Tumor volume significantly decreased in BALB/c nude mice bearing a subcutaneous xenograft of HT-29 cells treated with NPs compared to free PheoA	
GC iodinated	Ce6 (core)	Therapy	In vivo (intravenous injection)	Iodinated NPs enhanced $^1O_2$ photogeneration compared to their non-iodinated counterpart	[147]
				NPs exhibit high tumor targeting capability in BALB/c nude mice bearing a subcutaneous xenograft of SCC-7 squamous carcinoma cells	

CS cross-linked with GA	ICG (core)/ AuNR	Combined PTT/PDT	In vivo (intravenous injection)	Superior antitumor effect of PDT/PTT combination compared to separate treatments in male ICR mice bearing a subcutaneous xenograft of H22 hepatocellular carcinoma cells	[148]
HSA	IR780 (core)	Combined PTT/PDT	In vivo (intravenous injection)	NPs have the ability to target tumor and can simultaneously generate heat and ROS after laser irradiation at 808 nm	[153]
Apoferritin	MB (core)	Therapy	In vitro	MB is encapsulated successfully within apoferritin nanocages through a pH-controlled dissociation and reassembly process	[158]
				The nanocomposites effectively generate $^1O_2$ following irradiation	
PEG-GEL or PEG-GEL/PLA	HB or CHA2HB (core)	Therapy	In vivo (intravenous injection)	Cellular uptake of HB is increased in Daltons Lymphoma Ascites (DLA) cells	[156, 157]
				Improved PDT response in a xenograft model of DLA after photoirradiation	

complexes, improving the performance of the base material [139]. CS NPs can be created by emulsion cross-linking, emulsion-solvent extraction, emulsification solvent diffusion, emulsion droplet coalescence, ionotropic gelation, complex coacervation, reverse microemulsion techniques, and self-assembly [140]. Physiologically, lysozyme is the primary degrading enzyme and CS degradation rate is dependent on the degree of acetylation and crystallinity [141].

Because of cationic surface, CS-based NPs are especially suited to entrap hydrophilic PSs with the final aim to improve their cell uptake as demonstrated for the hydrophilic *meso*-tetra(*N*-methyl-4-pyridyl) porphine tetra tosylate (TMP) [142]. NPs of 560 nm in diameter were endocytosed into HCT116 colorectal carcinoma cells and elicited a more potent photocytotoxic effect than the free drug. To improve NP specificity toward cancer cells, surface-conjugation of an antibody to DR5, a cell surface apoptosis-inducing receptor up-regulated in various types of cancer, was demonstrated to enhance uptake and cytotoxicity further.

Recently, several hydrophobically modified CS derivatives able to self-assemble in NPs with a positively-charged shell entangling hydrophilic negatively-charged PSs and a hydrophobic core accommodating poorly soluble drugs have been reported [143]. It is envisaged that this feature can allow efficient delivery of multiple drugs with different physicochemical properties. When loading chlorin E6 (Ce6) in NPs fabricated from CS modified with ursodeoxycholic acid, fluorescence quenching was observed in aqueous solution. Surprisingly, Ce6 uptake into HuCC-T1 cholangiocarcinoma cells, phototoxicity and ROS generation were enhanced compared to free Ce6 [144], suggesting Ce6 photoactivation only takes place in a biological environment.

In another example, the importance of premature drug leaching from hydrophobically modified CS NPs on in vivo performance has been demonstrated [145]. In a comparative study, Ce6 was loaded into the hydrophobically-modified glycol CS-5-beta-cholanic acid conjugate (HGC-Ce6) or conjugated to glycol CS (GC-Ce6) to form NPs with similar average diameters (300–350 nm), similar in vitro $^1O_2$ generation, and rapid uptake in SCC-7 squamous-cell carcinoma cells. When intravenously injected into tumor-bearing mice, HGC-Ce6 did not accumulate efficiently in tumor tissue because of premature Ce6 release, although GC-Ce6 showed a prolonged circulation profile, a more efficient tumor accumulation, and high therapeutic efficacy.

Implementation of CS NPs was attempted with the aim of attaining responsiveness to a reductive tumor environment. Self-assembling NPs made of glycol CS (GC) with reducible disulfide bonds conjugated with pheophorbide A (GC-SS-PheoA) were designed [146]. As shown in Fig. 3, the photoactivity of NPs in an aqueous environment was greatly suppressed by the self-quenching effect, which enabled the PheoA-SS-CC NPs to remain photo-inactive. NPs were internalized in HT-29 human colorectal adenocarcinoma cells and dissociated instantaneously by reductive cleavage of the disulfide linkers. The following efficient dequenching process resulted in effective photodynamic activity on HT-29 cells. In subcutaneous tumor-bearing mice, NPs presented prolonged blood circulation, demonstrating

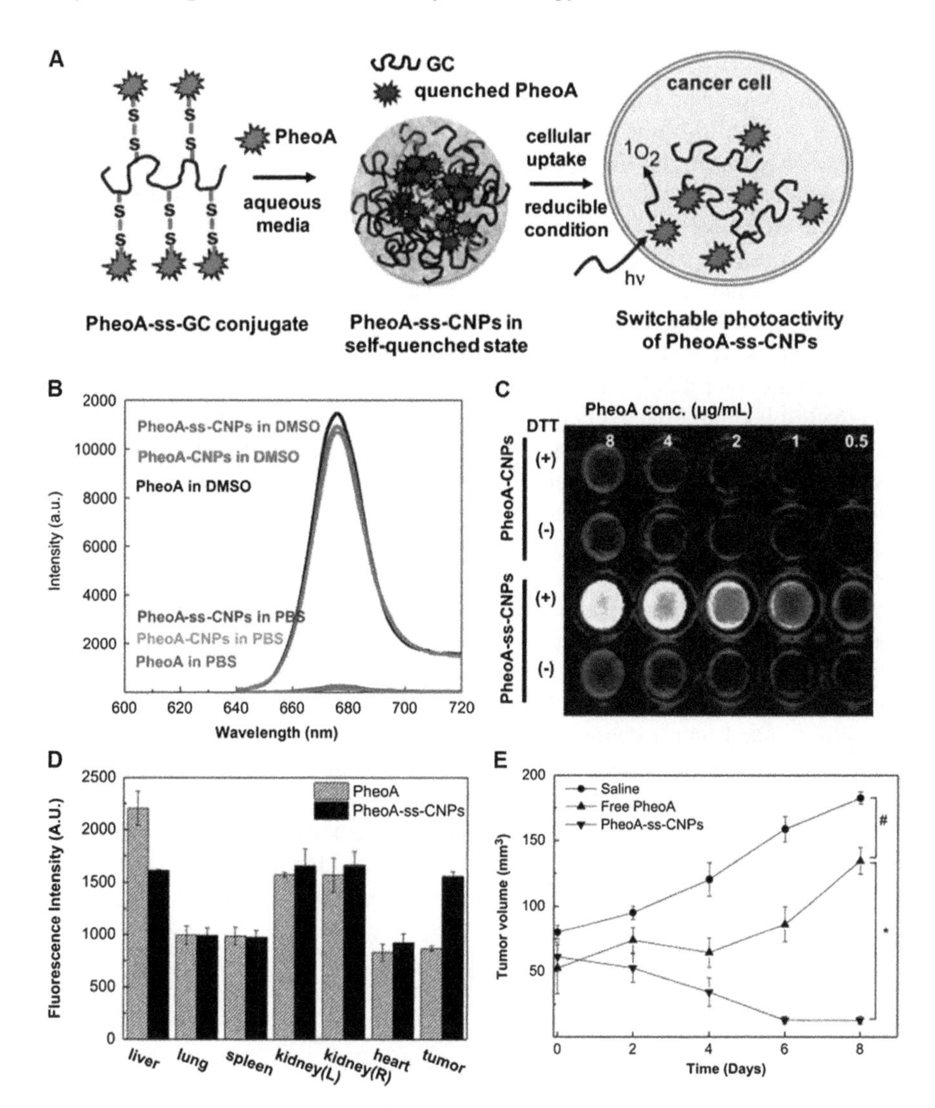

Fig. 3 Bioreducible chitosan NPs for switchable photoactivity of PheoAA. (**A**) PheoA is conjugated to glycol chitosan (GC) through reducible disulfide bonds (PheoA-ss-GC). (**B**) Self-quenching and dequenching of free PheoA, PheoA–NPs, and PheoA-ss-NPs in different solvents. (**C**) NIR images of PheoA–NPs and PheoA-ss-NPs in PBS with (+) or without (−) DTT solution. (**D**) Ex vivo fluorescence photon counts of tumor and organs. (**E**) Tumor growth of HT-29 tumor-bearing mice treated with PheoA and PheoA-ss-CNPs under irradiation. (#,*$p < 0.01$). Adapted from [146]

enhanced tumor specific targeting behavior through the EPR effect, and superior antitumor effects compared to free PheoA.

It has been demonstrated that iodine-concentrated nanoformulations can enhance the $^1O_2$ generation efficiency because of the intraparticle heavy-atom effect that facilitates intersystem crossing of the photoexcited PS from the singlet state to the long-lived triplet state. On this basis, a CS densely conjugated with diatrizoic acid (3,5-bis(acetamido)-2,4,6-triiodobenzoic acid) as an iodine-rich hydrophobic pendant and Ce6 (GC-I-Ce6) was synthesized and used to fabricate self-assembled polymeric NPs [147]. Actual improvement in the photodynamic efficacy of MDA-MB-231 human breast cancer cells demonstrated the potential of the hybrid bioconjugate approach in therapeutic applications.

Multifunctional hybrid NPs made of a glutaraldehyde-cross-linked CS entrapping indocyanine green (ICG) and gold nanorods (AuNR) were successfully prepared and used for combined PDT/PTT with a single irradiation [148]. It was found that the hybrid NPs with a spherical size of 180 nm and a broad adsorption from 650 to 900 nm effectively entrapped ICG and protected it from rapid hydrolysis. In vivo NIR imaging and biodistribution demonstrated that ICG and AuNR could be delivered to the tumor site with high accumulation. With the irradiation by 808 nm laser, CS hybrid nanospheres were able to produce simultaneously sufficient hyperthermia and ROS to kill cancer cells at irradiation sites, resulting in complete tumor disappearance in most tumor-bearing mice.

Among polysaccharides, alginates are linear polyanionic block copolymers composed of 1-4-linked β-D-mannuronic acid and α-L-guluronic acid with recognized biocompatibility and bioresorption properties. Because of ionic interactions with divalent ions, they can form cross-linked hydrated NPs. Alginate-docusate NPs cross-linked with calcium ions and encapsulating MB through electrostatic interaction have been developed [149, 150]. It was demonstrated that these NPs facilitate charge transfer and Type I reaction which is less sensitive to environmental oxygen concentration. NPs led to an increased production of ROS under both normoxic and hypoxic conditions and were able to eliminate cancer stem cells under hypoxic conditions, an important aim of current cancer therapy [151]

4.2 Protein NPs

The presence of multiple sites able to accommodate hydrophobic drugs has pointed to human serum albumin (HSA) as an interesting option to deliver hydrophobic PSs. HSA is an abundant plasma protein that is positively-charged, acidic, and multifunctional [152]. It is produced by extraction from plasma as an amphoteric, globular protein which maintains its structure in the pH range of 4–9, is soluble in 40% ethanol (an important parameter which is of great importance to albumin production processes such as cold ethanol fractionation), and can resist denaturation when heated at 60 °C for over 10 h [140]. Albumin NPs can be prepared by pH-induced desolvation which can include cross-linking by glutaraldehyde molecules, thermal and chemical treatments under emulsification, and self-assembly. NP albumin-bound (Nab)-technology has also been applied to create Abraxane® (albumin-bound paclitaxel NPs) which is currently used in clinics to deliver paclitaxel in breast, pancreatic, and lung cancers. Some examples of nanoPDT systems based on HSA, gelatin, and apoferritin are listed in Table 3.

IR780 iodide, a NIR dye for cancer imaging, PDT, and PTT were loaded into HSA NPs through protein self-assembly [153]. Compared to free IR-780, the solubility of HSA-IR780 NPs was greatly increased (1,000-fold), although a 10-fold decreased toxicity was observed. As illustrated in Fig. 4, both PTT and PDT could be observed in HSA-IR780 NPs, as determined by increased temperature and enhanced generation of $^1O_2$ after laser irradiation at a wavelength of 808 nm. In vivo studies also showed a great tumor inhibition in mice bearing a subcutaneous xenograft of CT26 colon adenocarcinoma cells.

Gelatin is another natural polymer tested for PS delivery. Gelatin is the result of acid or base catalyzed hydrolysis of collagen and its physiochemical properties depend upon the hydrolysis method employed. One of the most important properties of gelatin is its ability to form a thermally reversible gel in water under a variety of pH, temperature, and/or solute conditions [154]. This property is readily utilized by the largest part of the gelatin-based NP encapsulation methods. Gelatin NPs can be created by the water-in-oil emulsification process, the desolvation process, and the two-step desolvation process [140]. The presence of free amino groups on their surface is advantageous for surface modification with target molecules [155].

NPs formulated from biodegradable and natural gelatin were investigated for their potential to enable efficient delivery and enhanced efficacy of a well-known photodynamic agent, Hypocrellin B (HB) [156]. The HB-loaded PEG-conjugated gelatin NPs (HB-PEG-GNP), prepared by a modified two-step desolvation method, exhibited near-spherical shape, with particle size around 300 nm, and demonstrated characteristic optical properties for PDT. NPs tested for cell uptake on Daltons' Lymphoma Ascites (DLA) cells demonstrated dose-dependent phototoxicity upon visible light treatment, and induced mitochondrial damage leading to apoptotic cell death. Biodistribution measurements in solid tumor-bearing mice revealed that NPs reduce liver uptake and increase tumor uptake with time. In vivo PDT studies showed markedly significant regression for HB-PEG-GNP treated mice in contrast to those treated with free HB. In a subsequent study, polylactic acid (PLA) was added to PEGylated gelatin with the aim of better controlling release features of an HB derivative (cyclohexane-1,2-diamino hypocrellin B, CHA2HB) [157]. PS release was observed in normal conditions, whereas enzyme assistance resulted in a relatively fast release because of partial disintegration of CHA2HB-loaded PEG-GEL/PLA NPs. In vitro experiments indicated that NPs were efficiently taken up not only by Dalton's lymphoma cells but also by MCF-7 human breast adenocarcinoma and AGS human gastric sarcoma. Interestingly, PDT effectiveness was different for the different cell type studied and induced both apoptotic and necrotic cell death as a result of photoirradiation.

Another protein, apoferritin, has been proposed as a natural nanocage for PSs. Taking advantage of the fact that apoferritin nanocages can be disassociated into subunits at low pH (2.0) and the subunits reconstitute in a high pH (8.5) environment, a novel encapsulation approach has been proposed. As a model, MB was successfully encapsulated in apoferritin via a dissociation-reassembly process

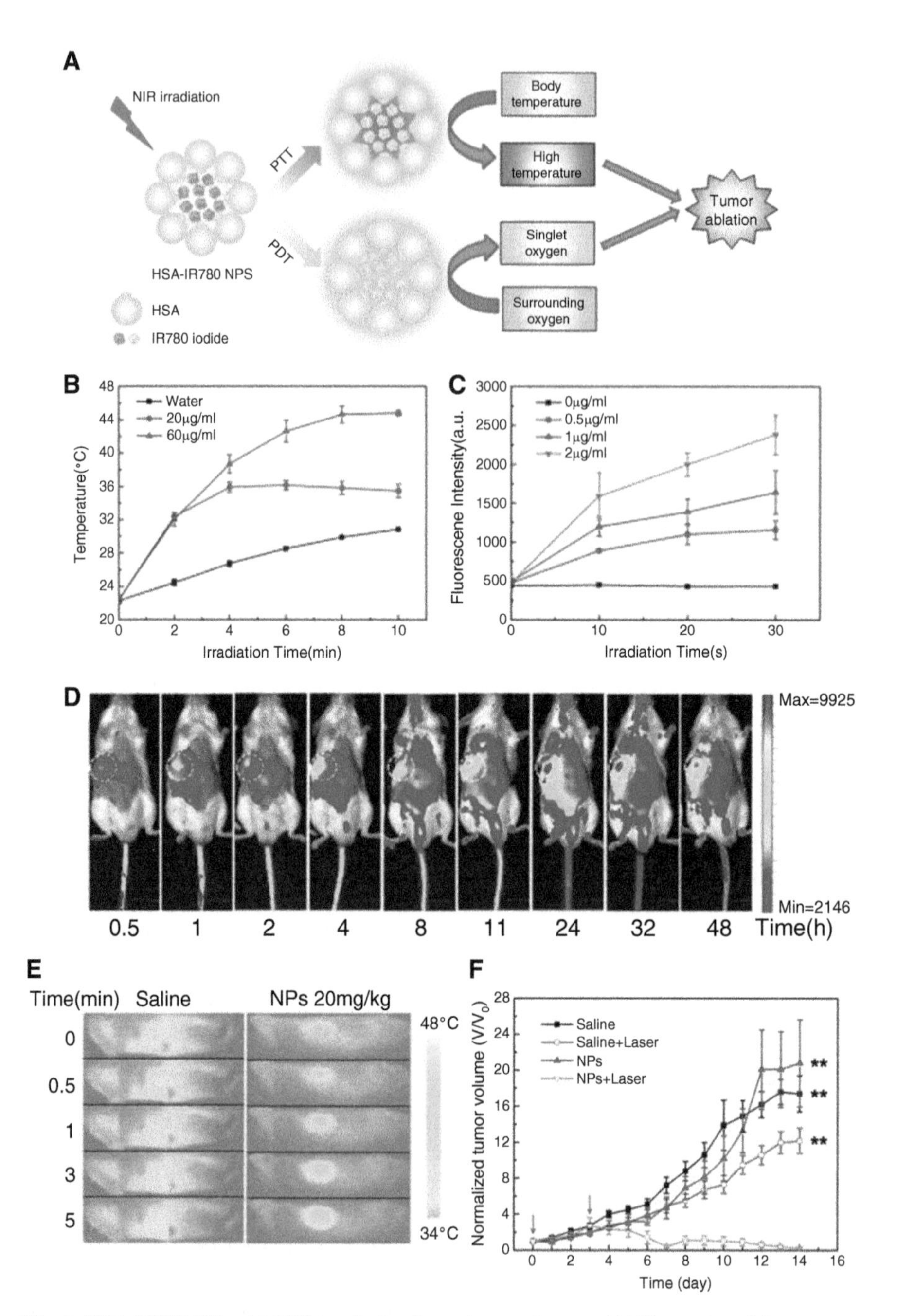

Fig. 4 HSA-IR780 NPs with NIR irradiation for antitumor therapy. (**A**) Illustration of the concept of multifunctional NPs. (**B**) Heating curves of water and HSA-IR780 NPs solutions. (**C**) Fluorescence intensity of $^1O_2$ sensor green (SOSG) combined with HSA-IR780 NPs solutions at different concentrations exposed to 808 nm laser irradiation (1 W cm^{-2}). (**D**) NIR images and (**E**) IR thermal images of tumor-bearing mice intravenously administered with HSA-IR780 NPs and saline. Mice were exposed to an 808-nm laser (1 W cm^{-2}) at 24 h post-injection. (**F**) Tumor

controlled by pH [158]. The resulting MB-containing apoferritin nanocages showed a positive effect on $^1O_2$ production, and cytotoxic effects on MCF-7 human breast adenocarcinoma cells when irradiated at the appropriate wavelength.

4.3 Polyester NPs

In the last 30 years, particular attention has been focused on nanocarriers based on biodegradable polyesters such as poly(ε-caprolactone) (PCL), PLA from D- and/or L-lactic acid monomers (PLLA, PDLLA) and copolymers of lactic acid with glycolic acid (PLGA) because of their better safety profile (degradation products are water and carbon dioxide). Their use has been approved by the regulatory agencies in implantable devices and in injectable products (implants, microspheres). As core-forming polymers, polyesters are appropriate to form NPs with sustained delivery features of the incorporated drug. Indeed, polymer molecular weight and crystallinity, along with the presence of more or less hydrophobic monomers, are the key properties to control both encapsulation efficiency and biodegradation rate, which in turn allows a fine tuning of drug delivery rate [159, 160]. A molecule entrapped in a PLA or PLGA matrix is protected from inactivation occurring in the biological environment and is slowly released in the milieu as a function of diffusion through matrix micropores and degradation of the polymer itself. A drug burst followed by a slow diffusion phase and a fast erosion phase in the time-window of months is observed. By regulating polymer features in term of monomer composition and molecular weight, a large variety of materials with different degradability can be obtained.

Biodegradable polymeric NPs have received tremendous attention for delivering PSs because of their excellent biodegradability, capacity of high drug loading, the possibility of controlling drug release rate, and the existence of a large variety of derivatives (Table 4).

The simplest type of polyester-based NPs developed for PDT application consists in PLGA or PCL hydrophobic NPs entrapping hydrophobic PSs. Early studies highlighted that PDT response in both in vitro and in vivo cancer models is strictly related to NP size [161, 162], cellular internalization pathways [163, 164], and timing/dosing of light exposure [165]. The time interval between NP administration and light irradiation is a determinant for therapeutic efficacy, because it is related to the time needed for NPs to biodistribute in the body. As an example, PLGA NPs delivering SL052, a hypocrellin-based photosensitizer, induced a higher tumor cure rates in syngenic C3H/HeN mice bearing a subcutaneous xenograft of SCC-7 squamous carcinoma cells with a drug-light interval of 4 h compared to 1 h

Fig. 4 (continued) growth of mice bearing CT26 tumor after various treatments as indicated. $**P < 0.01$, compared to the NPs with laser group. Adapted from [153]

Table 4 NPs made of polyesters

Polymer	PS/2nd drug/imaging agent	Intended use	Stage of development	Main finding	Ref.
PLGA	ZnPc (core)	Therapy	In vivo (intratumoral injection)	Smaller mean tumor volume, increased tumor growth delay and longer survival in mice bearing a subcutaneous xenograft of Ehrlich's Ascites Carcinoma cells for ZnPc-loaded NPs compared with free ZnPc	[166]
PLGA	SL052 (core)	Therapy	In vivo (intravenous injection)	NPs show a stronger PDT efficacy compared to a liposome formulation in syngenic C3H/HeN mice bearing subcutaneous SCCVII squamous cell carcinoma	[165]
				Drug-light interval of 4 h produces tumor cure rates higher compared to 1 h interval	
PLGA	TCPP (core)	Therapy	In vivo (intravenous injection)	NPs are internalized by SW480 colon cancer cells through clathrin-mediated endocytosis and induce effective PDT in SW480 xenografts	[163]
c(RGDfK)-PLGA	MB (core)	Therapy	In vivo (intravenous injection)	Targeted NPs are selectively taken up by cancer cells and generate $^{1}O_2$, with shell rupture and PS release	[186]
				In glioblastoma bearing mice, tumor growth is completely inhibited and the tumor eliminated after 7 days of treatment	
PLGA@lecithin-PEG	ICG (core) Lipophilic/DOX	Combined drug therapy	In vivo/intratumoral injection	The combined treatment of NPs with laser irradiation synergistically induced the apoptosis and death of DOX-sensitive MCF-7 and DOX-resistant MCF-7/ADR human breast cancer cells	[148]
				Suppressed MCF-7 and MCF-7/ADR tumor growth also in vivo with no tumor recurrence after only a single dose of NPs	

PLGA@HA	$TPPS_4$ (shell)/DTX (core)	Therapy	In vitro	$TPPS_4$ completely aggregates when associated to NPs	[188]
				$TPPS_4$ uptake is greatly increased in MDA-MB-231 human breast cancer cells overexpressing CD44 receptor	
				Improved cytotoxicity of DTX/$TPPS_4$-NPs compared to single drugs in the same cell line	
PCL	ZnPc (core)	Therapy	In vitro	Concentration, light dose and time-dependent response in A549 human lung adenocarcinoma cells	[164]
PEG-PLA	THPP (core)	Therapy	In vitro	Micelles exhibit fluorescence and photodynamic activity against head and neck cancer cells in vitro depending on formulation parameters and PS actual loading	[172, 173]
PEG-PDLLA	HMME (core)/DOX	Combined drug therapy	In vitro	Nanovesicles show a strong synergistic cytotoxic effect against HepG2 hepatic cancer cells through apoptotic mechanisms	[180, 189]
PEG-PLGA	*m*-THPC (core)	Therapy	In vivo (intravenous injection)	PEGylation significantly increased release of *m*-THPC monomers and reduced the uptake in U937 cells	[176]
				Dark cytotoxicity of *m*-THPC delivered by NPs is less than that of *m*-THPC in the commercial formulation	
PEG-PLGA	ZnPc (core)/CpG (shell)/ gold NPs (shell)	Combined drug therapy	In vitro	Treatment of mouse bone marrow-derived dendritic cells with hybrid NPs showed that the combination of PDT with immunostimulant CpG results in a synergistic immune response, useful for the treatment of metastatic breast cancer	[182]
PEG-PCL	Pc4 (core)	Therapy	In vitro	PS encapsulated in micelles is internalized in MCF-7 human breast cancer cells and co-localized in mitochondria and lysosomes thus inducing cell death through apoptosis	[174]
PEG-PCL	PheoA (core)	Therapy	In vitro	PS is monomeric in micelle core and able to generate $^1O_2$ which may diffuse outside and then induce effective cytotoxicity in MCF-7 human breast cancer cells	[175]

(continued)

Table 4 (continued)

Polymer	PS/2nd drug/imaging agent	Intended use	Stage of development	Main finding	Ref.
PEG-PCL	ZnPc (core)/DTX(core)	Combined drug therapy	In vivo (intravenous injection)	NPs were able to slowly release the chemotherapeutic drug and to produce $^1O_2$ inducing a synergic anticancer effect in an animal model of orthotopic amelanotic melanoma	[179]
GE11 peptide-PEG-PCL	Pc4 (core)	Therapy	In vivo (intravenous injection)	EGFR-targeted micelles are selectively uptaken by A431 epidermoid carcinoma cells thus causing cell death depending on specific photoirradiation parameters	[184, 232]
				High intratumoral NP concentration and post-PDT significant response is found in head and neck SCC15 xenografts models	
PEG-PLA	PpIX (core)	Therapy	In vitro	Micelles with lower PpIX loading density show brighter fluorescence and higher $^1O_2$ yield than those with higher PpIX loading density whereas PDT efficacy in H2009 lung cancer cells shows an opposite trend	[177]
PEG-PLA-PpIX					
PLGA-Ce6/PEG-PLGA	Ce6 (core)/Iron oxide	Theranostic	In vivo (intravenous injection)	NPs exhibit an improved in vivo MRI luminescence imaging and PDT into female nude mice bearing a subcutaneous xenograft of KB human nasopharyngeal epidermal carcinoma cells	[181]
PEG-PCL-Chlorine	Chlorine (core)/SN38	Combined drug therapy	In vivo (intravenous injection)	Prolonged plasma residence time of micelles allowing increased tumor accumulation	[178]
				After light activation, micelles synergistically inhibit tumor growth in a subcutaneous xenograft of HT-29 human colorectal adenocarcinoma cells, resulting in up to 60% complete regression after three treatments	
				Decrease of microvessel density and cell proliferation within the a subcutaneous xenograft	

[165]. When injecting zinc(II) phthalocyanine (ZnPc)-loaded PLGA NPs intratumorally into mice bearing an Ehrlich's Ascites Carcinoma, good PDT outcome was observed in term of tumor growth and survival compared to free ZnPc [166].

PEG is a hydrophilic polymer which can be chemically conjugated to polyester forming ABCs able to provide a vast variety of nanostructures (micelles, NPs, polymersomes, filomicelles) and to entrap hydrophobic and hydrophilic drugs [167]. The opportunity to tune the length of the single chains, the chemical composition of polyester segments, and the arrangement of PEG-polyester segments (diblock, triblock, star-shape) has allowed the building of a wide range of nanocarriers designed with specific delivery requirements. As far as spontaneous self-assembly is concerned, hydrophilic/lipophilic balance, copolymer molecular weight, and properties of the core (crystallinity, hydrophobicity) strongly affect critical micelle concentration, thereby controlling micelle disassembly in biological media [168]. When employing PEGylated polyesters to form core/shell NPs, the conformation of PEG on the surface is dictated by PEG molecular weight, architecture, and surface density [110, 112, 169].

PEGylated polyesters nowadays represent one of the most promising classes of copolymers to translate nanoncologicals in a clinical setting because of excellent biocompatibility and chemical versatility [170, 171]. Polymeric micelles of PDLLA-PEG (Genexol-PM) represent the first polyester system for passive targeting of taxanes approved in Korea in 2006 as a first-line therapy for metastatic breast and non-small cell lung cancer (Phase III) and are currently being evaluated in the USA in a Phase II study on metastatic pancreatic cancer.

Several types of NPs made of PEGylated polyesters, such as PEG-PCL, PEG-PLGA, and PEG-PLA, have been tested in nanoPDT as delivery system for hydrophobic PSs (Table 3) based on the concept that PEGylated NPs exhibited therapeutically favorable tissue distribution compared to non-PEGylated counterparts in delivering PSs in vivo [172–176].

Different entrapment strategies to incorporate PSs into NPs can severely affect photochemical profile. For instance, Ding et al. [177] demonstrated that PpIX loaded in the core of PEG-PLA micelle nanocarriers were monomeric, dimeric, and aggregated depending on the method of encapsulation (physical entrapment or chemical conjugation to the copolymer). The obvious consequence was different photochemical behavior in terms of $^1O_2$ generation and PDT activity, with the highest PDT efficacy in the case of conjugates micelles. Along this line, chlorin-core star block PEG-PLA micelles loaded with SN-38 as second anticancer drug were found to improve significantly the cytotoxicity of SN-38 in HT-29 human colorectal adenocarcinoma cells after irradiation [178]. Micelles exhibited a prolonged plasma residence time in mice bearing subcutaneous xenografts of HT-29 human colon adenocarcinoma cells, as well as increased tumor accumulation which improved antitumor activity.

Dual drug delivery of ZnPc and the anticancer drug docetaxel from PEG-PCL was recently demonstrated by our group [179]. These systems showed superior antitumor activity compared to the free drugs in an orthotopic mice model of amelanotic melanoma. Similar synergic effects were demonstrated for PEG-PDLLA nanovesicles loaded with Hp/doxorubicin (DOX) against HepG2 human hepatocellular carcinoma cells through apoptotic cell death pathways [180].

The versatility of PEGylated polyesters and the ability to employ different polymer combinations allow the fabrication of hybrid NPs for theranostic applications. Multifunctional NPs based on PEG-PLGA mixed with PLGA-Ce6, and loaded with superparamagnetic iron oxide nanoparticles (SPIONs) for luminescence/magnetic resonance imaging and PDT, have been developed [181]. It was demonstrated that, depending on the amount of PLGA conjugated to the PS, NPs exhibited a different ability to produce $^1O_2$ because of concentration-dependent aggregation of Ce6 in NPs when encountering a biological environment.

Combined PDT nanosystems have also been tested to evaluate the ability of PS in eliciting immune response against tumor. An ideal cancer treatment should not only cause tumor regression and eradication but also induce a systemic antitumor immunity controlling metastasis formation and long-term tumor resistance. Marrache et al. [182] formulated PLGA-PEG NPs loaded with ZnPc and modified on the surface with gold NPs (AuNPs) by using non-covalent interactions. For immune stimulation, the surface of the AuNPs was utilized to introduce 5′-purine-purine/T-CpG-pyrimidine-pyrimidine-3′-oligodeoxynucleotides (CpG-ODN) as a potent dendritic cell activating agent. In vitro cytotoxicity on 4T1 metastatic mouse breast carcinoma cells showed significant photocytotoxicity of NPs and the treatment of mouse bone marrow-derived dendritic cells with the PDT-killed 4T1 cell lysate highlighted the immunostimulant activity of PDT through involvement of several cytokines.

As an alternative to preparing NPs from preformed PEGylated polyesters, surface coating of PLGA NPs with PEGylated lecithin has recently been reported [183]. NPs loaded with both ICG and DOX were prepared in one step for combination of chemotherapy with PTT NPs showed excellent temperature response, faster DOX release under laser irradiation, and longer retention time in mice bearing a subcutaneous xenograft of MCF-7 human breast cancer cells compared with free ICG. NPs induced the apoptosis and death of both DOX-sensitive MCF-7 and DOX-resistant MCF-7/ADR cells and suppressed MCF-7 and MCF- 7/ADR tumor growth in vivo. Notably, no tumor recurrence was observed after only a single dose of NPs with laser irradiation.

Despite the obvious promises shown, PEGylated nanoncologicals are poorly prone to be uptaken inside cells (PEG dilemma) [176] and remain entangled in tumor matrix, forming an extracellular drug depot releasing drug cargo. To encourage internalization of PSs in cancer cells, the most diffused strategy lies in nanocarrier surface decoration with ligands, typical or overexpressed in tumor microenvironments, which can promote nanocarrier transport through receptor-mediated endocytosis. As an example, Master et al. built PEG-PCL micelles targeted to cancers overexpressing epidermal growth factor receptor (EGRF), such as head and neck cancers, and delivering silicon phthalocyanine [184, 185]. Analogously, micelles decorated with the 12-amino acid EGFR-targeting peptide GE11 (i.e., GE11-PEG-PCL) showed high uptake in targeted

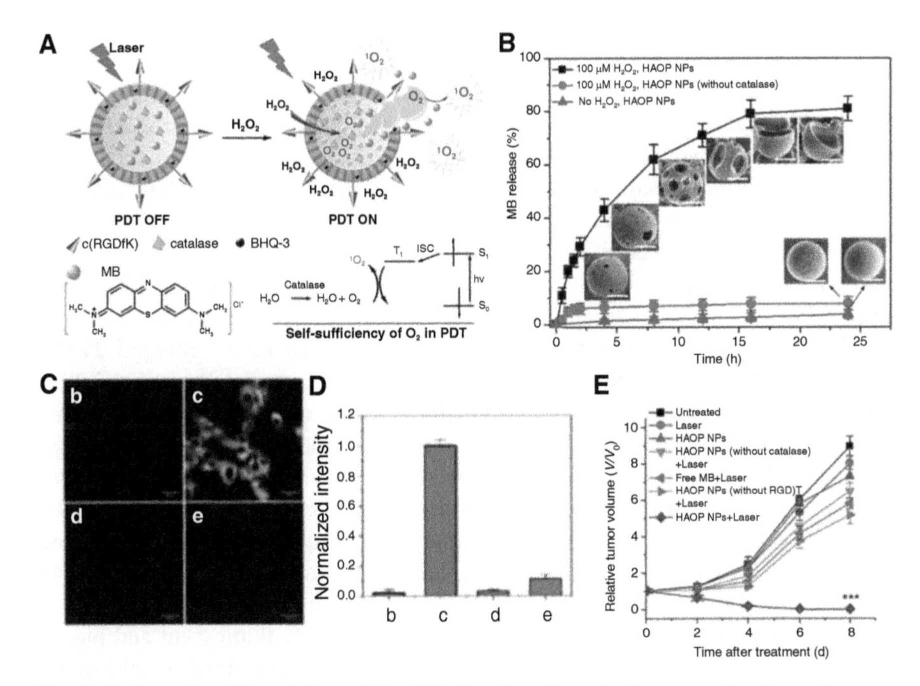

Fig. 5 H_2O_2-activatable and O_2-evolving NPs for photodynamic therapy against hypoxic tumor cells. (**A**) Mechanism of H_2O_2-controllable release of photosensitizer and O_2 to implement PDT. (**B**) In vitro release profiles of MB from NPs (with/without catalase) in the presence or absence of 100 μM H_2O_2. *Insets*: SEM micrographs of NPs incubated with 100 μM H_2O_2 (*scale bars*: 100 nm). (**C**) Confocal fluorescence images of U87-MG cells after 635 nm irradiation in the presence of $^1O_2$ sensor green (SOSG) (*b*) SOSG only; (*c*) NPs + SOSG; (*d*) NAC + NPs + SOSG; (*e*) NPs (without catalase) + SOSG. (**D**) Normalized average intracellular fluorescence intensity of cells in **C**. (**E**) Change of relative tumor volume (V/V_0) in U87-MG tumor-bearing mice injected with NPs. After 24 h, PDT treatment was performed on groups 4–7 by irradiating the tumor region with a 635-nm laser at a power of 100 mW cm^{-2} for 5 min. Adapted from [186]

tumor cells and, as a consequence, a strong PDT response, depending on photoirradiation parameters. H_2O_2-activatable and O_2-evolving NPs targeted to α,β integrin receptor exploiting a clever strategy to treat hypoxic tumor area—where PDT effect is expected to be poor—have recently been reported (Fig. 5) [186]. NPs bear MB and catalase in the aqueous core, a black hole quencher in the polymeric shell, and are functionalized with a tumor targeting peptide c(RGDfK) to be selectively taken up by α,β integrin-rich tumor cells. In the intracellular compartment, H_2O_2 penetrates into NP core and generate O_2, which is the substrate for $^1O_2$ production under light irradiation, through catalase activity. Following shell rupture and release of PS activate local PDT. In vivo studies in glioblastoma bearing mice showed that tumor growth is completely inhibited, and the tumor eliminated after 7 days of treatment.

Non-covalent approaches useful to modify NP surface have recently been investigated as an alternative to chemical functionalization of copolymers. On this basis, overwhelming interest has been taken in layer-by-layer (LbL) NPs engineered for cancer therapy [187]. Consecutive deposition through electrostatic interactions of ionized polymers with opposite charge onto a nanotemplate results in ultrathin multilayers of polymer chains. By dictating and controlling type, composition, number of alternating layers surrounding the core, and final layer thickness, multifunctional core-shell nanostructures can be obtained where multiple drugs and magnetic or luminescent layers can be formed. In this context, our group developed double coated NPs (dcNPs) targeted to CD44 receptor, which is overexpressed in several cancers [188, 189]. Negatively-charged DTX-loaded NPs of PLGA were sequentially decorated through electrostatic interactions with a polycationic shell of polyethyleneimine entangling negatively-charged $TPPS_4$ and a final layer of hyaluronan (HA), a CD44 ligand. dcNPs bears $TPPS_4$ completely aggregated and photochemically inactive at their surface and they release the active monomer after cell internalization, which is higher in MDA-MB231 breast cancer cells overexpressing CD44 receptor. This aspect is of relevance in view of in vivo application because dcNPs are expected to be nonfluorescent and non-photoactive in non-target organs, thereby strongly reducing phototoxicity of carried PS. Nevertheless, taking advantage of targeting to CD44 receptor, dcNPs can localize in tumor tissue, where the PDT component is disassembled from the dcNPs surface and becomes highly fluorescent and phototoxic. The concerted delivery of DTX and $TPPS_4$ resulted in a higher uptake of the hydrophilic PS and tremendous improvement of single drug activity.

Modification of polyesters with pH-sensitive segments allows the building of NPs responsive to tumor acidic microenvironments. pH-responsive micelles made of poly(2-ethyl-2-oxazoline)-*b*-poly(D-L-lactide) entrapping *m*-THPC deliver PS at pH around 5 and suppresses release at pH 7.4. Nevertheless, micelles exhibited in vivo PDT activity similar to that of free *m*-THPC while strongly attenuating skin phototoxicity [190]. Analogously, PEG-*b*-poly(beta-aminoesters) entrapping PpIX displayed in vivo PDT activity and tumor specificity in mice bearing SCC-7 squamous carcinoma, causing complete tumor ablation [191].

4.4 Polyacrylamide NPs

Polyacrylamide (PAA) NPs are hydrogel-like nanostructures prepared by polymerization of a nanoemulsion template widely studied in PS delivery (Table 5). Hydrophilic PAA NPs can be prepared from biodegradable or non-biodegradable cross-linkers and decorated on the surface with targeting ligands.

The simplest strategy to load a PS inside PAA NPs is either encapsulation or post-loading, which can have a different impact on PDT efficacy depending on PS water solubility. When employed for hydrophilic PSs such as MB and its derivatives, conjugation gave PAA NPs with higher $^1O_2$ production yield and enhanced cell mortality under PDT [192, 193]. When those NPs were surface-decorated with F3 peptide to attain specific targeting to 9L rat gliosarcoma cells, MDA-MB-435

Table 5 NPs made of acrylic polymers

Polymer	PS/2nd drug/ imaging agent	Intended use	Stage of development	Main finding	Ref.
AFPAA	HPPH (core)	Therapy	In vivo (intravenous injection)	HPPH leaching is negligible when encapsulated, conjugated and post-loaded	[194]
				The highest $^1O_2$ production is achieved by the post-loaded formulation, which caused the highest phototoxicity in Colon 26 cells	
				Post-loaded NPs induce a similar tumor response compared to that of free HPPH at an equivalent dose in BALB/cAnNCr mice bearing a subcutaneous xenograft of Colon 26 cells	
AFPAA	HPPH (shell)/cyanine dye (shell)	Theranostic	In vivo (intravenous injection)	Undesirable FRET between HPPH and cyanine dye can be controlled playing on loading ratio during post-loading	[195]
				NPs at optimal ratios allow efficient imaging and improved survival after PDT compared with free HPPH in BALB/c mice bearing a subcutaneous xenograft of Colon 26 cells	
AFPMMA	$TPPS_4$ (shell)/NO photodonor	Therapy	In vitro	NPs are well tolerated by B78H1 melanoma cells in the dark and exhibit strongly amplified cell mortality under visible light excitation because of the combined action of $^1O_2$ and nitric oxide	[197]
F3 peptide-PAA-MB	MB derivatives (core)	Therapy	In vitro	MB-conjugated NPs show a higher $^1O_2$ production compared to MB-encapsulated NPs	[192]
				Targeted NPs killed MDA-MB-435 human breast cancer cells more effectively than non-targeted NPs	

(continued)

Table 5 (continued)

Polymer	PS/2nd drug/ imaging agent	Intended use	Stage of development	Main finding	Ref.
F3 peptide - PEG-PAA-MB	MB (core)	Therapy	In vitro	Targeted NPs show a large enhancement of PDT efficacy compared to the non-targeted NPs and free MB in 9L (rat gliosarcoma), MDA-MB-435 (breast) and F98 (rat glioma) cells	[193]

human breast cancer cells, and F98 rat glioma cells, excellent PDT efficacy, increasing with the NP dose and irradiation time, was found [192, 193].

In the case of a hydrophobic PSs such as HPPH and its derivatives, post-loading in preformed amine functionalized polyacrylamide (AFPAA) gives the highest $^1O_2$ production and phototoxicity in vitro compared to encapsulation and conjugation [194]. NPs, tested in a mice colon carcinoma xenograft, enabled fluorescence imaging of the tumor and produced a photodynamic response similar to that of free HPPH at an equivalent dose [194].

When developing multifunctional nanoplatforms loaded with more than one drug, confinement of multiple absorbing species in NPs can have a strong impact on in vitro and in vivo PDT outcomes. To operate in parallel, the two photoresponsive agents should not interfere with each other when in close proximity in the same polymeric scaffold. This aspect was clearly shown with HPPH and a tailor-made cyanine dye for PDT and fluorescence imaging [195]. By playing on HPPH/cyanine dye ratio, the undesirable quenching of the HPPH because of Förster Resonance Energy Transfer between the two molecules was minimized. NPs at optimal ratio resulted in an excellent tumor-imaging (NIR fluorescence) and PDT efficacy in mice bearing a subcutaneous xenograft of Colon 26 cells.

Besides $^1O_2$, light can trigger the release of cytotoxic species such as nitric oxide (NO) exerting antitumor cooperative effects. Indeed, cytotoxic effects induced by NO [196] combined with PDT represent a very promising strategy in view of a multimodal cancer treatment because of the ability to attack biological substrates of different natures, to avoid MDR, and to improve selectivity of therapy. Finally, as NO release is independent of O_2 availability, it can potentially very well complement PDT at the onset of hypoxic conditions. A nanoplatform releasing $^1O_2$ and nitric oxide was thus prepared by the electrostatic entangling of two anionic photoactivable components ($TPPS_4$ and a tailored nitro-aniline derivative) onto the cationic shell of NPs [197]. Photochemical characterization of the nanoplatform clearly showed that the drugs operate in parallel under the exclusive control of light, providing a combinatory effect of the two photogenerated cytotoxic species in B78H1 melanoma cells.

4.5 Pluronic Micelles

Among a variety of triblock copolymers, Pluronics (also termed poloxamers) have achieved the most noticeable interest in pharmaceutics because of their versatility and biocompatibility [198, 199]. Pluronics are commercial FDA-approved material, consisting of a central poly(propylene oxide) flanked by two poly(ethylene oxide) blocks, available in different molecular weights and block lengths and thus characterized by different hydrophilic-lipophilic balances. Pluronics form spontaneously nanosized micelles in aqueous media. Because of their amphiphilic character, these copolymers display surfactant properties including the ability to interact with hydrophobic surfaces and biological membranes. These systems avoid MPS uptake, increase drug solubility, and improve its circulation time and passive tumor targeting by EPR effects [198]. Previously thought to be "inert," Pluronics display a unique set of biological activities and have been shown to be potent sensitizers of MDR cancer cells in vitro and in vivo [198, 200]. The key attribute for the biological activity of Pluronics is their ability to incorporate into membranes followed by subsequent translocation into the cells where they affect various cellular functions, such as mitochondrial respiration, ATP synthesis, activity of drug efflux transporters, apoptotic signal transduction, and gene expression. As a result, Pluronics cause drastic sensitization of MDR tumors to various anticancer agents, enhance drug transport across the blood brain and intestinal barriers, and cause transcriptional activation of gene expression both in vitro and in vivo [200].

The first example of Pluronic use as carriers for PDT agents was the delivery of Verteporfin® derivatives in monomeric form [201, 202]. Thereafter, Pluronic micelles have been considered a promising vehicle for the delivery of other PDT agents such as porphyrins, chlorins, phthalocyanines, chlorophylls, and xanthene derivatives [203–206]. Sobczyński and collaborators recently evaluated the influence of Pluronics on the photocytotoxicity and cytolocalization of four porphyrin-based PSs, i.e., tetraphenyl porphyrins 4-substituted on the phenyl groups with trimethylamine (TAPP), hydroxyl (THPP), sulfonate (TSPP), and carboxyl (TCPP), in WiDr colon adenocarcinoma cell line [205]. Pluronics were found to deaggregate the PSs and improve PS solubility efficiently. Moderate to profound effects on intracellular localization of the PSs and cellular sensitivity to photoinactivation were found. P123 and F127 strongly attenuated the uptake and photocytotoxicity of THPP and redirected the cellular uptake to endocytosis, while P123 stimulated translocation of TAPP from endocytic vesicles to a cytosolic and nuclear localization followed by an enhanced phototoxicity. P123 and F127 lowered the fraction of TCCP in endocytic vesicles followed by a reduced sensitivity to photoinactivation. F68 had only moderate effects on intracellular localization of the evaluated PSs with the exception of a higher endocytic accumulation of TCPP and lowered photocytotoxicity of TCPP and THPP.

To enhance $^1O_2$ generation, chlorine e6 was encapsulated in heavy-atomic NPs based on Pluronic F127 and in vitro PDT efficacy was evaluated in MDA-MB-231 human breast-cancer cell line [207]. Pluronic F68 was used to increase the intracellular level of ALA in human cholangiocarcinoma cells, resulting in enhanced PpIX formation and phototoxicity [208]. Pluronics P123 and F127 improved the delivery of Photofrin® overcoming MDR in MCF-7/WT human breast and SKOV-3 ovarian cancer cells inducing cell apoptosis through photodynamic effects [209].

A few pieces of work have reported in vivo results with Pluronic-based nanoPDT. Park and Na [210] conjugated chlorine e6 in F127 micelles and observed higher internalization rates, tumor-specific distribution, and in vivo tumor growth inhibition after intravenous injection into mice bearing a colon tumor (CT-26) when compared with free PS. In subsequent work these authors introduced DOX in the formulations for combined photodynamics and chemotherapy overcoming drug resistance in drug-resistant cancer cells [211]. Furthermore, the natural PS chlorophyll (Chl) extracted from vegetables was encapsulated into Pluronic F68 micelles for in vivo cancer imaging and therapy. Results showed that, after intravenous injection and laser irradiation, the growth of melanoma cells and mouse xenograft (A375) were effectively inhibited by laser-triggered PTT and PDT synergistic effects.

A simple and biocompatible nanocomplex of MB in a combination of Pluronic F68 and oleic acid (named NanoMB) driven by the dual (electrostatic and hydrophobic) interactions between the ternary constituents was developed [212]. The nanocomplexed MB showed greatly enhanced cell internalization in different cancer cell lines while keeping the photosensitization efficiency as high as free MB, leading to distinctive phototoxicity toward cancer cells. When administered to human breast cancer xenograft mice by peritumoral injection, nanocomplexed MB was capable of facile penetration into the tumor followed by cancer cell accumulation. After five PDT treatments consisting in a combination of peritumorally injected nanocomplexed MB and selective laser irradiation, tumor volume was significantly decreased, demonstrating potential for adjuvant locoregional cancer treatment.

Additionally, Pluronics have been associated with graphene oxide and gold NPs for in vivo combined PDT and PTT. The complex graphene oxide sheet, Pluronic F127 and MB showed high tumor accumulation after intravenous injection into tumor-bearing mice, causing total ablation of tumor tissue exposed to NIR light [213]. A Pluronic-based nanogel was combined with both gold nanorods as a PTT agent and Ce6 as a PS for PDT. In both in vitro cell culture and in vivo tumor-bearing mice experiments in SCC-7 squamous carcinoma cells or NIH/3T3 fibroblast cells, a remarkably enhanced tumor ablation was observed with treatment by PDT (red laser) followed by PTT (NIR laser) as compared with separate treatments [214].

4.6 Other Systems

Polymer-PS conjugates can spontaneously form different nanostructures. The simplest example among them is represented by PEGylated HpD assembled into NPs and loaded with DOX to achieve a synergistic effect of chemotherapy and PDT [215]. This approach is useful to form core-shell-structured bioreducible self-quenched NPs that dissociate under intracellular reductive conditions, triggering the rapid release of PS in a photoactive form [216]. More complex concepts have been reported recently. NPs based on a star shaped 4-arm PEG functionalized with biotin as targeting unit and a chlorambucil-coumarin fluorophore were synthesized for site-specific and image guided treatment of cancer cells [217]. Telodendrimers, a novel class of hybrid amphiphilic polymers comprised of linear PEG and dendritic oligomers of PheoA and cholic acid (CA) formed NPs by self-assembly (Nanoporphyrins), which greatly increased the imaging sensitivity for tumor detection through background suppression in blood, as well as preferential accumulation and signal amplification in tumors. Nanoporphyrins also functioned as multiphase nanotransducers that can efficiently convert light to heat inside tumors for PTT, and light to $^1O_2$ for PDT.

The Kataoka group studied polyion complex micelles of PEG-poly(L-lysine) block copolymers with an anionic dendrimer Pc (dPc) [218]. The PDT effect of DPc was two orders of magnitude higher than the free DPc with the same irradiation time. In vivo PDT efficacy of dPc-loaded micelles [219] was higher than free dPc and Photofrin®. Furthermore, the skin phototoxicity of the DPc-loaded micelles was significantly reduced after white light irradiation. Lu et al. have loaded in PEG-poly(L-lysine) micelles entrapping dPc and also DOX [220]. In vivo studies carried out in a xenograft model of breast cancer highlighted that the internalized DPc micelles showed unique PCI properties inside the cells and thereby facilitated DOX release from the endo-lysosomes to nuclei after photoirradiation, thus reversing MDR.

To implement multimodality of a cancer therapy, very promising results have been obtained by employing light-activated supramolecular nanoassemblies based on cyclodextrin branched polymers for simultaneous imaging and therapy [221–225]. In these systems, beside the photochemical independence of the two chromogenic centers, a high association constant between the two units was a key prerequisite to avoid displacement in a biological environment if their association is not sufficiently strong. Bichromophoric NPs exploiting TPE fluorescence have been obtained by the self-assembly of a NO photodonor, ZnPc, and a β-CD polymer [226]. The macromolecular assembly delivered its photoresponsive cargo of active molecules not only within the cytoplasm but also in human skin as exemplified in ex vivo experiments.

In the attempt to apply layer-by-layer fabrication technology to theranostic nanocapsules for cancer PDT combined with magnetic resonance imaging (MRI), a platform entrapping dendrimer porphyrin (DP) in the shells and SPIONs in the core was designed [227]. SPIONs-embedded polystyrene NPs were used as a template to build up multilayered nanocapsules through sequential poly(allylamine hydrochloride)/DP deposition. NCs exhibited typical superparamagnetic behavior, and cell viability study (HeLa) upon light irradiation revealed that NCs can successfully work in PS formulation for PDT.

5 Conclusions

PDT is gaining momentum as an alternative or complementary treatment of solid tumors. Potentiating PDT effects and coupling PDT to other treatment modalities are considered promising strategies to fight tumors. Polymeric NPs are supportive of this evolution, offering a wide variety of options to engineering multimodal systems useful for therapeutic, diagnostic and theranostic purposes. NP design needs to be established ab initio and should be driven by biologically-oriented design rules to accumulate drug cargo at the pharmacological target and by specific requirements to preserve/optimize photochemical properties of delivered PS. Rational combination of building elements in a single nanoplatform can also couple PDT to other treatment modalities (conventional chemotherapy, PPT, radio-therapy) or imaging (MRI, fluorescence) propelling the application of PDT to the forefront of diagnosis and therapy of cancer.

Acknowledgments The authors wish to thank the Italian Ministry of University and Research (PRIN 2010H834LS) and Italian Association for Cancer Research (IG2014 15764).

References

1. Holohan C, Van Schaeybroeck S, Longley DB, Johnston PG (2013) Nat Rev Cancer 13:714–726
2. Wu Q, Yang Z, Nie Y, Shi Y, Fan DE (2014) Cancer Lett 347:159–166
3. Jain KK (2010) BMC Med 8:83
4. Jain RK, Stylianopoulos T (2010) Nat Rev Clin Oncol 7:653–664
5. Wang AZ, Langer R, Farokhzad OC (2012) Annu Rev Med 63:185–198
6. Schroeder A, Heller DA, Winslow MM, Dahlman JE, Pratt GW, Langer R, Jacks T, Anderson DG (2012) Nat Rev Cancer 12:39–50
7. Markman JL, Rekechenetskiy A, Holler E, Ljubimova JY (2013) Adv Drug Deliv Rev 65:1866–1879
8. Palakurthi S, Yellepeddi VK, Vangara KK (2012) Expert Opin Drug Deliv 9:287–301
9. Iyer AK, Singh A, Ganta S, Amiji MM (2013) Adv Drug Deliv Rev 65:1784–1802
10. Webster DM, Sundaram P, Byrne ME (2013) Eur J Pharm Biopharm 84:1–20
11. Moghimi SM, Hunter AC, Andresen TL (2012) Annu Rev Pharmacol Toxicol 52:481–503
12. Bertrand N, Leroux JC (2012) J Control Release 161:152–163
13. Alexis F, Pridgen E, Molnar LK, Farokhzad OC (2008) Mol Pharm 5:505–515
14. Fang J, Nakamura H, Maeda H (2011) Adv Drug Deliv Rev 63:136–151
15. Maeda H (2012) J Control Release 164:138–144

16. Huynh NT, Roger E, Lautram N, Benoit JP, Passirani C (2010) Nanomedicine (Lond) 5:1415–1433
17. Bertrand N, Wu J, Xu X, Kamaly N, Farokhzad OC (2014) Adv Drug Deliv Rev 66:2–25
18. Kim CS, Duncan B, Creran B, Rotello VM (2013) Nano Today 8:439–447
19. Torchilin VP (2014) Nat Rev Drug Discov 13:813–827
20. Mura S, Nicolas J, Couvreur P (2013) Nat Mater 12:991–1003
21. Gao W, Chan JM, Farokhzad OC (2010) Mol Pharm 7:1913–1920
22. Tian L, Bae YH (2012) Colloids Surf B Biointerfaces 99:116–126
23. Du JZ, Mao CQ, Yuan YY, Yang XZ, Wang J (2013) Biotechnol Adv 32:789–803
24. Nowag S, Haag R (2014) Angew Chem Int Ed Engl 53:49–51
25. Abulateefeh SR, Spain SG, Aylott JW, Chan WC, Garnett MC, Alexander C (2011) Macromol Biosci 11:1722–1734
26. Dolmans DEJG, Fukumura D, Jain RK (2003) Nat Rev Cancer 3:380–387
27. Agostinis P, Berg K, Cengel KA, Foster TH, Girotti AW, Gollnick SO, Hahn SM, Hamblin MR, Juzeniene A, Kessel D, Korbelik M, Moan J, Mroz P, Nowis D, Piette J, Wilson BC, Golab J (2011) CA Cancer J Clin 61:250–281
28. Allison RR (2014) Future Oncol 10:123–142
29. Kawczyk-Krupka A, Bugaj AM, Latos W, Zaremba K, Wawrzyniec K, Sieron A (2015) Photodiagnosis Photodyn Ther 12:545–553
30. Huang Z (2005) Technol Cancer Res Treat 4:283–293
31. Avci P, Erdem SS, Hamblin MR (2014) J Biomed Nanotechnol 10:1937–1952
32. Lim CK, Heo J, Shin S, Jeong K, Seo YH, Jang WD, Park CR, Park SY, Kim S, Kwon IC (2013) Cancer Lett 334:176–187
33. Master A, Livingston M, Sen Gupta A (2013) J Control Release 168:88–102
34. Anand S, Ortel BJ, Pereira SP, Hasan T, Maytin EV (2012) Cancer Lett 326:8–16
35. Mohamed S, Parayath NN, Taurin S, Greish K (2014) Ther Deliv 5:1101–1121
36. Kamaly N, Xiao Z, Valencia PM, Radovic-Moreno AF, Farokhzad OC (2012) Chem Soc Rev 41:2971–3010
37. Kuimova MK, Yahioglu G, Ogilby PR (2009) J Am Chem Soc 131:332–340
38. Robertson CA, Evans DH, Abrahamse H (2009) J Photochem Photobiol B 96:1–8
39. Ormond AB, Freeman HS (2013) Materials 6:817–840
40. O'Connor AE, Gallagher WM, Byrne AT (2009) Photochem Photobiol 85:1053–1074
41. Allison RR, Bagnato VS, Sibata CH (2010) Future Oncol 6:929–940
42. Nokes B, Apel M, Jones C, Brown G, Lang JE (2013) J Surg Res 181:262–271
43. Fukuda H, Casas A, Batlle A (2006) J Environ Pathol Toxicol Oncol 25:127–143
44. Calzavara-Pinton PG, Venturini M, Sala R (2007) J Eur Acad Dermatol Venereol 21:293–302
45. Kreimer-Birnbaum M (1989) Semin Hematol 26:157–173
46. Garland MJ, Cassidy CM, Woolfson D, Donnelly RF (2009) Future Med Chem 1:667–691
47. Staneloudi C, Smith KA, Hudson R, Malatesti N, Savoie H, Boyle RW, Greenman J (2007) Immunology 120:512–517
48. Nowis D, Makowski M, Stoklosa T, Legat M, Issat T, Golab J (2005) Acta Biochim Pol 52:339–352
49. Brancaleon L, Moseley H (2002) Lasers Med Sci 17:173–186
50. Sandell JL, Zhu TC (2011) J Biophotonics 4:773–787
51. Lee YE, Kopelman R (2011) Methods Mol Biol 726:151–178
52. Barolet D (2008) Semin Cutan Med Surg 27:227–238
53. Vicente MG (2001) Curr Med Chem Anticancer Agents 1:175–194
54. Chowdhary RK, Sharif I, Chansarkar N, Dolphin D, Ratkay L, Delaney S, Meadows H (2003) J Pharm Pharm Sci 6:198–204
55. Jori G, Reddi E (1993) Int J Biochem 25:1369–1375
56. Woodburn KW, Vardaxis NJ, Hill JS, Kaye AH, Phillips DR (1991) Photochem Photobiol 54:725–732
57. Malik Z, Amit I, Rothmann C (1997) Photochem Photobiol 65:389–396

58. Reiners JJ Jr, Agostinis P, Berg K, Oleinick NL, Kessel D (2010) Autophagy 6:7–18
59. Castano AP, Mroz P, Hamblin MR (2006) Nat Rev Cancer 6:535–545
60. Krammer B (2001) Anticancer Res 21:4271–4277
61. Fingar VH, Wieman TJ, Wiehle SA, Cerrito PB (1992) Cancer Res 52:4914–4921
62. Fingar VH (1996) J Clin Laser Med Surg 14:323–328
63. Bhuvaneswari R, Gan YY, Soo KC, Olivo M (2009) Cell Mol Life Sci 66:2275–2283
64. Mitra S, Cassar SE, Niles DJ, Puskas JA, Frelinger JG, Foster TH (2006) Mol Cancer Ther 5:3268–3274
65. Pizova K, Tomankova K, Daskova A, Binder S, Bajgar R, Kolarova H (2012) Biomed Pap Med Fac Univ Palacky Olomouc Czech Repub 156:93–102
66. Gollnick SO, Owczarczak B, Maier P (2006) Lasers Surg Med 38:509–515
67. Patel G, Armstrong AW, Eisen DB (2014) JAMA Dermatol 150:1281–1288
68. Clark CM, Furniss M, Mackay-Wiggan JM (2014) Am J Clin Dermatol 15:197–216
69. Lansbury L, Bath-Hextall F, Perkins W, Stanton W, Leonardi-Bee J (2013) BMJ 347:f6153
70. Biel MA (2010) Methods Mol Biol 635:281–293
71. Kostron H (2010) Methods Mol Biol 635:261–280
72. Minnich DJ, Bryant AS, Dooley A, Cerfolio RJ (2010) Ann Thorac Surg 89:1744–1748
73. Simone CB, Friedberg JS, Glatstein E, Stevenson JP, Sterman DH, Hahn SM, Cengel KA (2012) J Thorac Dis 4:63–75
74. Wiedmann MW, Caca K (2004) Curr Pharm Biotechnol 5:397–408
75. Ortner MA (2011) Lasers Surg Med 43:776–780
76. Keane MG, Bramis K, Pereira SP, Fusai GK (2014) World J Gastroenterol 20:2267–2278
77. Bown SG, Rogowska AZ, Whitelaw DE, Lees WR, Lovat LB, Ripley P, Jones L, Wyld P, Gillams A, Hatfield AW (2002) Gut 50:549–557
78. Tejeda-Maldonado J, Garcia-Juarez I, Aguirre-Valadez J, Gonzalez-Aguirre A, Vilatoba-Chapa M, Armengol-Alonso A, Escobar-Penagos F, Torre A, Sanchez-Avila JF, Carrillo-Perez DL (2015) World J Hepatol 7:362–376
79. Marien A, Gill I, Ukimura O, Betrouni N, Villers A (2014) Urol Oncol 32:912–923
80. Cheung G, Sahai A, Billia M, Dasgupta P, Khan MS (2013) BMC Med 11:13. doi:10.1186/1741-7015-11-13
81. Estel R, Hackethal A, Kalder M, Munstedt K (2011) Arch Gynecol Obstet 284:1277–1282
82. Celli JP, Spring BQ, Rizvi I, Evans CL, Samkoe KS, Verma S, Pogue BW, Hasan T (2010) Chem Rev 110:2795–2838
83. Zuluaga MF, Lange N (2008) Curr Med Chem 15:1655–1673
84. Lehar J, Krueger AS, Avery W, Heilbut AM, Johansen LM, Price ER, Rickles RJ, Short GF III, Staunton JE, Jin X, Lee MS, Zimmermann GR, Borisy AA (2009) Nat Biotechnol 27:659–666
85. Kessel D, Erickson C (1992) Photochem Photobiol 55:397–399
86. Norum OJ, Selbo PK, Weyergang A, Giercksky KE, Berg K (2009) J Photochem Photobiol B 96:83–92
87. Selbo PK, Weyergang A, Hogset A, Norum OJ, Berstad MB, Vikdal M, Berg K (2010) J Control Release 148:2–12
88. Berg K, Berstad M, Prasmickaite L, Weyergang A, Selbo PK, Hedfors I, Hogset A (2010) Nucleic Acid Transfect 296:251–281
89. Rajendran L, Knolker HJ, Simons K (2010) Nat Rev Drug Discov 9:29–42
90. Allison RR, Mota HC, Sibata CH (2004) Photodiagnosis Photodyn Ther 1:263–277
91. Gelderblom H, Verweij J, Nooter K, Sparreboom A (2001) Eur J Cancer 37:1590–1598
92. Aggarwal LPF, Borissevitch IE (2006) Spectrochim Acta A Mol Biomol Spectrosc 63:227–233
93. Gabrielli D, Belisle E, Severino D, Kowaltowski AJ, Baptista MS (2004) Photochem Photobiol 79:227–232
94. Gao D, Agayan RR, Xu H, Philbert MA, Kopelman R (2006) Nano Lett 6:2383–2386
95. Chatterjee DK, Fong LS, Zhang Y (2008) Adv Drug Deliv Rev 60:1627–1637

96. Davis ME, Chen ZG, Shin DM (2008) Nat Rev Drug Discov 7:771–782
97. Parhi P, Mohanty C, Sahoo SK (2012) Drug Discov Today 17:1044–1052
98. Oh J, Yoon H, Park JH (2013) Biomed Eng Lett 3:67–73
99. Zhang D, Wu M, Zeng Y, Wu L, Wang Q, Han X, Liu X, Liu J (2015) ACS Appl Mater Interfaces 7:8176–8187
100. Mallidi S, Spring BQ, Chang S, Vakoc B, Hasan T (2015) Cancer J 21:194–205
101. Taratula O, Schumann C, Duong T, Taylor KL, Taratula O (2015) Nanoscale 7:3888–3902
102. Nichols JW, Bae YH (2014) J Control Release 190:451–464. doi:10.1016/j.jconrel.2014.03.057, Epub@2014 Apr 30
103. Monopoli MP, Aberg C, Salvati A, Dawson KA (2012) Nat Nanotechnol 7:779–786
104. Karmali PP, Simberg D (2011) Expert Opin Drug Deliv 8:343–357
105. Euliss LE, DuPont JA, Gratton S, DeSimone J (2006) Chem Soc Rev 35:1095–1104
106. Decuzzi P, Pasqualini R, Arap W, Ferrari M (2009) Pharm Res 26:235–243
107. Geng Y, Dalhaimer P, Cai S, Tsai R, Tewari M, Minko T, Discher DE (2007) Nat Nanotechnol 2:249–255
108. Fujimori K, Covell DG, Fletcher JE, Weinstein JN (1989) Cancer Res 49:5656–5663
109. Ernsting MJ, Murakami M, Roy A, Li SD (2013) J Control Release 172:782–794
110. Owens DE III, Peppas NA (2006) Int J Pharm 307:93–102
111. Amoozgar Z, Yeo Y (2012) Wiley Interdiscip Rev Nanomed Nanobiotechnol 4:219–233
112. Gref R, Domb A, Quellec P, Blunk T, Muller RH, Verbavatz JM, Langer R (2012) Adv Drug Deliv Rev 64:316–326
113. Kettler K, Veltman K, van de Meent D, van Wezel A, Hendriks AJ (2014) Environ Toxicol Chem 33:481–492
114. Lee H, Fonge H, Hoang B, Reilly RM, Allen C (2010) Mol Pharm 7:1195–1208
115. Kim JA, Aberg C, Salvati A, Dawson KA (2012) Nat Nanotechnol 7:62–68
116. Mahon E, Salvati A, Baldelli Bombelli F, Lynch I, Dawson KA (2012) J Control Release 161:164–174
117. Mukherjee S, Ghosh RN, Maxfield FR (1997) Physiol Rev 77:759–803
118. Fleige E, Quadir MA, Haag R (2012) Adv Drug Deliv Rev 64:866–884
119. Bae YH, Huh KM, Kim Y, Park KH (2000) J Control Release 64:3–13
120. Xiong XB, Binkhathlan Z, Molavi O, Lavasanifar A (2012) Acta Biomater 8:2017–2033
121. Adams ML, Lavasanifar A, Kwon GS (2003) J Pharm Sci 92:1343–1355
122. Letchford K, Burt H (2007) Eur J Pharm Biopharm 65:259–269
123. Aliabadi HM, Lavasanifar A (2006) Expert Opin Drug Deliv 3:139–162
124. Grubbs RB, Sun Z (2013) Chem Soc Rev 42:7436–7445
125. Sutton D, Nasongkla N, Blanco E, Gao J (2007) Pharm Res 24:1029–1046
126. Hoffman AS (2013) Adv Drug Deliv Rev 65:10–16
127. Pasquier E, Kavallaris M, Andre N (2010) Nat Rev Clin Oncol 7:455–465
128. Li J, Wang Y, Zhu Y, Oupický D (2013) J Control Release 172:10
129. Tang W, Xu H, Kopelman R, Philbert MA (2005) Photochem Photobiol 81:242–249
130. Muthu MS, Wilson B (2012) Nanomedicine 7:307–309
131. Mahapatro A, Singh DK (2011) J Nanobiotechnol 9:55–55
132. Saraf S (2009) Expert Opin Drug Deliv 6:187–196
133. Anton N, Benoit JP, Saulnier P (2008) J Control Release 128:185–199
134. Quaglia F, Ostacolo L, De Rosa G, La Rotonda MI, Ammendola M, Nese G, Maglio G, Palumbo R, Vauthier C (2006) Int J Pharm 324:56–66
135. Venkataraman S, Hedrick JL, Ong ZY, Yang C, Ee PL, Hammond PT, Yang YY (2011) Adv Drug Deliv Rev 63:1228–1246
136. Valencia PM, Hanewich-Hollatz MH, Gao WW, Karim F, Langer R, Karnik R, Farokhzad OC (2011) Biomaterials 32:6226–6233
137. Holzer M, Vogel V, Mantele W, Schwartz D, Haase W, Langer K (2009) Eur J Pharm Biopharm 72:428–437
138. Wu L, Zhang J, Watanabe W (2011) Adv Drug Deliv Rev 63:456–469

139. Jayakumar R, Prabaharan M, Nair SV, Tamura H (2010) Biotechnol Adv 28:142–150
140. Hudson D, Margaritis A (2013) Crit Rev Biotechnol 34:161–179
141. Ren D, Yi H, Wang W, Ma X (2005) Carbohydr Res 340:2403–2410
142. Abdelghany SM, Schmid D, Deacon J, Jaworski J, Fay F, McLaughlin KM, Gormley JA, Burrows JF, Longley DB, Donnelly RF, Scott CJ (2013) Biomacromolecules 14:302–310
143. Philippova OE, Korchagina EV (2012) Polym Sci Ser A 54:552–572
144. Lee HM, Jeong YI, Kim dH, Kwak TW, Chung CW, Kim CH, Kang DH (2013) Int J Pharm 454:74–81
145. Lee SJ, Koo H, Jeong H, Huh MS, Choi Y, Jeong SY, Byun Y, Choi K, Kim K, Kwon IC (2011) J Control Release 152:21–29
146. Oh IH, Min HS, Li L, Tran TH, Lee YK, Kwon IC, Choi K, Kim K, Huh KM (2013) Biomaterials 34:6454–6463
147. Lim CK, Shin J, Kwon IC, Jeong SY, Kim S (2012) Bioconjug Chem 23:1022–1028
148. Chen R, Wang X, Yao X, Zheng X, Wang J, Jiang X (2013) Biomaterials 34:8314–8322
149. Khdair A, Gerard B, Handa H, Mao G, Shekhar MP, Panyam J (2008) Mol Pharm 5:795–807
150. Khdair A, Chen D, Patil Y, Ma LN, Dou QP, Shekhar MPV, Panyam J (2010) J Control Release 141:137–144
151. Usacheva M, Swaminathan SK, Kirtane AR, Panyam J (2014) Mol Pharm 11:3186–3195
152. Kratz F, Elsadek B (2012) J Control Release 161:429–445
153. Jiang C, Cheng H, Yuan A, Tang X, Wu J, Hu Y (2015) Acta Biomater 14:61–69
154. Fakirov S (2007) Handbook of engineering biopolymers. Hanser, München, pp 417–464
155. Jain SK, Gupta Y, Jain A, Saxena AR, Khare P, Jain A (2008) Nanomed Nanotechnol Biol Med 4:41–48
156. Babu A, Jeyasubramanian K, Gunasekaran P, Murugesan R (2012) J Biomed Nanotechnol 8:43–56
157. Babu A, Periasamy J, Gunasekaran A, Kumaresan G, Naicker S, Gunasekaran P, Murugesan R (2013) J Biomed Nanotechnol 9:177–192
158. Yan F, Zhang Y, Kim KS, Yuan HK, Vo-Dinh T (2010) Photochem Photobiol 86:662–666
159. Danhier F, Ansorena E, Silva JM, Coco R, Le BA, Preat V (2012) J Control Release 161:505–522
160. Sah H, Thoma LA, Desu HR, Sah E, Wood GC (2013) Int J Nanomedicine 8:747–765
161. Vargas A, Eid M, Fanchaouy M, Gurny R, Delie F (2008) Eur J Pharm Biopharm 69:43–53
162. Vargas A, Lange N, Arvinte T, Cerny R, Gurny R, Delie F (2009) J Drug Target 17:599–609
163. Hu Z, Pan Y, Wang J, Chen J, Li J, Ren L (2009) Biomed Pharmacother 63:155–164
164. da Volta SM, Oliveira MR, dos Santos EP, de Brito GL, Barbosa GM, Quaresma CH, Ricci-Junior E (2011) Int J Nanomedicine 6:227–238
165. Korbelik M, Madiyalakan R, Woo T, Haddadi A (2012) Photochem Photobiol 88:188–193
166. Fadel M, Kassab K, Fadeel DA (2010) Lasers Med Sci 25:283–292
167. Conte C, D'Angelo I, Miro A, Ungaro F, Quaglia F (2014) Curr Top Med Chem 14:1097–1114
168. Mikhail AS, Allen C (2009) J Control Release 138:214–223
169. Perry JL, Reuter KG, Kai MP, Herlihy KP, Jones SW, Luft JC, Napier M, Bear JE, DeSimone JM (2012) Nano Lett 12:5304–5310
170. Pridgen EM, Langer R, Farokhzad OC (2007) Nanomedicine 2:669–680
171. Dinarvand R, Sepehri N, Manoochehri S, Rouhani H, Atyabi F (2011) Int J Nanomedicine 6:877–895
172. Cohen EM, Ding H, Kessinger CW, Khemtong C, Gao J, Sumer BD (2010) Otolaryngol Head Neck Surg 143:109–115
173. Ding H, Mora R, Gao J, Sumer BD (2011) Otolaryngol Head Neck Surg 145:612–617
174. Master AM, Rodriguez ME, Kenney ME, Oleinick NL, Gupta AS (2010) J Pharm Sci 99:2386–2398
175. Knop K, Mingotaud AF, El-Akra N, Violleau F, Souchard JP (2009) Photochem Photobiol Sci 8:396–404

176. Rojnik M, Kocbek P, Moret F, Compagnin C, Celotti L, Bovis MJ, Woodhams JH, MacRobert AJ, Scheglmann D, Helfrich W, Verkaik MJ, Papini E, Reddi E, Kos J (2012) Nanomedicine 7:663–677
177. Ding H, Sumer BD, Kessinger CW, Dong Y, Huang G, Boothman DA, Gao J (2011) J Control Release 151:271–277
178. Peng CL, Lai PS, Lin FH, Yueh-Hsiu WS, Shieh MJ (2009) Biomaterials 30:3614–3625
179. Conte C, Ungaro F, Maglio G, Tirino P, Siracusano G, Sciortino MT, Leone N, Palma G, Barbieri A, Arra C, Mazzaglia A, Quaglia F (2013) J Control Release 167:40–52
180. Xiang GH, Hong GB, Wang Y, Cheng D, Zhou JX, Shuai XT (2013) Int J Nanomedicine 8:4613–4622
181. Lee DJ, Park GY, Oh KT, Oh NM, Kwag DS, Youn YS, Oh YT, Park JW, Lee ES (2012) Int J Pharm 434:257–263
182. Marrache S, Choi JH, Tundup S, Zaver D, Harn DA, Dhar S (2013) Integr Biol (Camb) 5:215–223
183. Zheng M, Yue C, Ma Y, Gong P, Zhao P, Zheng C, Sheng Z, Zhang P, Wang Z, Cai L (2013) ACS Nano 7:2056–2067
184. Master AM, Qi Y, Oleinick NL, Gupta AS (2012) Nanomedicine 8:655–664
185. Master A, Malamas A, Solanki R, Clausen DM, Eiseman JL, Sen GA (2013) Mol Pharm 10:1988–1997
186. Chen H, Tian J, He W, Guo Z (2015) J Am Chem Soc 137:1539–1547
187. Yan Y, Such GK, Johnston APR, Lomas H, Caruso F (2011) ACS Nano 5:4252–4257
188. Maiolino S, Moret F, Conte C, Fraix A, Tirino P, Ungaro F, Sortino S, Reddi E, Quaglia F (2015) Nanoscale 7:5643–5653
189. Maiolino S, Russo A, Pagliara V, Conte C, Ungaro F, Russo G, Quaglia F (2015) J Nanobiotechnol 13:29
190. Shieh MJ, Peng CL, Chiang WL, Wang CH, Hsu CY, Wang SJJ, Lai PS (2010) Mol Pharm 7:1244–1253
191. Koo H, Lee H, Lee S, Min KH, Kim MS, Lee DS, Choi Y, Kwon IC, Kim K, Jeong SY (2010) Chem Commun (Camb) 46:5668–5670
192. Qin M, Hah HJ, Kim G, Nie G, Lee YE, Kopelman R (2011) Photochem Photobiol Sci 10:832–841
193. Hah HJ, Kim G, Lee YE, Orringer DA, Sagher O, Philbert MA, Kopelman R (2011) Macromol Biosci 11:90–99
194. Wang S, Fan W, Kim G, Hah HJ, Lee YE, Kopelman R, Ethirajan M, Gupta A, Goswami LN, Pera P, Morgan J, Pandey RK (2011) Lasers Surg Med 43:686–695
195. Gupta A, Wang S, Pera P, Rao KV, Patel N, Ohulchanskyy TY, Missert J, Morgan J, Koo-Lee YE, Kopelman R, Pandey RK (2012) Nanomedicine 8:941–950
196. Fukumura D, Kashiwagi S, Jain RK (2006) Nat Rev Cancer 6:521–534
197. Fraix A, Manet I, Ballestri M, Guerrini A, Dambruoso P, Sotgiu G, Varchi G, Camerin M, Coppellotti O, Sortino S (2015) J Mater Chem B 3:3001–3010
198. Batrakova EV, Kabanov AV (2008) J Control Release 130:98–106
199. Jung YW, Lee H, Kim JY, Koo EJ, Oh KS, Yuk SH (2013) Curr Med Chem 20:3488–3499
200. Alakhova DY, Kabanov AV (2014) Mol Pharm 11:2566–2578
201. Chowdhary RK, Chansarkar N, Sharif I, Hioka N, Dolphin D (2003) Photochem Photobiol 77:299–303
202. Hioka N, Chowdhary RK, Chansarkar N, Delmarre D, Sternberg E, Dolphin D (2002) Can J Chem 80:1321–1326
203. Pellosi DS, Estevão BM, Semensato J, Severino D, Baptista MS, Politi MJ, Hioka N, Caetano W (2012) J Photochem Photobiol A Chem 247:8–15
204. Chu M, Li H, Wu Q, Wo F, Shi D (2014) Biomaterials 35:8357–8373
205. Sobczynski J, Kristensen S, Berg K (2014) Photochem Photobiol Sci 13:8–22
206. Vilsinski BH, Gerola AP, Enumo JA, Campanholi KSS, Pereira PCS, Braga G, Hioka N, Kimura E, Tessaro AL, Caetano W (2015) Photochem Photobiol 91:518–525

207. Lim CK, Shin J, Lee YD, Kim J, Park H, Kwon IC, Kim S (2011) Small 7:112–118
208. Chung CW, Kim CH, Choi KH, Yoo JJ, Kim DH, Chung KD, Jeong YI, Kang DH (2012) Eur J Pharm Biopharm 80:453–458
209. Lamch L, Bazylinska U, Kulbacka J, Pietkiewicz J, Biezunska-Kusiak K, Wilk KA (2014) Photodiagnosis Photodyn Ther 11:570–585
210. Park H, Na K (2013) Biomaterials 34:6992–7000
211. Park H, Park W, Na K (2014) Biomaterials 35:7963–7969
212. Lee YD, Cho HJ, Choi MH, Park H, Bang J, Lee S, Kwon IC, Kim S (2015) J Control Release 209:12–19
213. Sahu A, Choi WI, Lee JH, Tae G (2013) Biomaterials 34:6239–6248
214. Kim TH, Song C, Han YS, Jang JD, Choi MC (2014) Soft Matter 10:484–490
215. Ren Y, Wang R, Liu Y, Guo H, Zhou X, Yuan X, Liu C, Tian J, Yin H, Wang Y, Zhang N (2014) Biomaterials 35:2462–2470
216. Kim WL, Cho H, Li L, Kang HC, Huh KM (2014) Biomacromolecules 15:2224–2234
217. Gangopadhyay M, Singh T, Behara KK, Karwa S, Ghosh SK, Singh NDP (2015) Photochem Photobiol Sci 14:1329–1336
218. Jang WD, Nakagishi Y, Nishiyama N, Kawauchi S, Morimoto Y, Kikuchi M, Kataoka K (2006) J Control Release 113:73–79
219. Nishiyama N, Nakagishi Y, Morimoto Y, Lai PS, Miyazaki K, Urano K, Horie S, Kumagai M, Fukushima S, Cheng Y, Jang WD, Kikuchi M, Kataoka K (2009) J Control Release 133:245–251
220. Lu HL, Syu WJ, Nishiyama N, Kataoka K, Lai PS (2011) J Control Release 155:458–464
221. Fraix A, Kandoth N, Manet I, Cardile V, Graziano AC, Gref R, Sortino S (2013) Chem Commun (Camb) 49:4459–4461
222. Kandoth N, Vittorino E, Sciortino MT, Parisi T, Colao I, Mazzaglia A, Sortino S (2012) Chemistry 18:1684–1690
223. Sortino S (2010) Chem Soc Rev 39:2903–2913
224. Fraix A, Goncalves AR, Cardile V, Graziano AC, Theodossiou TA, Yannakopoulou K, Sortino S (2013) Chem Asian J 8:2634–2641
225. Swaminathan S, Garcia-Amoros J, Fraix A, Kandoth N, Sortino S, Raymo FM (2013) Chem Soc Rev 43:4167–4178
226. Kandoth N, Kirejev V, Monti S, Gref R, Ericson MB, Sortino S (2014) Biomacromolecules 15:1768–1776
227. Yoon HJ, Lim TG, Kim JH, Cho YM, Kim YS, Chung US, Kim JH, Choi BW, Koh WG, Jang WD (2014) Biomacromolecules 15:1382–1389
228. Thambi T, Park JH (2014) J Biomed Nanotechnol 10:1841–1862
229. Kaneda Y (2010) Expert Opin Drug Deliv 7:1079–1093
230. Huang Z, Yang Y, Jiang Y, Shao J, Sun X, Chen J, Dong L, Zhang J (2013) Biomaterials 34:746–755
231. Hammond P (2013) ACS Nano 7:3733–3735
232. Master AM, Livingston M, Oleinick NL, Sen GA (2012) Mol Pharm 9:2331–2338

Top Curr Chem (2016) 370: 113–134
DOI: 10.1007/978-3-319-22942-3_4
© Springer International Publishing Switzerland 2016

Inorganic Nanoparticles for Photodynamic Therapy

L. Colombeau, S. Acherar, F. Baros, P. Arnoux, A. Mohd Gazzali, K. Zaghdoudi, M. Toussaint, R. Vanderesse, and C. Frochot

Abstract Photodynamic therapy (PDT) is a well-established technique employed to treat aged macular degeneration and certain types of cancer, or to kill microbes by using a photoactivatable molecule (a photosensitizer, PS) combined with light of an appropriate wavelength and oxygen. Many PSs are used against cancer but none of them are highly specific. Moreover, most are hydrophobic, so are poorly soluble in aqueous media. To improve both the transportation of the compounds and the selectivity of the treatment, nanoparticles (NPs) have been designed. Thanks to

---

L. Colombeau and K. Zaghdoudi
LRGP, CNRS UMR-UL 7274, Université de Lorraine, ENSIC, 1 rue Grandville, BP 20451 - 54001 Nancy Cedex, France

LCPM, UMR CNRS-UL 7375, Université de Lorraine, ENSIC, 1 rue Grandville, BP 20451 - 54001 Nancy Cedex, France

S. Acherar and A.M. Gazzali
LCPM, UMR CNRS-UL 7375, Université de Lorraine, ENSIC, 1 rue Grandville, BP 20451 - 54001 Nancy Cedex, France

F. Baros, P. Arnoux, and C. Frochot (✉)
LRGP, CNRS UMR-UL 7274, Université de Lorraine, ENSIC, 1 rue Grandville, BP 20451 - 54001 Nancy Cedex, France

GDR3049 CNRS “Médicaments Photoactivables – Photochimiothérapie (PHOTOMED)”, 31062 Toulouse cedex 9, France
e-mail: Celine.frochot@univ-lorraine.fr

M. Toussaint
CRAN, CNRS UMR-UL 7039, Université de Lorraine, Campus Sciences, BP 70239 - 54506 Vandœuvre Cedex, France

R. Vanderesse
LCPM, UMR CNRS-UL 7375, Université de Lorraine, ENSIC, 1 rue Grandville, BP 20451 - 54001 Nancy Cedex, France

GDR3049 CNRS “Médicaments Photoactivables – Photochimiothérapie (PHOTOMED)”, 31062 Toulouse cedex 9, France

their small size, these can accumulate in a tumor because of the well-known enhanced permeability effect. By changing the composition of the nanoparticles it is also possible to achieve other goals, such as (1) targeting receptors that are over-expressed on tumoral cells or neovessels, (2) making them able to absorb two photons (upconversion or biphoton), and (3) improving singlet oxygen generation by the surface plasmon resonance effect (gold nanoparticles). In this chapter we describe recent developments with inorganic NPs in the PDT domain. Pertinent examples selected from the literature are used to illustrate advances in the field. We do not consider either polymeric nanoparticles or quantum dots, as these are developed in other chapters.

Keywords Inorganic nanoparticle • Photodynamic therapy • Singlet oxygen

Contents

Abbreviations

@	Coated
5-ALA	5-Aminolevulinic acid
APBA	Aminophenylboronic acid
AuNPs	Gold nanoparticles
C60	Fullerene
Capt	Captopril

Ce6	Chlorin e6
DNA	Deoxyribonucleic acid
DPBF	1,3-Diphenylisobenzofuran
Fmp	Fibronectin-mimetic peptide
FRET	Förster resonance energy transfer
GA	Gambogic acid
HA	Hyaluronic acid
HER-2	Human epidermal growth factor receptor 2
HNSCC	Head and neck squamous cell carcinomas
IC_{50}	Median growth inhibitory concentration
ICG	Indocyanine green
IO	Iron oxide
LDH	Layered double hydroxide
MB	Methylene blue
MRI	Magnetic resonance imaging
MSN	Mesoporous silica nanoparticles
MTAP	*meso*-tetra(*o*-Amino phenyl) porphyrin
MWNTs	Multiwalled carbon nanotubes
NPs	Nanoparticles
ORMOSIL	Organically modified silica
PAH	Poly(allylamine hydrochloride)
Pc	Phthalocyanine
Pc 4	Silicon phthalocyanine 4
PDT	Photodynamic therapy
PEG	Polyethylene glycol
PET	Phenylethanedithiol
PFHA	Perfluoroheptanoic acid
pHLIP	pH low insertion peptide
PpIX	Protoporphyrin IX
PS	Photosensitizer
PSS	Poly-sodium-4-styrenesulfonate
PTT	Photothermal therapy
RB	Rose bengal
RGD	Arginyl-glycyl-aspartic acid
RNA	Ribonucleic acid
ROS	Reactive oxygen species
SiNPs	Silica nanoparticles
SLPDT	Self-lighting PDT
SWNTs	Single walled carbon nanotubes
TAT	Trans-activator of transcription
UCN	Upconverting nanoparticles
ZnPc	Zinc(II) phthalocyanine

1 Introduction

Photodynamic therapy (PDT) is a well-established means for the treatment of aged macular degeneration, some types of cancer, and infectious microorganisms by using a photoactivatable molecule (a photosensitizer, PS) together with light of an appropriate wavelength and molecular oxygen. After excitation of the photosensitizer and energy transfer to the substrate or surrounding oxygen, reactive oxygen species (ROS) are produced, such as superoxide, hydroxyl radicals, and hydroperoxides for Type 1 pathways and singlet oxygen ($^1O_2$) for Type II pathways, which lead to photoreactions. Many photosensitizers are used in clinical applications, but none of them are very specific. To improve the PDT efficiency, it is crucial that the photosensitizers accumulate in diseased cells and avoid normal ones. Moreover, most photosensitizers belong to the porphyrins family, so are quite hydrophobic and thus poorly soluble in physiological liquids. This lack of solubility can lead to aggregation, which is disastrous for singlet oxygen formation. To improve both the transportation of the compounds and the selectivity of the treatment, nanoparticles (NPs) can be designed. Because of their small size, they can accumulate preferentially in tumors by the well-known enhanced permeability effect. By changing the composition of NPs, it is also possible to achieve others goals, such as targeting receptors that are over-expressed in tumoral cells or neovessels, allowing the NPs to absorb two photons (upconversion or biphoton), improving singlet oxygen generation by the surface plasmon resonance effect (gold nanoparticles), etc. Here we describe recent developments in the field of PDT with inorganic NPs. Compared to organic NPs (liposomes, micelles, microspheres, etc.) these provide the clear advantages of being photo-activatable without the need to release the photosensitizer. Another promising strategy is the release of the PS into cells, triggered by an acidic pH or by enzymes specific to tumoral zones.

In this chapter we describe the progress on leading NPs based on their interest for researchers in terms of publications, as presented in Fig. 1. We first describe

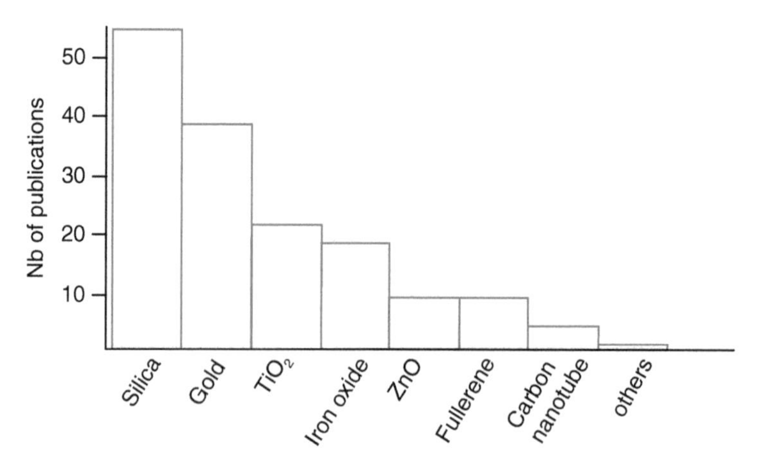

Fig. 1 Number of overall publications related to the type of NP developed for PDT applications

silica-based NPs then gold NPs, those that produce ROS after light excitation (except quantum dots), those that can be excited by sources other than one photon light, and those that can be used for theranostic tests.

2 Silica-Based Nanoparticles (SiNPs)

Silica NPs present several advantages such as controllable size and shape, high stability, and easy functionalization, which make them soluble in aqueous media and provide a high biocompatibility. Prasad [1] and Kopelman [2] have been pioneers in the elaboration of silica nanoparticles for PDT applications. Different synthetic methods can be used to elaborate various silica-based NPs such as hollow silica, organically modified silicate (ORMOSIL), mesoporous silica NPs (MSN), and other nanoparticles made by sol-gel methods. In 2010, we described the different methods for the synthesis of silica-based nanoparticles from pioneering works in the field to the latest achievements [3]. In another book chapter we detailed the main NPs possessing a three-dimensional rigid matrix that serve for PDT applications [4].

Photosensitizers can be encapsulated or covalently linked onto the surface of the silica NPs. In the case of internally grafted PSs, and to achieve a good PDT efficiency, molecular oxygen and $^1O_2$ should diffuse easily in and out of the silica nanoparticle (Fig. 2).

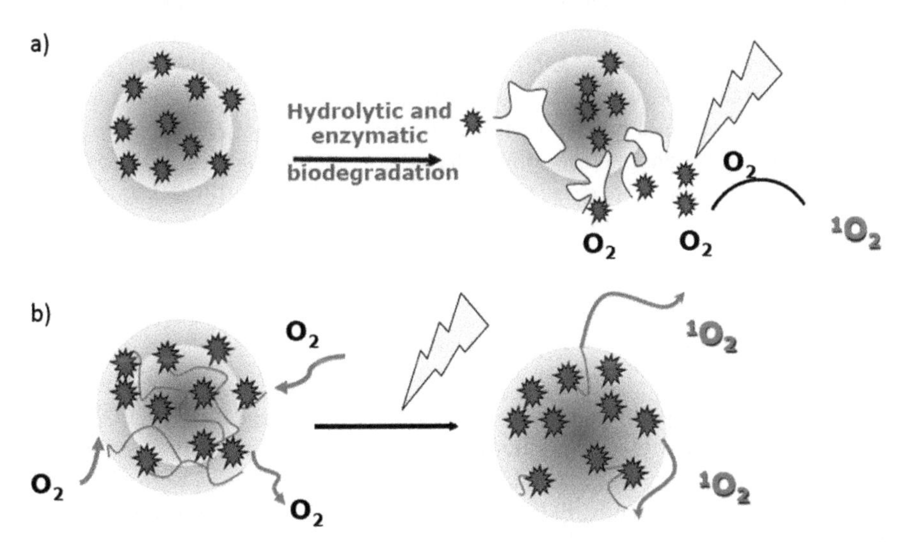

Fig. 2 (**a**) $^1O_2$ formation after release of the PS encapsulated into an NP. (**b**) $^1O_2$ formation after excitation of the PS covalently grafted into the matrix, diffusion of O_2 in the NP and $^1O_2$ out of the NP

To study the influence of the silica matrix on drug efficacy, Chu et al. [5, 6] designed two types of nanoparticles and evaluated the variation in efficacy. A loose SiO_2 nanoparticle with embedded methylene blue (SiO_2-MB) was elaborated that allowed good diffusion of both O_2 and methylene blue (MB). MB was also embedded in the surface layer of dense silica outside a gold nanorod Au@(SiO_2-Au). As expected, loose SiO_2 NPs showed enhanced ROS generation compared to dense SiO_2 NPs. Nevertheless, because of dimer formation of MB inside the NPs, the formation of $^1O_2$ was lower for MB inside the nanoparticle NPs than for MB in solution. Apoptosis was dominant in cells treated with free MB, loose and dense SiO_2-MB, whereas Au@(SiO_2-Au) induced necrosis.

2.1 *ORMOSIL*

The name ORMOSIL is shorthand for ORganically Modified SILica or Organically MOdified SILicate. These can be used for drug, DNA, or gene delivery. Most of the time they are elaborated in a sol-gel process by adding silane to silica-derived gel. In the PDT field, in 2003, Prasad's group prepared ORMOSIL NPs to entrap 2-devinyl-2-(1-hexyloxyethyl)pyropheophorbide (HPPH) [1]. A microemulsion technique was used, with vinyltriethoxysilane and aminopropyltriethoxysilane as reagents for the gelation procedure. The NPs killed HeLa and UCI-107 cancer cells very efficiently after irradiation at 650 nm. The same year, this same group pioneered the biological application of this kind of nanoparticle [7].

Tang et al. [2] used ORMOSIL NPs to encapsulate methylene blue. These authors compared the Stöber procedure to immobilize PS inside pure silica, with ORMOSIL nanoparticles prepared by methyltrimethoxysilane and phenyltrimethoxysilane as precursors. A higher PS loading was observed with the Stöber procedure but the production of $^1O_2$ after irradiation of the nanoparticles at 650 nm showed a higher kinetic rate with ORMOSIL than with Stöber NPs.

Since then, different photosensitizers have been coupled or encapsulated in this kind of silica NPs and more details can be found in Bechet et al. [8] and Vanderesse et al. [4].

2.2 *Hollow Silica*

Other types of silica particles have been designed to encapsulate PS. In 2008, Zhou et al. [9] prepared porous hollow silica nanospheres and embedded hypocrellin A in them. The particles presented good light and thermal stabilities and were efficiently incorporated into HeLa cells. PDT efficiency was better for embedded hypocrellin A than for free hypocrellin A because of higher $^1O_2$ generation.

Very recently, Liu et al. [10] performed in vivo experiments in nude mice for liver cancer. They compared the efficiency of free Photosan-II and Photosan-II-loaded hollow silica nanospheres of 37.8 nm mean diameter. They observed a better therapeutic efficiency with the NPs both in vitro and in vivo, with apoptosis involving a death-receptor exogenous pathway as well as a mitochondria endogenous pathway.

2.3 Mesoporous SiNPs (MSNs)

MSN are of great interest because of their particular properties such as large surface area, pore volume, and good biocompatibility. Moreover, it is easy to adjust their size and modify their surface. The first studies using MSN for PDT applications were carried out in 2009 by Mou's and Durand's teams. Mou and collaborators [11] coupled protoporphyrin IX (PpIX) to MSN. No dark cytotoxicity was observed up to 80 μg/mL. Uptake of the PpIX-MSNs by HeLa cells was high and both concentration and irradiation time-dependent cell viability were observed. A few weeks later, Brevet et al. [12] synthesized MSN nanoparticles coupled with water-soluble PS and mannose units attached to the surface as targeting units. The involvement of mannose receptors in MDA-MB-231 cell line experiments with these mannose-functionalized MSN was demonstrated as well as phototoxicity of the nano-objects. More recently, the same group developed two-photon excitation MSN [13] and further details of this are given in Sect. 5.2.

In 2014, Pan et al. [14] elaborated sophisticated MSN with two targeting units: RGD peptide to target integrin $\alpha_\nu\beta_3$ over-expressed on neovessels and the trans-activator of transcription (TAT) peptide for nuclear targeting. The photosensitizer (chlorin e6, noted Ce6) was strongly adsorbed onto the silica wall by electrostatic interaction. Although the formation of $^1O_2$ was lower with MSN-RGD/TAT than for Ce6 in solution, the nanoplatforms proved to be very efficient in vitro on HeLa cells and in vivo at very low power density (0.02 W/cm^2 for 5 min, light dose = 6 J/cm^2). Comparing these NPs to those without RGD and/or TAT, the authors proved that adding both a tumor vasculature targeting agent and a TAT peptide, which induces nuclear accumulation and DNA oxidation after irradiation, strongly enhanced PDT efficiency. Another way to improve the therapeutic index of PDT is to combine PDT with another treatment such as radiotherapy or photothermal therapy (PTT). For PTT, Zhao et al. [15] reported the synthesis of MSN coupled to tetra-substituted carboxyl aluminum phthalocyanines and covered by Pd nanosheets of different sizes. After incubation in HeLa cells of NPs with PS alone, NP with Pd sheet alone, or NPs with both phthalocyanine and Pd sheet, and excitation with a single laser at 660 nm, they concluded that there was a co-operative therapeutic effect of radiotherapy and PTT.

2.4 Sol-Gel Method

A very interesting approach has been described recently by Vivero-Escoto and Vega [16]. These authors synthesized stimuli-responsive silica NPs in which PpIX was attached to the surface of the nanoparticles through a redox-responsive (RR) linker. The PpIX stays attached to the nanoparticles via a disulfide bond which is cleaved when the NP reaches the cancer cells, because of the reducing environment. PpIX is released in its monomeric form, avoiding aggregation, and allowing high $^1O_2$ formation. They compared the behavior of this RR-PpIX-SiNP to a control NP in which the link between PpIX and the surface is a silane ligand (PpIX-SiNP). $^1O_2$ production was lower for PpIX attached to both NPs than PpIX in solution, probably because of the aggregation of PpIX into the NPs. When dithiothreitol was used as a reducing agent in solution, 90% of PpIX was released from RR-PpIX-SiNP after 6 days of incubation, whereas only 25% of PpIX was released from PpIX-SiNP after 10 days of incubation. In HeLa cells, a better phototoxicity was observed for RR-PpIX-SiNP than for PpIX-SiNP and the NPs were transported by endososomal pathway. However, the best phototoxicity was observed for PpIX in solution.

3 Gold Nanoparticles (AuNPs)

Among all the metal nanoparticles, gold NPs (AuNPs) have received particular attention, because of a combination of unique properties which lend them to multiple applications such as labeling, delivery, heating, and sensing. For the first time, in 2002 Russell's team [17] compared free phthalocyanine (Pc), Pc-coated AuNPs, and the effect of adding tetraoctylammonium bromide phase transfer reagent. They found an increased formation of $^1O_2$ of around 50% with the three-component Pc-coated AuNPs-phase transfer agent. Since this first study, around 40 papers have been published which describe the use of AuNPs for PDT applications. The same team [18] improved their systems and reported the elaboration of AuNPs functionalized with jacalin or anti-HER-2 antibodies, which target the over-expressed HER-2 receptor. Following PDT treatment, the destruction of SK-BR-3 cells is greater than that of HT-29 cells. A clear advantage of anti-HER-2 compared to jacalin is the lower levels of dark toxicity after incubation with either HT-29 or SK-BR-3 cells. Both types of nanoparticle conjugate were internalized through receptor mediated endocytosis. The next step is the combination loading of both targeting ligands onto a single NP for dual targeting.

Another recent example, published in 2014 by Prasanna et al. [19], described the preparation of gold–Rose bengal Nanoparticles (AuNP-RB), which are nanocomplexes with electrostatic and covalent bonding. The in vitro PDT efficiency of the free RB and the AuNP-RB was studied in Vero and HeLa cell lines. The results demonstrated that the nanocomplexes are more photo-cytotoxic than free RB, and that the covalent complex is more toxic than the electrostatic complex. It also appears that the PDT efficiency is higher in HeLa than in Vero cells.

Kawasaki et al. [20] reported the synthesis of organic-soluble $Au_{25}(PET)_{18}$ (PET = phenylethanethiol) and water-soluble $Au_{25}(Capt)_{18}$ (Capt = captopril) clusters. They showed that $^1O_2$ is generated by organic-soluble $Au_{25}(PET)_{18}$ and water-soluble $Au_{25}(Capt)_{18}$ clusters under visible/near infrared irradiation (532, 650, and 808 nm) in the absence of organic photosensitizers. In human serum, $Au_{25}(Capt)_{18}$ clusters produced more $^1O_2$ than new methylene blue. Cytotoxicity tests and in vitro cellular experiments were carried out in HeLa cells using water-soluble $Au_{25}(Capt)_{18}$ clusters. It was found that the size of clusters influence $^1O_2$ generation. It was also reported that the energy transfer to $^3O_2$ rather than to $^1O_2$ is enabled by a larger optical gap of $Au_{25}(PET)_{18}$ or $Au_{25}(Capt)_{18}$ clusters. Finally, the water-soluble clusters are advantageous for penetration into human tissue for PDT applications.

Some authors have studied the effect of the light dose. Shi et al. [21] analyzed the high/low PDT mode using methoxypoly(ethylene glycol) thiol modified gold nanorods coupled to the Al(III) phthalocyanine chloride tetrasulfonic acid (AlPcS4) photosensitizer. The polyelectrolyte-coated gold nanorods present multilayers of electrostatically coated negatively charged poly-sodium-4-styrenesulfonate (PSS) and positively charged branched polyethyleneimine. The complexes were synthesized by combining the negatively charged AlPcS4 with the positively charged NPs. The high/low PDT mode (high dose for short duration/low dose for long duration) was investigated in MCF-7 cell lines. First, the high power density light allowed the release of photosensitizers from the surface of NPs combined with the photothermal effect of NPs. The low density light achieved PDT efficiency because of the release of the photosensitizer, leading to the death of 90% of the tumor cells. This high/low PDT mode presented a very good efficiency and a risk-free PDT method.

Several publications report that the excitation efficiency of the photosensitizer can be increased, because of localized surface plasmon resonances. For example, Benito et al. [22] observed an increase in ROS formation by AuNPs coupled to 5-ALA when compared with 5-ALA alone and concluded that this may be caused by a plasmon effect. The same conclusion was reached by Chu et al. [5] who confined MB into a silica matrix in the close vicinity of gold nanorods and found that more hydroxyl radical and superoxide were generated than by both free MB and SiO_2-MB NPs. Nevertheless, free MB was found to produce the most $^1O_2$ among the three species.

Li et al. [23] reported the synthesis of a nanoplatform composed of mesoporous silica-coated AuNP (AuNR@SiO_2-ICG) incorporating indocyanine green (ICG) to increase PDT efficiency based on the surface plasmon resonance of AuNP. Compared to free ICG, the production of $^1O_2$ by AuNP@SiO_2-ICG is enhanced upon laser excitation. Li and co-workers also described the substantial in vitro and in vivo cellular uptake in the tumor region of ICG provided by the gold nanoplatform. They showed that $^1O_2$ generation allows the photodynamic destruction of the MDA-MB-231 human breast carcinoma cells. They also investigated the effects of PTT and

PDT, confirming that photothermal heating with low laser irradiation causes a decrease of tumor growth.

Indeed, much research has focused on the possibility of coupling PDT and PTT. Trinidad et al. [24] described the use of rat alveolar macrophages loaded with gold nanoshells as the drug delivery system for an in vitro study on head and neck squamous cell carcinoma using both PDT and PTT treatments. The combination of the two treatments (irradiation at $\lambda = 670$ nm for PDT with disulfonated aluminum phthalocyanine and $\lambda = 810$ nm for PTT) showed synergy and led to an increase of 40% cell death. No toxicity was observed for empty macrophages. Nanocomposites, made up of gold nanorod core and a hematoporphyrin-doped mesoporous silica shell on the surface, were used by Terentyuk and co-workers [25]. An in vivo study was carried out on alveolaris liver cancer PC-1 cells in rats using two wavelengths of irradiation ($\lambda = 633$ nm for PDT and 808 nm for PTT). When PDT was used alone, no conclusive result was found. In contrast, a drastic decline of tumor volume during the first 3–5 days was observed upon combining the two therapies ($\lambda = 633$ and 808 nm). This concept could be interesting for clinical use after optimization of irradiation conditions (light delivery, power density, and doses).

In an attempt to increase the accumulation of photosensitizer and hyperthermia agents in cancer cells, for an application of bimodal PDT/PTT therapy, Wang et al. [26] designed a new NP. This contains a pH-low insertion peptide (pHLIP) bound by a disulfide bridge, which confers extracellular pH driven targeting ability. Exposed to a reducing agent, such as DTT, the disulfide bond is cleaved, resulting in the liberation of Ce6. The authors clearly showed that the $^1O_2$ generation in the conjugate is better with than without DTT. These results demonstrated that Ce6 is phototoxic and that the conjugate is a good candidate for PDT. The photothermal analysis showed that the AuNP conjugates and the AuNPs alone are affected by laser irradiation with an increased temperature from 20 to 50 °C for AuNRs conjugates and 49 °C for AuNPs alone. These results suggest that the conjugate is a good hyperthermia agent for tumor therapy. The cytotoxicity studies in human lung cancer 95-C cell lines showed that the accumulation of Ce6-pHLIP-AuNPs in the cells is better at pH 6.2 than at pH 7.4, suggesting the utility of the bimodal PDT/PTT strategy for cancer therapy and potential clinical applications.

Vankayala and co-workers [27] were the first to show that AuNPs can produce $^1O_2$ upon NIR light activation alone in the absence of a photosensitizer. This team demonstrated that, by changing the wavelength of NIR irradiation, AuNPs can move from a PDT to PTT effect, or even a combination of the two. In vitro $^1O_2$ formation in HeLa cells was demonstrated by detecting $^1O_2$ phosphorescence emission. PDT and PTT were performed using two wavelengths ($\lambda = 550$ nm for PTT and 940 nm for PDT). Cell death generated by 940-nm LED light was ten times higher than that with 550-nm light. Concerning an in vivo study on B16F0 melanoma tumors in mice, the authors demonstrated that these tumors are completely destroyed when AuNPs are irradiated under very low LED/laser doses of single photon NIR (915 nm, <130 mW/cm^2), because of a PDT effect with the generation of $^1O_2$.

Another strategy that has begun to be developed is the use of triggered linker. Cheng et al. [28] coupled AuNPs to the silicon phthalocyanine Pc 4 drug via an Au–S covalent bond. The interesting property of this ligand is that it is photosensitive and induces hemolytic photocleavage of the Si–C bond after NIR irradiation ($\lambda = 660$ nm) that generates free Pc 4 once the ligand exchange with water has occurred on the central Si atom. Using HeLa cells, the authors showed that PDT efficacy is similar to that obtained with non-covalently bound Pc 4, but that this covalent linkage provides a better targeting of the Pc 4 drug release, because it is quenched by the AuNP during transport.

In 2014, Huh's group [29] described the synthesis, photoactivity, and in vitro and in vivo bioactivity of iron oxide/gold nanoparticles. These NPs are decorated on their surface by thiolated heparin-pheophorbide via an Au–S covalent bond. The originality of these multifunctional carriers is their self-quenching caused by FRET between pheophorbide a and gold, rendering the NPs photo-inactive during circulation. The photoactivity of the photosensitizer is regenerated after internalization in tumor cells because of the reductive cleavage of the gold–sulfur bond by glutathione. The quenching and dequenching were evaluated by fluorescence of the PS in the presence or absence of glutathione. The intensity of fluorescence at 670 nm after 60 min incubation was ninefold higher than the control without glutathione. The generation of $^1O_2$ was clearly shown with 9,10-dimethylanthracene used as a $^1O_2$ trap. The cellular uptake by A549 cells was determined by confocal laser scanning microscopy showing a strong PS signal of the internalized NPs, indicating the effect of cytoplasmic glutathione. Under light exposure, the cell viability decreased similarly whether the PS was free or released from NPs. Finally, a comparison of free PS and NPs in an animal model (nude mice xenografted with A459 cells) showed that NPs exhibit better tumor selectivity than free pheophorbide a and a prolonged time in the tumor, the latter being rapidly excreted from the body. Tumor-bearing mice were treated every 2 days by saline (control), free PS or NPs, and were exposed to light. The tumor volume and weight were measured and it could be clearly seen that NP treatment was 2.5-fold better than that of free PS.

4 Nanoparticles that Can Produce ROS by Themselves

4.1 ZnO

Zinc oxide NPs (ZnO) are also good candidates for biomedicine because of their low toxicity and unique optical and electronic properties. ZnO has a wide band gap (3.37 eV), high bond strength, and high exciton binding energy (60 meV) at room temperature. ZnO nanostructures typically have a near-band-edge emission in the UV region and a defect-related visible emission in the blue to green part of the spectrum. Upon UV excitation in aqueous solution, ZnO can generate ROS such as hydroxyl radical, hydrogen peroxide, and superoxide.

The first report of the use of ZnO NPs for PDT was presented in 2008 by Liu et al. [30]. They described the potential applications of ZnO NPs for cancer treatment. ZnO NPs were conjugated to *meso*-tetra(*o*-aminophenyl)porphyrin. After UV irradiation, FRET as high as 83% could be observed between ZnO and the PS. To evaluate the in vitro efficiency of these NPs, human carcinoma NIH:OVCAR-3 cells were exposed to UV (365 nm, 30 min) and ZnO, porphyrin, or ZnO–(L-cysteine)-porphyrin. The authors demonstrated that ZnO–(L-cysteine)-porphyrin conjugates allowed a high phototoxicity, compared to either ZnO or porphyrin alone, when exposed to UV. Since 2008, ten papers have reported the use of ZnO NPs for PDT application.

In 2014, Ismail et al. [31] described the effects of UV irradiation (320–400 nm) of ZnO NPs doped with Fe^{3+}, Ag^{+}, Pb^{2+}, or Co^{2+}, silica-coated ZnO-NPs, TiO_2-NPs, titanate nanotubes (TiO_2-NTs), and ZnO-NPs/TiO_2-NTs nanocomposites in HepG2 human liver adenocarcinoma cells. Doxorubicin was used as a standard. The median growth inhibitory concentration (IC_{50}) values for ZnO-NPs and metal-doped ZnO-NPs were, respectively, 42.6, 37.2, 45.1, 77.2, and 56.5 μg/mL as compared to doxorubicin (IC_{50} value 20.1 μg/mL), proving good antitumor activity. Silica-coated ZnO-NPs, TiO_2-NPs, titanate nanotubes (TiO_2-NTs), and ZnO-NPs/TiO_2-NTs nanocomposites did not demonstrate any antiproliferative activity. ZnO-NPs such as metal-doped ZnO-NPs (Fe^{3+}, Ag^{+}, Pb^{2+}, and Co^{2+}) showed significant antitumor activity in comparison to doxorubicin. In addition, high production of hydrogen peroxide, nitric oxide, and other free radicals was observed.

4.2 TiO_2

TiO_2 NPs have been considered as photosensitizing agents for PDT because of their unique phototoxic effect upon UV irradiation. They are novel photo-effecting materials with band gaps of 3.23 eV for anatase and 3.06 eV for rutile polymorph of TiO_2, respectively. Upon irradiation with UV light, electrons from the valence band are promoted to the conduction band. This results in the formation of photo-induced electron and hole pairs that have strong reduction and oxidation properties. In water, the photo-induced holes react with hydroxyl ions or water to produce hydroxyl radicals, perhydroxyl radicals, and superoxide anions, which allow the destruction of organic molecules. The first report of the use of TiO_2 nanoparticles for PDT application was in 1991 by Fujishima's team [32, 33]. Since then, around 20 papers have described the interest in TiO_2 NPs. We focus here on the more recent studies.

Zhang et al. [34] synthesized four types of TiO_2 nanofibers using anatase particles. In this study, single crystal TiO_2 (B) (calcination at 400 °C) is noted as T(B), mixed phase anatase/TiO_2 (calcination at 600 °C) is noted as T(AB)_T, and anatase (calcination at 700 °C) is noted T(A) nanofibers. Another core–shell structure is elaborated with mixed phase anatase/TiO_2 (B) nanofibers and is denoted as T(AB)_H. HeLa cells were incubated with the four types of fibers and irradiated with UV light for 30 min. Only 4% of the HeLa cells were killed by the UV light in the absence of TiO_2, 29% and 33% of the cells were killed by the pure TiO_2(B) and anatase nanofibers, and 41% and 46% of the cells were killed by T(AB)_T and T (AB)_H. These results demonstrate the photodynamic effect of the nanofibers produced by the active generation of ROS following UV light irradiation. The apoptosis ratios were measured and found to be 35%, 37%, 40%, and 43% for T(B), T(A), T(AB)_T, and T(AB)_H, respectively. The maximum results were obtained when mixed types of fibers were involved. The authors concluded that a good cell-killing efficiency of the mixed-phase nanofibers could be attributed to the band gap and stable interface between TiO_2(B) and anatase in the same nanofiber. This is a valid way of reducing the recombination of photogenerated electrons and holes. In this way, more ROS can be generated to kill cancer cells.

Li et al. [35] tested a novel nanocomposite, blending TiO_2 nanofibers with gambogic acid (GA), in vitro on human HepG2 liver adenocarcinoma cells under UV irradiation in aqueous media. GA is the major active component of gamboge resin secreted from *Garciniahanburyi* trees in Southeast Asia. A significant antitumor activity has been demonstrated for GA in vitro and in vivo. The Chinese FDA has approved a Phase II GA clinical trial. The mesoporosity of the fibers allows them to act as anticancer drug delivery carriers. Moreover, GA-TiO_2 nanocomposites have high pH sensitivity for efficient drug release into the targeted tumoral areas. The cytotoxicity of GA and GA-TiO_2 has been evaluated in HepG2 cancer cells. Cell inhibition depended on dose and time. The inhibitory concentration IC_{50} was lower with GA-TiO_2 than with GA, and much lower with GA-TiO_2 irradiated by UV light, showing a photodynamic effect. The morphological observation of the HepG2 cancer cells revealed that the presence of GA-TiO_2 irradiated with UV light led to the most serious damage and maximal apoptosis (59.3%). In summary, GA-TiO_2 nanocomposites irradiated by UV light show a time- and dose-dependent antiproliferative activity. Observations demonstrate that these NPs enhance apoptosis of the HepG2 cancer cells compared to GA or TiO_2 used separately.

4.3 Fullerene

Fullerenes are made of carbon and can present different forms such as hollow spheres, ellipsoids, or tubes. Under light irradiation they can produce ROS. In 2006, for the first time, Davydenko et al. [36] reported the sensibilization of C_{60} fullerene coupled to non-porous pyrogenic silica NPs modified by γ-aminopropyl groups. The fullerene generated ROS in solution, but neither in vitro nor in vivo experiments were performed. Only five publications about PDT and fullerene could be found dating from 2006 to 2013, but in 2014 alone, four papers describe the potential of these NPs in the PDT field.

Zhang's team [37] recently reported the elaboration of an NP composed of C_{60} fullerene and 5-aminolevulinic acid (C_{60}-5-ALA), the precursor of PpIX. In vivo phototoxicity experiments on mice with melanoma (B16-F10 cells) were performed with this new complex after irradiation at 630 nm. Induced PpIX was observed both in vitro and in vivo. The comparison with 5-ALA alone showed a better phototoxicity of the fullerene complex in vivo (about 50 vol.% volume for the tumor after 11 days of treatment) and in vitro (survival rate of 51.4 and 68.8% respectively for C_{60}-5-ALA and ALA alone).

Another well-designed nanoparticle made of fullerene, iron oxide (IO), PEG, and conjugated to folic acid (FA) was developed by Shi et al. [38]. No dark cytotoxicity could be observed in MCF-7 human breast cancer cells. After excitation at 532 nm, C_{60}-IONP-PEG presents a lower phototoxicity than the targeted C_{60}-IONP-PEG-FA. ROS production in vitro increased with the accumulation of the NPs. In vitro, the PDT effect was enhanced by the combination with photothermal therapy (PTT). In vivo, PDT and PTT treatments induced 62% and 37% apoptosis, respectively. Combined PTT and PDT increased this rate up to 96%.

Asada et al. [39] developed fullerene coupled to PEG. No dark toxicity was observed in human HT1080 fibrosarcoma cells after incubation with PEG-fullerene. The formation of radicals is 6.5 times greater under visible light irradiation using a halogen lamp than under natural light (outdoor) irradiation. Nuclear fragmentation and chromatin condensation were observed, indicating that fullerene was able to penetrate through the cell membrane and reach the cytoplasm.

4.4 Carbon Nanotubes and Graphene

Single walled (SWNTs) or multiwalled carbon nanotubes (MWNTs) can also be used as delivery agents for PDT photosensitizers. Carbon nanotubes offer many advantages, such as good internalization through endocytosis in mammalian cells, fast elimination, no significant cytotoxicity, and easy chemical design. Zhu et al. [40] linked non-covalently a single strand DNA aptamer with a chlorin e6 as photosensitizer and a SWNT. Chlorin e6 can produce $^1O_2$ upon irradiation. Because of π-stacking between the DNA aptamer and the SWNT, energy transfer occurs between the PS and the SWNT and this leads to 98% $^1O_2$ quenching. DNA aptamers are good linkers to biological targets, illustrated by the chlorin e6 fluorescence that increased up to 20-fold after the addition of 2.0 μM thrombin. $^1O_2$ production is restored when the construct probe is placed in the presence of thrombin and then irradiated at the maximum absorption wavelength of chlorin e6 (404 nm). The high specificity of aptamers towards one biological target and the quenching efficiency of the probe are two of many qualities that make an agent relevant for PDT.

5 Nanoparticles with Other Types of Excitation than One Photon

5.1 Upconverting Nanoparticles (UCNs): NPs Transduce Low-Energy Light to Higher Energy Emissions that Activate Associated PS

To treat deep tumors, in some cases it is necessary to excite the photosensitizer or the NPs in the red part of the light spectrum. Some photosensitizers can be designed to absorb at high wavelength, but another strategy is to develop upconverting nanoparticles (UCNs) and two-photon absorption NPs. In both cases, excitation light at a longer wavelength produces emission at a shorter wavelength and it becomes possible to excite the photosensitizers in the NIR. Upconversion is mainly observed in the rare earth elements, such as the lanthanide series, yttrium, and scandium. Three different upconversion luminescence mechanisms have been recognized: excited state absorption, photon avalanche, and energy transfer upconversion. This last mechanism is observed in the system of $NaYF^4$:$Yb^{3+}Er^{3+}$ UCNs frequently used in PDT.

A multifunctional nanoplatform was synthesized by Yin et al. [41] with a lanthanide core and a mesoporous shell. The core material is $NaGdF_4$:Yb 25%, Tm 0.3%, which allows energy transfer from the lanthanides luminescence to the UV absorption of TiO_2 by an upconversion process. The energy transfer was confirmed by observing a blue emission and the generation of hydroxyl radicals.

Doxorubicin was loaded as a guest molecule in the mesoporous TiO_2 and the pores were capped by hyaluronic acid (HA) specific to CD44, a receptor that is over-expressed at the surface of various tumor cells. In vitro tests were performed with MDA-MB-231 cells, which express CD44 abundantly. The authors demonstrated that, in these cells, HA-capped NPs were more internalized than non-targeted NPs and ROS formation was detected. No toxicity was encountered, either from direct irradiation alone or from NPs alone. Finally, MDA-MB-231 cells were incubated with increasing concentrations of the mesoporous TiO_2 upconverting nanoparticle-hyaluronic acid with or without doxorubicin. The therapeutic effect of mesoporous TiO_2 with doxorubicin was better with the combination of PDT and chemotherapy than with PDT alone.

A very sophisticated and clever approach to nano-object was developed by Wang et al. [42]. For the first time in the field of PDT, upconversion nanoparticles with two different targeting units were designed. One of the main drawbacks of fullerenes is that they have to be excited by UV/visible light, the penetration of such light into the tissues being limited. To overcome this problem, upconversion nanoparticle (UCN) can be used with excitation in the red part of the spectrum. The authors demonstrated FRET between the UCN and fullerene C_{60}. Moreover, to increase the selectivity for tumoral cells, they coupled two types of targeting ligands – aminophenylboronic acid (APBA), which is very specific for polysialic acid, and hyaluronic acid (HA), which targets over-expressed CD44 receptors on cancer cells.

PC12 cancer cells were incubated with UCPs-C_{60}, APBA-UCNs, and HA-C_{60}. No dark cytotoxicity was detected. A better cellular uptake was observed for APBA-UCNs-C_{60}-HA than for the single targeted probes APBA-UCNs and HA-C_{60}. The therapeutic efficacy was evaluated by further measurement of ROS in vitro after 980-nm excitation. The synergic targeting effect led to improved incorporation and better phototoxicity of the dual-targeting nanoparticles. In vivo experiments are in progress.

5.2 Two-Photon Absorption Nanoparticles

Two-photon absorption-induced excitation of photosensitizers is a promising approach to improve the efficiency of PDT by increasing the penetration of light penetration into tissues. After two-photon excitation, the photosensitizers absorb simultaneously two less-energetic photons at higher wavelength than one photon excitation. The excitation in the NIR region prevents absorbing or scattering by tissues. Very recently, Secret et al. [13] synthesized two-photon excited porphyrin-functionalized porous silicon NPs. Porous silicon NPs were coupled to mannose *via* two different linkers – either a phenyl squarate or a ketone moiety – to porphyrin, to both mannose and porphyrin, or neither porphyrin nor mannose. Nanoparticles with mannose units accumulated in MCF-7 human breast cancer cells through a mannose receptor-mediated endocytosis mechanism. Secret et al. compared PDT efficiency after one- (650 nm) or two-photon excitation (800 nm). The best system for inducing cell death by PDT was NPs containing both porphyrin and mannose, coupled via the ketone moiety. The phototoxicity was 2.3-fold better with two-photon PDT than one-photon PDT. Chen et al. [43] described the synthesis of gold nanorods coated with a mesoporous silica shell having PdTPPs as photosensitizers on their surface. In vivo and in vitro PDT studies after two-photon excitation (800 nm, 96 J ($\approx$250 J/cm^2)) were performed on MDA-MB-231 breast cancer cells in mice. The authors showed that the PdTPPs on the cancer cell surface could be excited by intra-particle plasmonic resonance energy transfer from the two-photon excited AuNP core that enables $^1O_2$ generation.

5.3 Self-Illuminating Nanoparticles: Scintillation on Excitation by X-Rays Activates Attached PS

A new concept combines the principle of radiotherapy and PDT. Luminescent NPs with attached photosensitizers are used as a new type of agent for PDT. Upon exposure to ionizing radiation such as X-rays, light is emitted from the

Fig. 3 Principle of self-lighting PDT

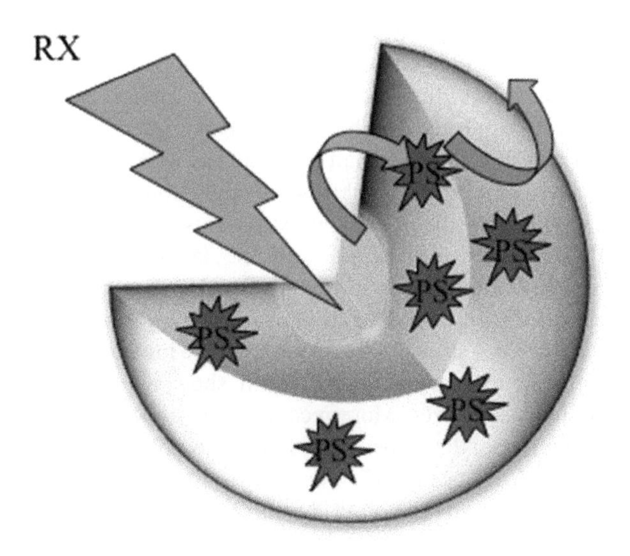

nanoparticles and activates the photosensitizers. As a consequence, $^1O_2$ is produced, which induces the destruction of malignant cells by ionizing radiation. With this novel therapeutic approach, no external light is necessary to activate the photosensitizing agent within the tumor tissue. This new modality is called nanoparticle Self-Lighting PDT (SLPDT) (Fig. 3). It is hoped that this original concept of cancer treatment is more efficient than classical concepts for deep tumors and requires a lower radiation dose than conventional radiotherapy. Few publications describe the potential of NPs that enable simultaneous radiation and PDT. Chen and Zhang have been the pioneers in this field [44, 45]. In 2006, they described the synthesis of LaF_3:Tb^{3+}-*meso*-tetra(4-carboxyphenyl)porphine NP conjugates and investigated the energy transfer as well as the formation of $^1O_2$ following X-ray irradiation [30].

Because X-ray absorption is driven by the square of the effective atomic number, heavy metal NPs are required to enhance the phenomenon. The same team also elaborated ZnO-*meso*-(*o*-aminophenylporphyrin) conjugate, in which the energy transfer is around 89%. In 2009, Morgan et al. [46] evaluated the physical parameters required for nanoscintillators to deliver cytotoxic levels of $^1O_2$ at therapeutic radiation doses drawing on the published literature from several disparate fields. They found that the most important parameters to release $^1O_2$ at therapeutic radiation doses are light yield to the NPs, efficiency of energy transfer between NPs, and the PS cellular uptake. Moreover, their calculations showed that combination therapy might be limited to X-ray energies below 300 keV.

In 2011, Scaffidi et al. [47] described the Y_2O_3 nanoscintillator, a fragment of the HIV-1 TAT peptide and psoralen. They used commercially available 12 nm diameter cubic-phase Y_2O_3 nanoscintillators, onto which they coupled TAT or psoralen-TAT. They observed a modest in vitro reduction in PC-3 human prostate cancer cells numbers after X-ray excitation (2 Gy at 160 or 320 keV). In the same year, Abliz et al. [48] described the excitation by X-rays of photofrin II incubated with gadolinium oxysulfide particles doped with terbium Gd_2O_2S:Tb (20 μm dimension) in glioblastoma cells. These authors observed a 90% decrease of cell viability produced by the concomitant presence of photofrin II and Gd_2O_2S:Tb whereas the NPs alone had no effect and even conferred protection against X-rays.

Our team recently published a study concerning X-ray-induced $^1O_2$ activation with nanoscintillator-coupled porphyrins [49]. A biocompatible nanoscintillator of Tb_2O_3 coated with a polysiloxane layer exhibited an appropriate pattern of biodistribution in vivo after injection. Using time-resolved laser spectroscopy and $^1O_2$ chemical probes, we demonstrated $^1O_2$ formation after X-ray irradiation of the nano-objects, as well as a decrease in the luminescence of the core and an increase in the fluorescence of the porphyrin, showing energy transfer between the core and the photosensitizer. More recently, Zou et al. [50] studied the potential of Ce^{3+}-doped LaF_3 NPs loaded with PpIX into poly-lactic-*co*-glycolic acid microspheres for X-ray induced PDT. They noticed that X-ray excitation (2 Gy) of the 2 μm diameter nanospheres in PC3 cells exposed to the microspheres during 24 h induced a decrease of the cell survival compared to control or microsphere without PpIX. Because of the large size of microspheres, the uptake is low but the nanocompounds are more efficient with less dose than previous work [44, 30].

6 Iron Oxide and Theranostic NPs

Theranostic NPs are a class of NPs that allow both imaging and therapy. In the field of PDT, they are mainly designed to allow magnetic resonance imaging signal enhancement (visualization of tumoral areas) and to serve for PDT treatment.

Narsireddy et al. [51] synthesized magnetite/gold nanoparticles by the hydrolysis of Fe_3O_4 and chitosan. The positive surface of these NPs was treated with gold particles followed by lipoic acid, which plays the role of a linker to attach 5,10,15,20-tetrakis(4-hydroxyphenyl)-21*H*,23*H*-porphyrin onto gold. For targeting, a human epidermal growth factor receptor specific peptide was grafted onto the chitosan with a peptide/NP ratio of 0.05/1 (w/w). It was shown that the generation of $^1O_2$ by free and bound PS was the same. As expected, cellular uptake was better for the NPs than free PS and even better with the peptide in SK-OV-3 cells. A xenograft tumor (SK-OV-3) model in nude mice was used to study the biodistribution of the PS and it appears that the peptide promotes accumulation of NPs in the tumor tissue compared to control and NPs without a targeting agent. The only organ which significantly accumulates the NPs is spleen. In vivo PDT experiments showed that the best treatments were obtained when the PS was at a high dose, or using the NP-peptide. The dose of NP-peptide needed to slow tumor growth is comparable to that of Foscan.

Recently, Fei's team [52] developed commercially available targeted amphiphilic polymer coated iron oxide nanoparticles (IO) for PDT and imaging of head and neck squamous cell carcinomas (HNSCC). The polymer was functionalized by carboxylic acid, which allowed bioconjugation with fibronectin-mimetic peptide (Fmp), an integrin β1 ligand. The phthalocyanine Pc 4 photosensitizer was encapsulated. In vitro experiments on four HNSCC cell lines (M4E, ME4-15, 686LN, and TU212) showed that, after laser irradiation, Fmp-IO-Pc 4 and IO-Pc 4 show the same efficacy as free Pc 4 with respect to cell survival. Binding tests on M4E at 4 °C showed that only Fmp-IO-Pc 4 was bound to cells in contrast to the untargeted NP. In vivo experiments in an HNSCC animal model indicated that the NPs slow the tumor growth, whereas Pc 4 alone slightly reduced the tumor size. This result was particularly significant with Fmp-IO-Pc 4 compared to non-targeted NP (a difference by a factor of 2.7). Biodistribution demonstrated the presence of Pc 4 essentially in the tumor, lung, liver, kidney and muscle. The fluorescence signal remains for more than 48 h for Fmp-IO-Pc 4, indicating a higher accumulation of the targeted NP in the tumor.

Yoon and coworkers [53] elaborated nanocapsules based on superparamagnetic iron oxide–polystyrene NPs on which they deposited alternatively multiple layers of positively charged poly(allylamine hydrochloride) (PAH) and negatively charged dendrimer porphyrin. Pegylation was then applied to increase NP stability. Until there are three layers of PAH and three layers of dendrimer porphyrin, the magnetic properties are compatible with an MRI application. The generation of ROS under laser light was detected in HeLa cells in the presence of nanocapsules with a 3.5-fold higher yield than that of the control (cells treated only with light). As expected, the results of the study clearly demonstrate that the more PAH/dendrimer porphyrin bilayers the nanocapsules contain, the more the cell viability decreases.

Our team is also involved in the design of multifunctional NPs. We analyzed particles with a gadolinium oxide core for MRI surrounded by a silica shell in which a photosensitizer is covalently coupled (porphyrin or chlorin), coated with polyethylene glycol to which a peptide targeting the neuropilin-1 receptor is covalently linked [54]. We showed that loading of the PS influences the photophysical properties of the nano-objects [55] and that increasing the concentration of PS reduces the fluorescence and singlet oxygen quantum yield because of short and long energy transfers. We demonstrated that these nanoparticles can be used both as MRI probes and as PDT agents [56] for vascular-targeted PDT of brain tumors guided by real-time [57]

7 Mg-Al Hydroxides

Layered double hydroxides (LDHs) are synthetic clay materials which form successive layers of metal hydroxides separated by layers of anions and water. A publication by Kantonis et al. [58] described the immobilization of PpIX on LDHs. The authors synthesized a nanohybrid of LDH coupled to perfluoroheptanoic acid (LDH-PFHA), a perfluorocarbon that can dissolve a large amount of oxygen. They evaluated the production of $^1O_2$ and the oxidation of various substrates for both PpIX and PpIX in LDH and showed that the reaction rate depends on both the catalyst and the type of substrate. For example, when the substrate is imidazole, free PpIX reacts five times faster than LDH-PpIX and LDH-PpIX-PFHA. In the case of hydrophobic substrates such as linoleic acid, LDH-PpIX-PFHA reacts four to seven times faster than LDH-PpIX or free PpIX.

8 Conclusion

Many types of inorganic NPs can be used to improve the PDT efficiency in term of vectorization, theranostics, or increased photosensitizer loading. Each type of NP presents some advantages. For example, for AuNPs it seems that their surface plasmonic effect could enhance singlet oxygen quantum yield. Moreover, they can also be used as phototherapy agents and AuNPs are already in Phase I studies for cancer treatment. With silica NPs, the diffusion of O_2 into the silica shell or the pores could increase the local concentration and efficiency of PDT. For graphene, some authors have claimed that the adsorption of MB onto graphene oxide enhanced $^1O_2$ quantum yield, compared to MB alone. Parameters such as chemical composition of the NPs, their size, and loading of the photosensitizers could be taken into account to evaluate the benefit of encapsulation/grafting on the formation and consumption of $^1O_2$ and PDT efficiency.

References

1. Roy I, Ohulchanskyy TY, Pudavar HE, Bergey EJ, Oseroff AR, Morgan J, Dougherty TJ, Prasad PN (2003) J Am Chem Soc 125:7860–7865
2. Tang W, Xu H, Kopelman R, Philbert MA (2005) Photochem Photobiol 81:242–249
3. Couleaud P, Morosini V, Frochot C, Richeter S, Raehm L, Durand JO (2010) Nanoscale 2:1083–1095
4. Vanderesse R, Frochot C, Barberi-Heyob M, Richeter S, Raehm L, Durand JO (2011) Nanoparticles for photodynamic therapy applications. In: Prokop A (ed) Intracellular delivery: fundamentals and applications, vol 5, Fundamental biomedical technologies. Springer, New York, pp 511–565
5. Chu Z, Yin C, Zhang S, Lin G, Li Q (2013) Nanoscale 5:3406–3411
6. Chu ZQ, Zhang SL, Yin C, Lin G, Li Q (2014) Biomater Sci 2:827–832
7. Prasad PN (2003) Introduction to biophotonics. Wiley, Hoboken
8. Bechet D, Couleaud P, Frochot C, Viriot ML, Guillemin F, Barberi-Heyob M (2008) Trends Biotechnol 26:612–621
9. Zhou J, Zhou L, Dong C, Feng Y, Wei S, Shen J, Wang X (2008) Mater Lett 62:2910–2913
10. Liu ZT, Xiong L, Liu ZP, Miao XY, Lin LW, Wen Y (2014) Nanoscale Res Lett 9:319
11. Tu HL, Lin YS, Lin HY, Hung Y, Lo LW, Chen YF, Mou CY (2009) Adv Mater 21:172–177
12. Brevet D, Gary-Bobo M, Raehm L, Richeter S, Hocine O, Amro K, Loock B, Couleaud P, Frochot C, Morère A, Maillard P, Garcia M, Durand JO (2009) Chem Commun 1475–1477
13. Secret E, Maynadier M, Gallud A, Chaix A, Bouffard E, Gary-Bobo M, Marcotte N, Mongin O, El Cheikh K, Hugues V, Auffan M, Frochot C, Morere A, Maillard P, Blanchard-Desce M, Sailor MJ, Garcia M, Durand JO, Cunin F (2014) Adv Mater 26:7643–7648

14. Pan LM, Liu JA, Shi JL (2014) Adv Funct Mater 24:7318–7327
15. Zhao ZX, Huang YZ, Shi SG, Tang SH, Li DH, Chen XL (2014) Nanotechnology 25:285701
16. Vivero-Escoto JL, Vega DL (2014) RSC Adv 4:14400–14407
17. Hone DC, Walker PI, Evans-Gowing R, FitzGerald S, Beeby A, Chambrier I, Cook MJ, Russell DA (2002) Langmuir 18:2985–2987
18. Obaid G, Chambrier I, Cook MJ, Russell DA (2015) Photochem Photobiol Sci. doi:10.1039/c4pp00312h
19. Prasanna SW, Poorani G, Kumar MS, Aruna P, Ganesan S (2014) Mater Express 4:359–366
20. Kawasaki H, Kumar S, Li G, Zeng CJ, Kauffman DR, Yoshimoto J, Iwasaki Y, Jin RC (2014) Chem Mater 26:2777–2788
21. Shi ZZ, Ren WZ, Gong A, Zhao XM, Zou YH, Brown EMB, Chen XY, Wu AG (2014) Biomaterials 35:7058–7067
22. Benito M, Martin V, Blanco MD, Teijon JM, Gomez C (2013) J Pharm Sci 102:2760–2769
23. Li YY, Wen T, Zhao RF, Liu XX, Ji TJ, Wang H, Shi XW, Shi J, Wei JY, Zhao YL, Wu XC, Nie GJ (2014) ACS Nano 8:11529–11542
24. Trinidad AJ, Hong SJ, Peng Q, Madsen SJ, Hirschberg H (2014) Lasers Surg Med 46:310–318
25. Terentyuk G, Panfilova E, Khanadeev V, Chumakov D, Genina E, Bashkatov A, Tuchin V, Bucharskaya A, Maslyakova G, Khlebtsov N, Khlebtsov B (2014) Nano Res 7:325–337
26. Wang NN, Zhao ZL, Lv YF, Fan HH, Bai HR, Meng HM, Long YQ, Fu T, Zhang XB, Tan WH (2014) Nano Res 7:1291–1301
27. Vankayala R, Huang YK, Kalluru P, Chiang CS, Hwang KC (2014) Small 10:1612–1622
28. Cheng Y, Doane TL, Chuang CH, Ziady A, Burda C (2014) Small 10:1799–1804
29. Li L, Nurunnabi M, Nafiujjaman M, Jeong YY, Lee YK, Huh KM (2014) J Mater Chem B 2:2929–2937
30. Liu Y, Zhang Y, Wang S, Pope C, Chen W (2008) Appl Phys Lett 92:043901
31. Ismail AFM, Ali MM, Ismail LFM (2014) J Photochem Photobiol B Biol 138:99–108
32. Cai R, Hashimoto K, Itoh K, Kubota Y, Fujishima A (1991) Bull Chem Soc Jpn 64:1268–1273
33. Cai R, Kubota Y, Shuin T, Sakai H, Hashimoto K, Fujishima A (1992) Cancer Res 52:2346–2348
34. Zhang SC, Yang DJ, Jing DW, Liu HW, Liu L, Jia Y, Gao MH, Guo LJ, Huo ZY (2014) Nano Res 7:1659–1669
35. Li JY, Wang XM, Shao YX, Lu XH, Chen BA (2014) Materials 7:6865–6878
36. Davydenko MO, Radchenko EO, Yashchuk VM, Dmitruk IM, Prylutskyy YI, Matishevska OP, Golub AA (2006) J Mol Liq 127:145–147
37. Li Z, Pan LL, Zhang FL, Zhu XL, Liu Y, Zhang ZZ (2014) Photochem Photobiol 90:1144–1149
38. Shi JJ, Wang L, Gao J, Liu Y, Zhang J, Ma R, Liu RY, Zhang ZZ (2014) Biomaterials 35:5771–5784
39. Asada R, Liao F, Saitoh Y, Miwa N (2014) Mol Cell Biochem 390:175–184
40. Zhu Z, Tang Z, Phillips JA, Yang R, Wang H, Tan W (2008) J Am Chem Soc 130:10856–10857
41. Yin ML, Ju EG, Chen ZW, Li ZH, Ren JS, Qu XG (2014) Chem Eur J 20:14012–14017
42. Wang X, Yang CX, Chen JT, Yan XP (2014) Anal Chem 86:3263–3267
43. Chen NT, Tang KC, Chung MF, Cheng SH, Huang CM, Chu CH, Chou PT, Souris JS, Chen CT, Mou CY, Lo LW (2014) Theranostics 4:798–807
44. Chen W (2008) J Biomed Nanotechnol 4:369–376
45. Chen W, Zhang J (2006) J Nanosci Nanotechnol 6:1159–1166
46. Morgan NY, Kramer-Marek G, Smith PD, Camphausen K, Capala J (2009) Radiat Res 171:236–244
47. Scaffidi JP, Gregas MK, Lauly B, Zhang Y, Vo-Dinh T (2011) ACS Nano 5:4679–4687
48. Abliz E, Collins JE, Bell H, Tata DB (2011) J Xray Sci Technol 19:521–530
49. Bulin AL, Truillett C, Chouikrat R, Lux F, Frochot C, Amans D, Ledoux G, Tillement O, Perriat P, Barberi-Heyob M, Dujardin C (2013) J Phys Chem C 117:21583–21589

50. Zou X, Yao M, Ma L, Hossu M, Han X, Juzenas P, Chen W (2014) Nanomedicine 9:2339–2351
51. Narsireddy A, Vijayashree K, Irudayaraj J, Manorama SV, Rao NM (2014) Int J Pharm 471:421–429
52. Wang DS, Fei BW, Halig LV, Qin XL, Hu ZL, Xu H, Wang YA, Chen ZJ, Kim S, Shin DM, Chen Z (2014) ACS Nano 8:6620–6632
53. Yoon HJ, Lim TG, Kim JH, Cho YM, Kim YS, Chung US, Kim JH, Choi BW, Koh WG, Jang WD (2014) Biomacromolecules 15:1382–1389
54. Couleaud P, Bechet D, Vanderesse R, Barberi-Heyob M, Faure AC, Roux S, Tillement O, Porhel S, Guillemin F, Frochot C (2011) Nanomedicine 6:995–1009
55. Seve A, Couleaud P, Lux F, Tillement O, Arnoux P, Andre JC, Frochot C (2012) Photochem Photobiol Sci 11:803–811
56. Benachour H, Seve A, Bastogne T, Frochot C, Vanderesse R, Jasniewski J, Miladi I, Billotey C, Tillement O, Lux F, Barberi-Heyob M (2012) Theranostics 2:889–904
57. Bechet D, Auger F, Couleaud P, Marty E, Ravasi L, Durieux N, Bonnet C, Plénat F, Frochot C, Mordon S, Tillement O, Vanderesse R, Lux F, Perriat P, Guillemin F, Barberi-Heyob M (2015) Nanomed Nanotechnol Biol Med. doi:10.1016/j.nano.2014.12.007
58. Kantonis G, Trikeriotis M, Ghanotakis DF (2007) J Photochem Photobiol A Chem 185:62–66

Top Curr Chem (2016) 370: 135–168
DOI: 10.1007/978-3-319-22942-3_5
© Springer International Publishing Switzerland 2016

Photoactivatable Nanostructured Surfaces for Biomedical Applications

Jiří Mosinger, Kamil Lang, and Pavel Kubát

Abstract This review aims to summarize the current status of photoactivatable nanostructured film and polymeric nanofiber surfaces used in biomedical applications with emphasis on their photoantimicrobial activity, oxygen-sensing in biological media, light-triggered release of drugs, and physical or structural transformations. Many light-responsive functions have been considered as novel ways to alter surfaces, i.e., in terms of their reactivities and structures. We describe the design of surfaces, nano/micro-fabrication, the properties affected by light, and the application principles. Additionally, we compare the various approaches reported in the literature.

Keywords Antibacterial • Antiviral • Film • Nanofiber • Photofunctional • Photorelease • Photosensitizer • Singlet oxygen

---

J. Mosinger (✉)
Faculty of Science, Charles University in Prague, Hlavova 2030, 128 43 Praha 2, Czech Republic

Institute of Inorganic Chemistry of the CAS, v. v. i., Husinec-Řež 1001, 250 68 Řež, Czech Republic
e-mail: mosinger@natur.cuni.cz

K. Lang
Institute of Inorganic Chemistry of the CAS, v. v. i., Husinec-Řež 1001, 250 68 Řež, Czech Republic
e-mail: lang@iic.cas.cz

P. Kubát
J. Heyrovský Institute of Physical Chemistry of the CAS, v. v. i., Dolejškova 3, 182 23 Praha 8, Czech Republic
e-mail: pavel.kubat@jh-inst.cas.cz

Contents

Abbreviations

CFU	Colony-forming unit
ECM	Extra cellular matrix
FRET	Förster resonance energy transfer
LB	Langmuir–Blodgett
LbL	Layer-by-layer
LDH	Layered double hydroxide
$O_2(^1\Delta_g)$	First singlet excited state of molecular oxygen
PdTPPC	Pd(II)-5,10,15,20-Tetrakis(4-carboxyphenyl)porphyrin
PLA	Poly(lactic acid)
PS	Photosensitizer
PU	Polyurethane
SAM	Self-assembled monolayer
SEM	Scanning electron microscopy
SODF	Singlet oxygen-sensitized delayed fluorescence
TMPyP	5,10,15,20-Tetrakis(1-methylpyridinium-4-yl)porphyrin
TPP	5,10,15,20-Tetraphenylporphyrin
TPPS	5,10,15,20-Tetrakis(4-sulfonatophenyl)porphyrin
WetSEM	Scanning transmission electron microscopy of wet samples
ZnPc	Zinc phthalocyanine

1 Introduction

Recent progress in nanotechnology has allowed the characterization, manipulation, and organization of matter on the nanometer scale, providing control not only of the size and shape but also of the surface of the resulting material. Materials with nanostructured surfaces can be roughly defined as substrates with typical dimensions in the region of one nanometer to several hundreds of nanometers.

In this review we focus on the design, synthesis, surface functionalization, and properties of photoactivatable nanostructured films (films with one nanoscale dimension), nanocomposites (materials consisting of nanofillers dispersed in a polymer matrix), and polymer nanofiber materials. These nanomaterials can be fabricated from hard (inorganic) and/or soft (organic and polymeric) materials. In contrast, nanoparticles are defined as having three nanoscale dimensions. The last wide-ranging topic is beyond the scope of this review.

The preparation of nanostructured materials with encapsulated or attached photoactive compounds allows for the changing/tuning of the surface properties using light and/or the controlled photorelease of active species. Light represents an elegant and accurate trigger to control three crucial factors that determine the photoactivity: the location, timing, and dosage. The utility of these materials has greatly expanded over the last decade, and they now serve as hosts for several sensors and as viable light-activated species/drug delivery vehicles for various biomedical applications. These photoresponsive systems are of special interest to scientists because of their unique advantages of quick responses without the addition of chemical stimulants or the production of a chemical contaminant.

2 Nanostructured Films

A wide range of nanostructured surfaces are used at the present time in medical applications, including polymers, metals, and ceramics. All the surfaces of these materials must meet the required specifications of biocompatibility, bio-functionality, biodurability, and biosafety in the short, medium, and long terms [1]. Light activation could be an effective trigger for the delivery of bioactive molecules from the surfaces and could be used to induce changes in the surface's physical properties or morphology. The surface nanoarchitecture may provide spatially and temporally resolved light-triggered responses of the material and may offer control over the behaviors of the active molecules, biomolecules, and cells at the solid-liquid interface [2].

Nanostructured building blocks on surfaces for biomedical applications can be fabricated using many processing methods. 'Bottom-up' methods, including layer-by-layer (LbL) methods [3–5], Langmuir–Blodgett (LB) assembly [6], and the formation of self-assembled monolayers (SAMs) [7] have been widely used to tailor surface properties. Inorganic nanoparticle–polymer nanocomposites have also emerged as potential multifunctional surfaces whose function can be triggered by light.

2.1 Photocatalytic Surfaces

The attachment of bacteria to various surfaces can result in colonization and infection [8]. Biofilms formed by bacteria on material surfaces are extremely difficult to remove and show strong resistance to many biocides. Thus, the prevention of biofilm formation is the best way to avoid the spread of disease [9] and material deterioration. Reactive oxygen species formed by the irradiation of materials can be effective in killing pathogenic microbial cells (Fig. 1) [10, 11].

Nanostructured metal oxides, such as TiO_2 and ZnO, have been extensively employed as self-cleaning coatings because of their wide band gaps and their hydrophilic and photocatalytic properties [12, 13]. The titanium surfaces of these implants can be modified to give different crystallinities, morphologies, porosities, etc., which can have a critical impact on the inherent photocatalytic activity of a TiO_2 film, resulting from oxidation. TiO_2 generates reactive oxygen species upon irradiation, such as OH· and O_2^- (Fig. 1), which impart antiviral, antibacterial, and fungicidal properties and effectively inhibit the formation of biofilms on the surfaces of implants [14].

Self-cleaning surfaces based on photogenerated reactive oxygen species are already making a significant impact as antimicrobial paints for buildings and hospitals [15]. One of the most popular self-cleaning coatings is based on UV light-activated photocatalytic TiO_2/ZnO thin films, the antimicrobial properties of which have been understood for over a decade [16, 17]. Self-cleaning reduces maintenance time, costs, and water and chemical use.

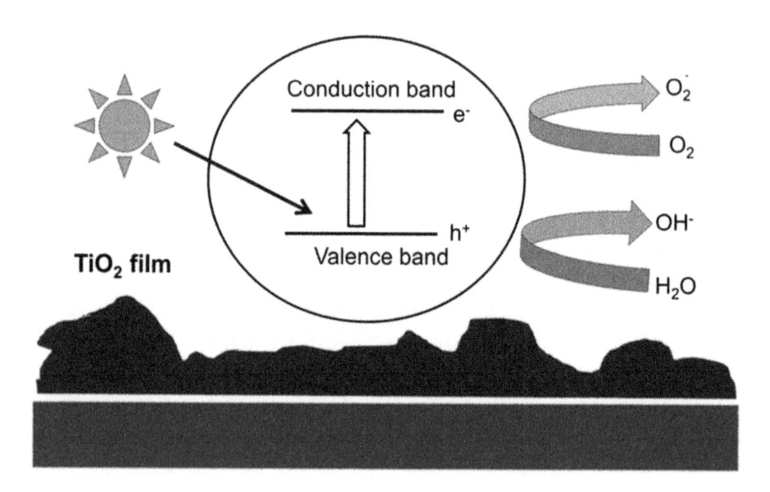

Fig. 1 Photocatalytic properties of a TiO_2 film. Processes occurring on TiO_2 surfaces are initiated by the absorption of ultraviolet radiation coupled with the generation of an electron–hole pair on the surface. The electron in the conduction band reduces O_2 to superoxide radical anions O_2^-, and the hole in the valence band can produce hydroxyl radicals OH· by reacting with water or hydroxyl ions on the surface

TiO_2 absorbs UV light; however, solar energy contains only 3–4% UV light ($\lambda < 380$ nm), which hinders its practical application in killing microbes and degrading environmental contaminants. To utilize solar energy and artificial light sources effectively, numerous efforts have been devoted to producing visible-light active TiO_2-based photocatalysts by doping [18–20].

TiO_2 nanofillers can be used to obtain poly(lactic acid) (PLA) nanocomposites for either short-term/single-use or long-term/durable applications [21]. The TiO_2 nanoparticles may play two roles: (1) blocking UV light below 350 nm, thus reducing the natural photodegradation of PLA (this effect has been achieved using TiO_2 nanowires) and (2) generating radicals produced by photo-induced electron/hole separation in the TiO_2 nanoparticles (Fig. 1). This latter approach leads to photodegradable PLA (this effect has been achieved by incorporating 1% TiO_2 nanoparticles into the nanocomposite).

2.2 Antibacterial Surfaces Photosensitizing Singlet Oxygen

The growing resistance against antibiotics and other chemotherapeutics has led to a search for novel antimicrobial treatments to which microorganisms find it difficult to develop resistance. In recent years, potential alternatives for the inactivation of bacteria include singlet oxygen, $O_2(^1\Delta_g)$, short-living, highly oxidative, and cytotoxic species generated in situ via energy transfer from an excited molecule of a photosensitizer (PS) to an oxygen molecule [22, 23]. Energy transfer from the excited triplet state (T_1) of the ^{3}PS to the ground state triplet oxygen $O_2\left(^3\Sigma_g^-\right)$ is a spin-allowed process, coupled with spin inversion of oxygen to singlet oxygen $O_2(^1\Delta_g)$ (Fig. 2):

$$^3\mathrm{PS} + O_2\left(^3\Sigma_g^-\right) \rightarrow {}^1\mathrm{PS} + O_2\left(^1\Delta_g\right) \tag{1}$$

Many materials, e.g., polymers containing PS, can generate $O_2(^1\Delta_g)$. There are numerous PSs absorbing light from UV to near-IR region, and among which porphyrins and phthalocyanines are some of the most frequently used. The utilization of PSs is advantageous because they affect the microorganisms in their vicinity without site specificity, which would result in bacterial resistance. The potential of porphyrin PSs and related compounds for the photodynamic inactivation of bacteria has been reviewed recently [23].

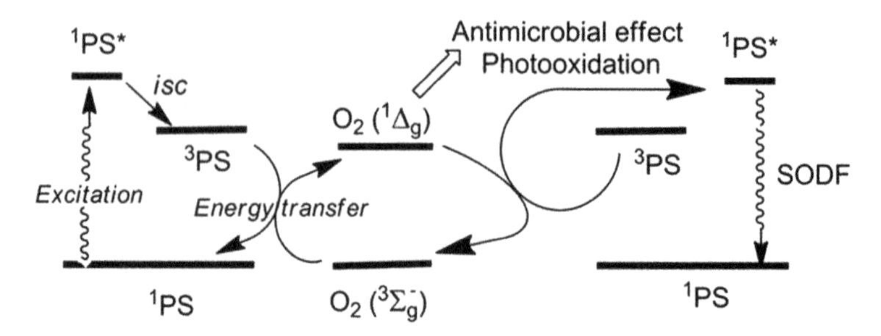

Fig. 2 Simplified energetic scheme of the photosensitized generation of $O_2(^1\Delta_g)$ and singlet oxygen sensitized delayed fluorescence (SODF): isc designates intersystem crossing, ^{1}PS* and ^{3}PS are excited singlet and triplet states of PS, respectively

2.2.1 Polymer Films with Photosensitizers

Polymer materials have the tendency to absorb exogenous proteins, grease, and microorganisms onto their surfaces via electrostatic interactions, which provides an opportunity for microorganisms to proliferate on medical devices and food packaging [24].

PSs can be incorporated into polymers during the polymerization process, or polymer surfaces can be coated with PSs during post-processing functionalization using covalent or non-covalent approaches (see Sect. 3.1.2). When the surface is exposed to light, photoproduced $O_2(^1\Delta_g)$ diffuses from the film with the potential to kill microorganisms on the surfaces or in the near proximity. The radial diffusion distance traveled by $O_2(^1\Delta_g)$, *d*, can be expressed as

$$d = (6\tau_\Delta D)^{1/2}, \tag{2}$$

where *D* is the diffusion coefficient of oxygen in the polymer. During the $O_2(^1\Delta_g)$ lifetime, τ_Δ, oxygen molecules can typically diffuse no more than tens to hundreds of nanometers [25, 26], indicating that the contribution of the $O_2(^1\Delta_g)$ formed deep below the surface to the antibacterial properties is negligible.

The LbL technique offers control over the chemical composition and architecture on the nanoscale and the flexibility of incorporating different materials or molecules, including PSs [27]. High-loading-capacity multilayers prepared via LbL overcome the main limitations of monolayer films, which have lower loading capacities for PSs and are thus less able to produce $O_2(^1\Delta_g)$ in high concentrations. Because they are applied to the surfaces of devices, they cannot affect the bulk mechanical properties of the devices to which they are applied. If the PS is not directly bound to the surface, it may leach into the solution/environment. Nevertheless, PSs incorporated into polymers have been found to be very effective at killing *Escherichia coli* and methicillin-resistant *Staphylococcus aureus* [28–30].

A photoactive surface coating formed by LbL that produces cytotoxic reactive oxygen species, including $O_2(^1\Delta_g)$, upon irradiation with near-infrared light has also been reported [28]. The coating consists of cross-linked hyaluronic acid and poly-L-lysine modified with the photoactive molecule pheophorbide A. In contrast to constantly active surface coatings (i.e., those not requiring photoactivation), this material is designed to allow cell attachment and, potentially, biointegration when not exposed to light.

5,10,15,20-Tetraphenylporphyrin (TPP) was used as a PS and immobilized onto a polyurethane (PU) film after being sprayed and polymerized as a thin layer onto a poly(methyl methacrylate) substrate [29]. PU is permeable to oxygen; thus, a sufficient number of oxygen molecules reach the PS in this coating. The biological experiments confirmed the antimicrobial effect of $O_2(^1\Delta_g)$ generated in the PU coating, which reached the bacterium *Staphylococcus aureus* via diffusion.

A flexible antibacterial film was prepared by depositing poly(terthiophene) with incorporated porphyrin (TPPS) onto a poly(ethylene terephthalate) sheet via a simple and rapid chemical polymerization method [30]. Under white light irradiation (400–800 nm), efficient Förster resonance energy transfer (FRET) from the poly(terthiophene) to the porphyrin occurs (Fig. 3). The excited porphyrin molecules convert to the triplet state by intersystem crossing followed by the generation of $O_2(^1\Delta_g)$. If bacteria adhere to the film surface, the microbial contamination can be eliminated.

Polymeric films formed via the electrochemical polymerization of 5,10,15,20-tetra(4-*N,N*-diphenylaminophenyl)porphyrin and its Pd(II) complex on optically transparent indium tin oxide have antibacterial properties towards the microorganisms *Escherichia coli* and *Candida albicans* [31]. In an alternative approach, methylene blue and toluidine blue were covalently bound to an activated silicone polymer via an amide condensation reaction [32]. The new polymers with the covalently attached dyes display significant bactericidal activities against *Escherichia coli* and *Staphylococcus epidermidis*.

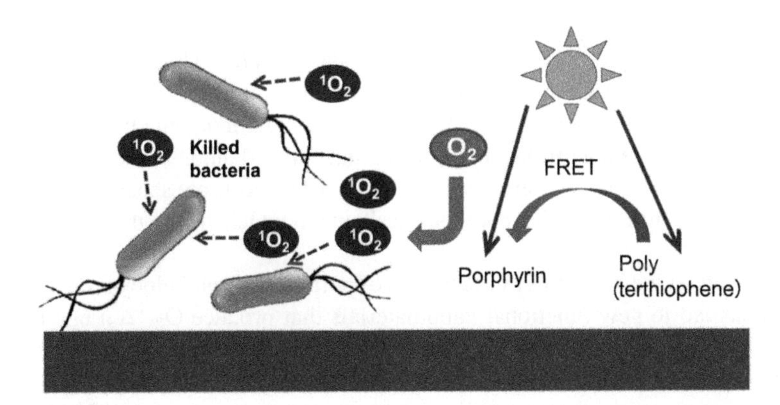

Fig. 3 Schematic of the antibacterial mechanism of the poly(terthiophene)/porphyrin film, including the formation of $O_2(^1\Delta_g)$ from the porphyrin triplets photogenerated by direct excitation of the porphyrin or photoexcitation of the poly(terthiophene) followed by FRET to the porphyrin moiety [30]

PSs externally bound to surfaces have weaknesses that may affect their photo-oxidation/antibacterial properties, such as the leaching of the PS molecules into solution [33].

2.2.2 Polymer Nanocomposites

Nanocomposite materials are comprised of two or more materials with significantly different physical or chemical properties, where at least one of these components has nanometric dimensions. Recently, inorganic nanoparticle–polymer nanocomposites have emerged as potential multifunctional platforms for the development of light-activated antimicrobial surfaces. There are two approaches to obtaining these surfaces: (1) a PS molecule is embedded in an inert inorganic nanocontainer with a suitable structure that protects this molecule from harsh interactions with the environment or (2) an inorganic photoactive nanofiller, a nanofiller with bound PS, or both a nanofiller and PS (with synergistic effects) are dispersed in a polymer film. The properties, functions, and applications of these nanomaterials are surveyed in this chapter.

Inorganic Nanocontainer

The incorporation of PSs into the galleries of two-dimensional inorganic layered structures can impart photosensitizing activity to solid materials [34, 35] which can be used for the fabrication of nanocomposites with photoactivatable surfaces. Layered double hydroxides (LDHs) are particularly promising building blocks in this context because of the versatility of the chemical compositions of the hydroxide layers and the exchangeable interlayer spacing together with their stability and biocompatibility [36]. The general formula for LDHs is $[M^{2+}{}_{1-x}M^{3+}{}_{x}(OH)_2]^{x+}[A^{n-}{}_{x/n}]^{x-}\cdot mH_2O$, where M^{2+} and M^{3+} represent metal cations octahedrally coordinated by hydroxyl groups and A^{n-} represents an n-valent intercalated anion, which can vary from a simple inorganic species to a targeted PS anion that introduces functionality to the organic/inorganic hybrid [37]. The distance between the inorganic hydroxide layers depends on the size and arrangement of the intercalated anions. The PS molecules located in the expandable interlayer space of the LDH hosts are separated from their surroundings, preserving the photophysical and photochemical properties of the introduced PS.

Research on the host–guest interactions of porphyrin or phthalocyanine PSs with LDHs has led to new functional nanomaterials that produce $O_2(^1\Delta_g)$ upon irradiation with visible light [34–38] and exhibit potential for the fabrication of LDH-porphyrin/polymer nanocomposites for antibacterial coatings [39–41], anti-cancer photodynamic therapy [38], and photooxidation reactions [42]. A supramolecular PS designed for photodynamic therapy was fabricated via the incorporation of zinc phthalocyanine (ZnPc) into the gallery of LDH [38]. The composite nanomaterial possessed uniform particle sizes (~120 nm), and the host–guest interactions resulted in a high dispersion of the ZnPc molecules in the monomeric state and a high $O_2(^1\Delta_g)$ production efficiency.

The porphyrins TPPS and PdTPPC were intercalated into LDH hosts via a co-precipitation procedure and then used as fillers in polyurethane and poly(butylene succinate), two eco-friendly polymers [39]. Both X-ray diffraction and transmission electron microscopy measurements indicated that the porphyrin-LDH fillers were well dispersed in the polymer matrices and that the porphyrin molecules remained intercalated within the LDH hydroxide layers (Fig. 4). LDH/poly(butyl methacrylate) nanocomposites were prepared using LDHs partially intercalated with TPPS [40]. An investigation of the kinetics of the quenching of the porphyrin triplet states by oxygen indicated limited diffusion of oxygen within the composite. Because the polymer matrix restricts the diffusion of oxygen (Fig. 5) and partially quenches $O_2(^1\Delta_g)$, the antibacterial effect on the nanocomposite surface results from the $O_2(^1\Delta_g)$ produced within a narrow surface layer. All these nanocomposites produced $O_2(^1\Delta_g)$ upon irradiation (Fig. 5) and were proposed as potential platforms for the fabrication of antibacterial surfaces activated by visible light.

The photostability of polyurethane composite films was studied to establish their applicability as new photodynamic surfaces [41]. It was shown that the LDH host enhanced the chemical stability of the porphyrin PSs by minimizing photobleaching and aggregation effects and that $O_2(^1\Delta_g)$ has no detrimental effects on the microstructure and viscoelastic properties of the nanocomposite. In vitro antimicrobial tests showed that *Staphylococcus aureus* growth on the composite surfaces is inhibited by white light irradiation. The complete inhibition of *Pseudomonas*

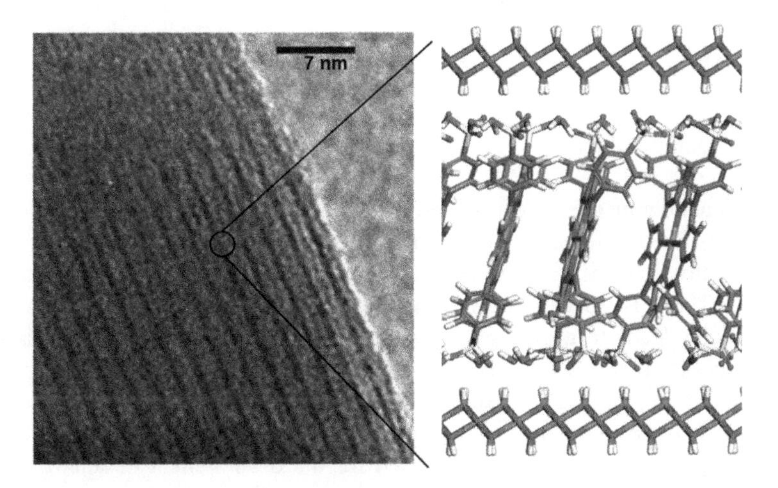

Fig. 4 LDHs are suitable hosts for porphyrin PSs. The interlayer space of an LDH is filled with nearly parallel, slightly inclined porphyrin units. The arrangement of intercalated porphyrin molecules was investigated using molecular dynamics simulations (*right*). The hydroxide sheets composed of Zn^{2+} and Al^{3+} octahedrally coordinated by OH groups are visualized as *dark parallel lines* in the high-resolution electron microscopy image (*left*). Reprinted with permission from [43]. Copyright 2010, American Chemical Society

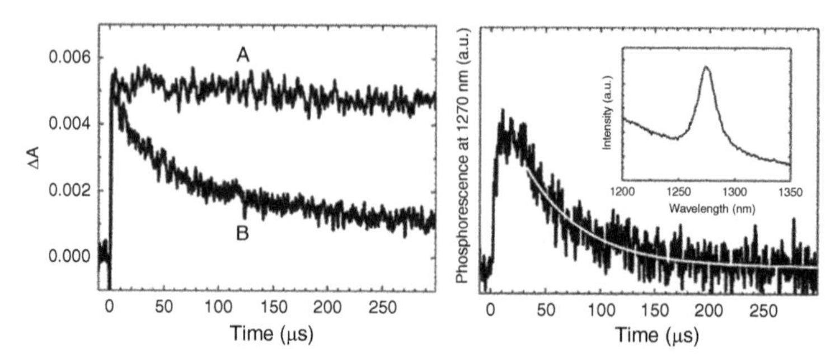

Fig. 5 The excited state dynamics of TPPS-based LDHs in PU nanocomposites (an LDH composed of Mg^{2+} and Al^{3+} was intercalated with TPPS and dispersed in PU) were probed using nanosecond laser flash photolysis. The decay curves of the triplet states of the TPPS in the nanocomposite in vacuum (*A*) and in an oxygen atmosphere (*B*) demonstrate the quenching of the triplet states by oxygen (*left*). The production of $O_2(^1\Delta_g)$ in the nanocomposite was evidenced by the appearance of an emission band peaking at 1275 nm (*right, inset*). The $O_2(^1\Delta_g)$ luminescence intensity decays monoexponentially (*right*). The *smoothed line* is a least-squares monoexponential fit. Conditions: excitation at 425 nm, triplet states recorded at 480 nm and $O_2(^1\Delta_g)$ at 1270 nm. Adapted from [39]. Reproduced with permission from The Royal Society of Chemistry

aeruginosa growth indicated the efficacy of the surface in preventing biofilm formation. As noted, the advantage is that the activity of the surface can be tuned by adjusting the amount of porphyrin-containing filler within it. The potential applications of such photodynamic surfaces, which are sterile under white light, are of great interest when it is desirable to maintain microbial populations at low levels.

Other Inorganic Fillers

Polymer nanocomposites encapsulated with methylene blue and gold nanoparticles exhibit enhanced light-activated antimicrobial activity against Gram-positive and Gram-negative bacteria relative to those encapsulated with methylene blue only [44, 45]. There is no definite explanation for why the presence of 2-nm gold nanoparticles leads to a synergistic enhancement of the photosensitized damage to the bacteria. It was speculated that this effect results from the reactive oxygen species produced on the irradiated methylene blue-nanogold embedded surface. Pre-clinical studies have indicated that silicone encapsulated with methylene blue and 2-nm gold nanoparticles can be used for the development of self-sterilizing polymers for medical devices for the effective reduction of *S. epidermidis* biofilm formation under laser irradiation [46].

2.3 Oxygen Sensing

The non-invasive, quantitative determination of oxygen concentration is essential for a variety of applications in fields ranging from the life sciences to the environmental sciences. The direct approach to obtaining oxygen nanosensors is to embed oxygen-sensitive molecules, generally luminescent transition metal complexes, inside an appropriate material. The nanosensors can be in the form of inorganic or polymer nanoparticles, nanometer-sized coatings, or nanostructured surfaces [47]. There are three common methods for immobilizing oxygen-sensitive probes in a polymer host: (1) homogeneous distribution in the host polymer; (2) adsorption on the surface of the micro- or nanoparticles incorporated in the host polymer; and (3) homogeneous distribution in micro/nanoparticles incorporated in the host polymer. White micro/nanoparticles (silica, TiO_2, etc.) can simultaneously serve as fillers and improve the intensity of the optical signal because of the improved scattering of both the exciting light and the luminescence. In contrast, black micro/nanoparticles can act as optical isolators.

Most sensors are based on the luminescence quenching of probes by oxygen. Regarding the probes, luminescent transition metal complexes offer high sensitivity to oxygen because their excited state lifetimes are usually on the microsecond timescale. Interactions between molecular oxygen in the ground state and the excited electronic state of a probe decreases the luminescence intensity and decay time. Such systems often generate $O_2(^1\Delta_g)$; however, for this application, the effectiveness is measured by the degree of luminescence quenching by oxygen, and the formation of $O_2(^1\Delta_g)$ can have detrimental effects (Fig. 2); see (1). The relationship between the luminescence intensity (or luminescence lifetime) and the concentration of oxygen ($[O_2]$) is described by the Stern–Volmer equation:

$$F_0/F = \tau_0/\tau = 1 + K_{SV}[O_2], \tag{3}$$

where F_0 (τ_0) and F (τ) are the intensities (lifetimes) of a probe in the absence and presence of oxygen, respectively, K_{SV} is the Stern–Volmer constant, and $[O_2]$ is the concentration of oxygen in the sample. In an ideal sensor system, there is a linear relationship between F_0/F (or τ/τ_0) and the oxygen concentration. The luminescence intensity also depends on the penetration of the oxygen into the material.

Lu and Winnik [48] concentrated on silica-polymer nanocomposites with high oxygen permeabilities and the effects of silica nanofillers on the performances of luminescent oxygen sensors. The silica phase serves as a carrier for the dye, improves the mechanical properties of the material, and may even reduce the photooxidation activity of the $O_2(^1\Delta_g)$ formed as a side product.

The most common nanofilms are those prepared by the Langmuir–Blodgett (LB) technique. LB layer-based sensors can be produced reproducibly and with well-defined thicknesses and have short response times. Self-assembled ionic nanofilms formed on top of fluorescent nanoparticles represent another type of nanoscale oxygen sensor [49]. This concept can be demonstrated with a tris

(4,7-diphenyl-1,10-phenanthroline)ruthenium(II) probe trapped in an ultrathin film deposited on fluorescent nanoparticles. The fluorescent nanoparticles act as scaffolds and provide an internal intensity reference. To demonstrate the feasibility of intracellular metabolic monitoring with such oxygen sensors, they were delivered into human dermal fibroblasts with no apparent loss in cell viability.

Polymer–carbon nanotubes and polymer–graphene-based nanocomposites have demonstrated their potential in a wide variety of biosensing applications [50]. These applications include the detection of glucose, NO, protons, and some proteins. The synergistic effect of the intrinsic properties of these carbon nanomaterials, such as fluorescence or fluorescence quenching, high electrical and thermal conductivity, chemical stability, and mechanical strength, with the tunable properties of polymers, makes these polymer nanocomposites ideal for the development of new types of chemical sensors.

2.4 Nanostructured Surfaces with Photocleavable Ligands

Photocleavable systems on nanostructured surfaces can be used to remove selected molecules and expose new functionalities, i.e., in photolithography. The immobilization of DNA molecules on a substrate is a crucial step in biochip research and related applications. A commonly used method for DNA immobilization is to functionalize it with a terminal reactive group that is selective for the surface of interest [51]. The photolithographic technique based on the photocleavage of linker-connected DNA strands showed potential for the repeated construction of different chemical and physical patterns on the same surface or for the fabrication of multifunctionalized DNA chip surfaces for genetic detection [52] or DNA computing [53].

Photoswitchable ligands on well-defined adhesive nanopatterned substrates were used to develop a method for analyzing collective cell migration via precisely tuned cell-substrate interactions [54]. In this method, gold nanoparticles are periodically arrayed on a glass substrate in a well-defined nanoscopic geometry and then functionalized with cell-adhesive extra cellular matrix (ECM) ligands. Because of the functionalization of the quasi-hexagonally arranged gold nanoparticles on the surface with photocleavable poly(ethylene glycol) (PEG), the ECM ligands only become accessible to cells when PEG is photoreleased.

The integration of photolabile protecting groups is also a promising strategy for the controlled release of bioactive compounds, e.g., drugs from nanostructured surfaces. The compounds are rendered inactive via the addition of a covalently bonded photosensitive conjugate. Light-induced processes cleave the photolabile protecting groups and restore the biological function of the released bioactive molecule. This approach has been used with nanoparticles [55] which can be directly delivered to tumors or other places of interest. An example of a surface decorated with photoswitchable ligands is a self-assembled monolayer of nitric oxide precursors on a gold surface. Light excitation ($\lambda > 380$ nm) quantitatively releases NO [56], which plays important roles in many physiological processes. The reaction is based on the photoreactivity of the anticancer drug flutamide which undergoes nitro-to-nitrite photorearrangement followed by the rupture of the O=N bond to generate a phenoxy radical and NO. The advances in utilizing photoactive approaches in macromolecules for prospective use in biomedical applications were recently reviewed [57].

2.5 *Photothermal Processes*

An important obstacle to the application of photo-triggered systems in biomedical therapeutic areas inside a human body is the requirement of UV light stimulation. Near-infrared light is preferable for triggering release in biological systems because it can pass through blood and tissue to depths of several centimeters [58]. However, very few suitable organic chromophores absorb in this region, and even fewer are capable of converting the absorbed energy into a chemical or thermal response that can be used to trigger drug release. A few years ago, gold nanostructures emerged as useful agents for photothermal therapy after they were shown to have strong absorption in the near-infrared region (four to five times higher than conventional photo-absorbing dyes) [59].

Photothermal mechanisms are advantageous because a broad range of materials are capable of collecting light to induce thermally activated processes. This method was used for dose- and spatially controlled drug release from photoactivated porous chitosan nanocomposite films [60]. Plasmonic gold nanorods are known to have extremely efficient photothermal conversion in the so-called therapeutic window between 700 and 1300 nm, where principal biological components (water, melanin, etc.) and common exogenous chromophores exhibit the highest optical transparency.

Graphene materials have been proposed to offer high therapeutic molecule loading capacities because of their large available surface areas and have been explored as potential drug delivery systems [61]. They play the dual role of versatile substrates for the temporary storage of drugs and transducers of near-infrared light into heat. Recently, the possibility of using light to trigger remotely the photothermal release of drugs from graphene-incorporated hybrids in a controlled manner was demonstrated [62]. Sada et al. [63] reported near-infrared-light-induced cell detachment using carbon-nanotube-coated substrates. A similar approach was used to deliver spatially and temporally defined mechanical forces to cells [64].

The other strategy is to trigger the release of photothermal drugs/biomolecules from multilayer nanostructured films using light [5]. This approach also offers a way to deliver biomolecules with high spatio-temporal resolution. The immobilization of DNA and gold nanoparticles in hyaluronic acid/poly-L-lysine (PLL) films is described in [65]. The thermal decomposition of the film surrounding the nanoparticles upon near-infrared light irradiation resulted in the weakening of the interactions between DNA and doped PLL and, as a result, the release of DNA from the film.

The photothermal effect can be used for the cultivation of cells and other biological processes. Viable mesenchymal stem cells were efficiently and selectively harvested using near-infrared light and a conductive polymer nanofilm consisting of poly(3,4-ethylenedioxythiophene) [66].

2.6 Light-Triggered Changes of the Physical Properties

The physical properties of some surfaces (wettability, hydrophobicity, lubrication, adhesiveness, etc.) can be changed by light. The application of such systems has been reported in biotribology, controlled drug release, and cell growth and separation, among other fields [67, 68].

Light provides an external handle by which the hydrophobic/hydrophilic balance and molecular weight (through photodegradation or crosslinking) of self-assembled copolymers can be influenced to manipulate nanoscale morphologies. Furthermore, the above-mentioned photo-responses can be tailored for specific wavelengths and intensities of light by the attachment of functional groups to the light-absorbing portions of macromolecules [69].

A variety of photoresponsive inorganic oxides (usually TiO_2 or ZnO) undergo the transition from hydrophobic to hydrophilic states upon UV irradiation, reverting to their original states in the dark (or when exposed to visible irradiation). These transitions can be repeated for several cycles. However, the recovery process is very slow, sometimes taking several days [70, 71]. A comprehensive overview of TiO_2 surfaces with switchable wettability driven by UV and visible light was published recently [12]. In addition to semiconductor inorganic oxides, some organic materials containing photochromic functional groups, such as azobenzenes, spiropyrans, dipyridylethylenes, stilbenes, and pyrimidines, also have the ability to undergo reversible conformation transitions triggered by UV/visible irradiation, leading to changes in their wetting properties [72].

The adhesion of cells to a solid substrate is strongly affected by the surface wettability [73]. Superhydrophilic patterns on superhydrophobic silicon nanowire surfaces were prepared using a standard optical lithography technique [74]. Exposure of the patterned surface to a suspension of *Bacillus cereus* spores in water led to their specific adsorption onto the superhydrophobic areas.

2.7 Surface Nanopatterning Using Light

There is an increasing interest in the use of micro/nano-patterned surfaces to organize and control the growth of cells on surfaces. Such surfaces offer new capabilities for the investigation of the dynamics of cell–environment interactions,

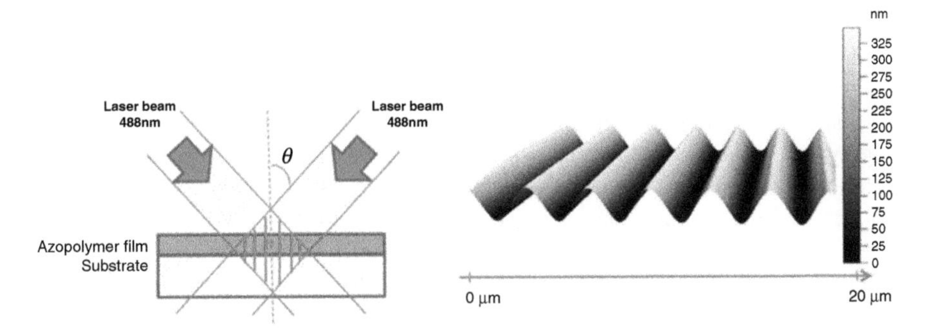

Fig. 6 Experimental set-up employed for light-induced nanopatterning [75]. The continuous 488-nm laser beam is absorbed by the azobenzene-based polymer. The beam is divided with a beam splitter, and the two resulting beams are superposed onto the film surface, which results in a sinusoidal interference pattern with a modulation axis given by the intersection of the incident plane of the beams and the film surface. An AFM image of a polymer film is shown on the *right*

cell-based biosensors, cell screening systems, and tissue engineering applications. In this respect, the modulation of the surface topography of photomechanically responsive polymers using light and the modulation of the chemical and physical properties using chemical modification may provide new opportunities.

Surface nanopatterning can be produced using the light-induced *trans–cis* isomerization of the azo-containing polymers that form the surface [75]. The isomerization induces a mechanical response and leads to mass transport in thin films of azo-polymers when exposed to light (Fig. 6). This method of preparing patterned surfaces on polymer thin films is advantageous over current methods. For example, the surfaces can be prepared quickly with practically no environmental limitations, and can be erased and patterned again. Thus, fabricated nanogrooves act as good scaffolds and provided directional cues to guide PC12 cell alignment and orientation [76]. The nanometric topography of surfaces has proven to be an important parameter for controlling neuronal behavior [77]. Azo-polysiloxanes also provided nanostructured surfaces and were successfully used to support cell adhesion and growth [78].

3 Polymer Nanofiber Materials

Since their initial development, polymeric nanofibers have rapidly become widely utilized nanostructured materials with high structural integrities. Because of their large specific surfaces, low weights, chemical specificities, and mechanical flexibilities, they have already been used in numerous areas, including drug delivery, filtration, sensing, engineering, nanoreinforcement, protective clothing and medicine [79–84]. Besides other known techniques, such as template synthesis, phase separation, drawing and self-assembly [85–89], the most extensively investigated method represents the electrospinning of polymers, which is a simple and low-cost method for the preparation of fibers with diameters in the range of nanometers to a few microns and lengths that can range from a few centimeters to kilometers.

The electrospinning equipment (spinneret) consists of a high-voltage power supply, a metallic capillary (or a set of capillaries) connected to the polymer solution supply as an electrode and, usually, a planar or cylindrical grounded collector as the counter electrode. The electric potentials on the electrodes at a distance of ca 5–50 cm can vary from a few to tens of kilovolts, resulting in an electric field of ca 10^5 V/m. Because the polymer droplets are electrically charged at the capillary tip, a fluid jet is ejected from the tip(s) of the capillary(ies) in the form of a so-called Taylor cone when the electrostatic force surpasses a threshold point [90–92]. The co-axial electrospinning technique is a modification of the traditional electrospinning technique in which two components are fed through different co-axial capillary channels and are integrated into core-shell structured composite nanofibers [93].

Recently, significant progress in electrospinning has led to the development of technologies that allow for the industrial-scale production of nanofiber materials [94–96]. For example, the capillary tip can be replaced by the charged surface of a roller spinning in a tank of polymer solution or a thin wire electrode coated with a polymer film from a moving tank of polymer solution. A thin material (e.g., textile) moves along a grounded counter-electrode and serves as a supporting material for the continuous nanofiber layer. The area weight of the nanofiber layer can easily be controlled by the linear velocity of the supporting material (Fig. 7).

The dimensions and morphologies of electrospun nanofibers are determined by many parameters: (1) the intrinsic properties of the polymer (chemical structure, solubility, and molecular weight); (2) the properties of the solvent (conductivity, viscosity, surface tension, vapor pressure, and dielectric constant); and (3) the processing parameters (electric voltage, distance between the electrodes, temperature, relative humidity, concentration of polymer and additional ions, number and structure of the inlet capillaries, shape and motion of the collector, and the presence

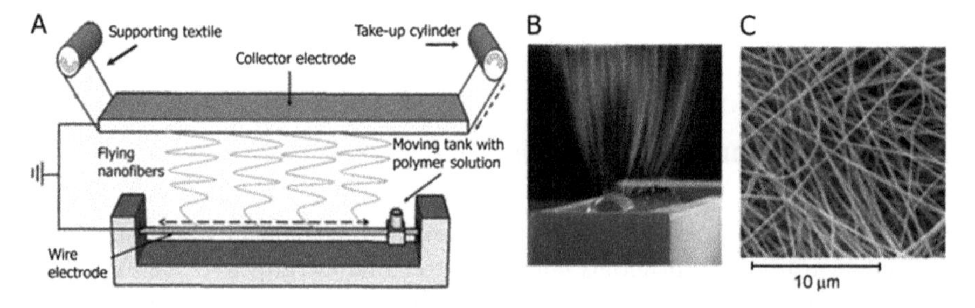

Fig. 7 (**A**) Scheme of the Nanospider™ industrial electrospinning device. (**B**) Electrospinning device consisting of a roller electrode immersed in a tank with the polymer solution. Flying nanofibers from a charged roller are placed on a grounded collector electrode. (**C**) SEM of the resulting nanofiber material. Adapted with permission from [97]. Copyright 2014, American Chemical Society

of a supporting material on the collector). Thus, the intrinsic and processing parameters have to be adjusted to enable electrospinning of each specific system and to control the size and morphology [98, 99]. Typically, randomly oriented electrospun nanofibers form materials (designated as membranes, textiles, or scaffolds) with high surface areas and nanoporous structures (Fig. 7C).

3.1 Nanofiber Materials with Light-Activated Antimicrobial Effects

Considerable research interest has been dedicated to the development of sterile surface materials/coatings for medical applications. The design and chemistry of various antimicrobial materials have been extensively reviewed [100–102]. These materials are usually based on a cationic polymer and contain functional group (s) (such as quaternary ammonium, quaternary pyridinium, phosphonium, and biguanide moieties) covalently attached to the matrix materials. However, only a limited number of studies have been dedicated to nanofibers prepared from antimicrobial polymers, despite the unique properties of nanofiber materials mentioned above [103, 104]. Besides the use of cationic polymers, antimicrobial activity can be introduced by doping with metal cations, metal nanoparticles (silver and gold), peroxides, antibiotics, antibacterial peptides, or other chemical compounds [105, 106]. These materials may have several disadvantages. For example, the release of low-molecular-weight antimicrobial agents from the surface of a nanofiber matrix can lead to a decrease in efficiency over time. The cytotoxicity of other agents, such as silver nanoparticles, hinders their usage in the treatment of infections.

At present, photoactive nanofiber materials are attracting growing interest, especially those whose antimicrobial character is induced by the photogeneration of reactive oxygen species after irradiation by visible light. The antimicrobial (also referred as photodynamic) character is mainly introduced by $O_2(^1\Delta_g)$ (see Sect. 2.2).

3.1.1 Nanofiber Materials with Encapsulated Photosensitizers

Recently, many studies have employed photoactive nanofibers electrospun from different polymers, including PU [26, 107–112], polystyrene [97, 113–117], polycaprolactone [26, 109, 118], and polyamide 6 [26, 119, 120], with homogenously distributed encapsulated porphyrin and/or phthalocyanine PSs. Similarly, pheophorbide and its poly-L-lysine conjugate were encapsulated in electrospun polycaprolactone nanofibers [121], and highly luminescent hexanuclear molybdenum clusters with high quantum yields of $O_2(^1\Delta_g)$ formation were encapsulated in PU nanofibers [122]. The fabrication of these photoactive nanofibers is straightforward; the PSs are dissolved in the polymer solution before electrospinning. This approach combines the properties of nanofiber materials comprised of the selected electrospinnable polymers (high flexibility, specific surface, transparency to light, oxygen permeability/diffusion, and nanoporous surface) with the properties of $O_2(^1\Delta_g)$ (cytotoxic effect and short lifetime) generated by the photosensitized reaction (Fig. 2); see (1). The concentration of nonpolar encapsulated porphyrinoid PSs with a broad absorption spectrum in the visible region usually ranges from 0.1 to 1 wt%, and the selected polymers are biocompatible. As a result, these materials do not leak PSs into aqueous media, exhibit low dark toxicity, and are activated by visible light.

The morphology, structure, and properties of these materials can be characterized using microscopic, steady-state, and time-resolved techniques. The transparency of the thin nanofiber materials enables the use of steady-state and time-resolved spectroscopies, even in transmission mode, to measure directly the kinetics of the transient species (PS excited states and $O_2(^1\Delta_g)$) (Fig. 8). The detailed photophysical characterization of these transients is crucial for understanding and monitoring the functionalities of these materials [108, 113].

The transient absorption spectra of porphyrinoid PSs embedded in polymer nanofiber materials exhibit characteristic features of the triplet state, such as an absorption band at approximately 460 nm and quenching by oxygen [26, 107, 108, 112, 113]. The quenching by oxygen (Fig. 8A) can be used to measure the oxygen permeability and diffusion coefficients of the polymers [26].

After the excitation of the embedded PS in oxygen-containing media, a new luminescence band appears at 1270 nm (Fig. 8B). The signals were attributed to $O_2(^1\Delta_g)$ based on the observed wavelength, and disappearance of the signal when the sample was purged with argon and the reappearance of the signal when the sample was purged with air or oxygen. Although the triplet state lifetimes ($\approx$ 20 μs) are not significantly affected by the medium in which the nanofiber material is

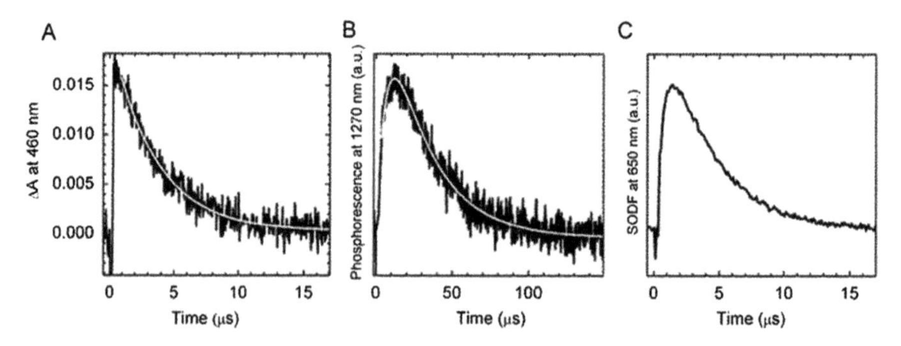

Fig. 8 Time-resolved detection of the triplet states of PS using transient absorption spectroscopy (**A**) and of $O_2(^1\Delta_g)$ (**B**) and SODF (**C**) using luminescence spectroscopy. The nanofiber material with the encapsulated TPP was excited by a nanosecond laser at 425 nm in an air atmosphere. The *solid lines* are least-square fits to the experimental data. Adapted with permission from [108]. Copyright 2010, American Chemical Society

immersed (except in the case of an oxygen-containing medium), the observed $O_2(^1\Delta_g)$ lifetimes indicate that $O_2(^1\Delta_g)$ is partially quenched by the polymer and partially diffuses outside the nanofibers [26, 107, 108, 112, 113].

Because of a combination of a high local concentration of photogenerated $O_2(^1\Delta_g)$ and a high concentration of fixed PS in the excited triplet state, a new phenomenon, singlet oxygen-sensitized delayed fluorescence (SODF), is observed. SODF is based on the repopulation of the excited singlet state (^{1}PS*) via the interaction of $O_2(^1\Delta_g)$ and the triplet state of another PS molecule (Figs. 2 and 8C); see (4). Thus, SODF can be used as a sensitive tool for the detection of oxygen, imaging of the distribution of oxygen, $O_2(^1\Delta_g)$, and PS molecules in polymer nanofibers, and the evaluation of the photosensitizing effectiveness of the nanofibers [108, 113].

$$^3\mathrm{PS} + \mathrm{O_2}\left(^1\Delta_g\right) \rightarrow {}^1\mathrm{PS}^* + \mathrm{O_2}\left(^3\Sigma_g^-\right) \tag{4}$$

Nanofiber materials with encapsulated PSs have photosensitizing properties, as proven by the photogeneration of $O_2(^1\Delta_g)$, SODF, and/or photooxidation properties under visible light illumination [26, 97, 107–114, 117, 118, 122]. Because $O_2(^1\Delta_g)$ oxidizes biological targets, e.g., proteins and cell membranes, the antibacterial effect on the surface of nanofiber materials demonstrated on *Escherichia coli*, *Staphylococcus aureus, and Pseudomonas aeruginosa* is very powerful (Fig. 9) [26, 97, 107, 109, 112, 114, 117].

Along with the photogeneration of short-lived $O_2(^1\Delta_g)$, the formation of H_2O_2, a well-known antibacterial agent, was proven on the surfaces of PU nanofibers [26]. Its formation was explained as a partial electron transfer process from the excited porphyrin PS to oxygen, yielding superoxide radical anions, O_2^-, followed

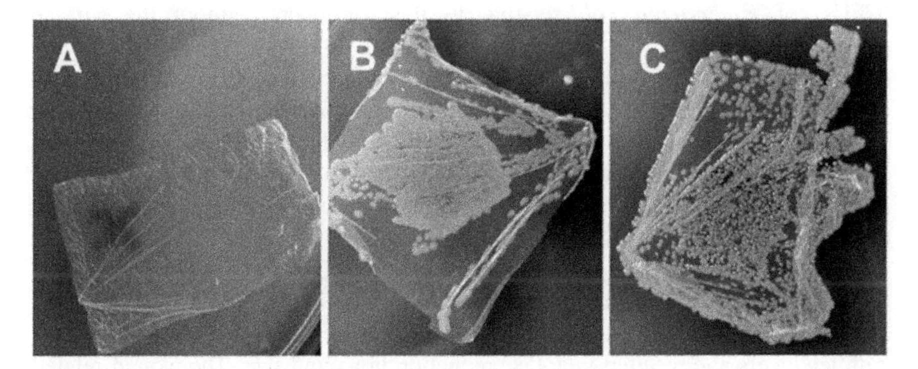

Fig. 9 Demonstration of the antibacterial effect of nanofiber materials with encapsulated TPP. (**A**) *Staphylococcus aureus* were inoculated on the hydrophilic surface of PU (Tecophilic™) nanofiber membranes encapsulated with 1 wt% TPP and illuminated by visible light. The bacteria colonies completely disappeared. (**B**) The same material illuminated but not doped with TPP. (**C**) The same material doped with TPP and kept in the dark. Reprinted with permission from [109]. Copyright 2012, John Wiley & Sons A/S

by their fast disproportionation to H_2O_2. Because of the presence of H_2O_2, these nanofiber materials exhibited prolonged antibacterial properties, even in the dark.

The antimicrobial (photodynamic) effect is not only predisposed by the intrinsic properties of the selected PS (spectral characteristics and quantum yield of $O_2(^1\Delta_g)$) but also influenced by the properties of the polymer. In addition to high transparency and low quenching ability toward the triplet states of PS and $O_2(^1\Delta_g)$, a key parameter determining the efficacy is the oxygen permeability/diffusion coefficient of the polymer. The use of a polymer with a high coefficient ensures the efficient quenching of the triplet states and an increase in the diffusion path of the oxygen and $O_2(^1\Delta_g)$ and thus controls the oxidative/cytotoxic range of the generated $O_2(^1\Delta_g)$ (Fig. 2).

Thus, only the surfaces of nanofiber materials made of polymers with high diffusion coefficients exhibit antimicrobial effects [26]. In addition, increasing the temperature from 5 to 32 °C significantly increases the diffusion length of oxygen, strengthening the antimicrobial effect [114]. Hence, the temperature of human skin can serve as a trigger of the antimicrobial effect, enabling possible applications of these materials in dermatology.

In addition to a high oxygen permeability/diffusivity, the surface hydrophilicity/wettability plays a crucial role in achieving efficient photooxidation and antimicrobial effects. As an example, post-processing modification of electrospun nanofibers, such as sulfonation, oxygen plasma treatment, and polydopamine coating, strongly increases the wettability of the originally hydrophobic polystyrene nanofibers without changing the inherent properties of the embedded TPP or the polymer nanofibers [97]. The high surface wettability results in effective contact between the surface and the biological target and significantly accelerates the antibacterial activity of the nanofibers in aqueous surroundings. For example, 20 min of irradiation of sulfonated polystyrene nanofibers with encapsulated TPP by simulated daylight significantly decreased the number of colony-forming units (CFUs) of *Escherichia coli* relative to the nanofibers without the surface modification.

Because of the short lifetime of $O_2(^1\Delta_g)$ and consequently relatively short diffusion pathway (hundreds of nanometers), the cytotoxic action of $O_2(^1\Delta_g)$ occurs on or very close to the surface of the nanofiber materials and does not limit medical applications. In this respect, it is advantageous that the most of the pathogens are detained on the surface because they cannot pass through the interconnected-pores of these materials [123, 124], as shown in Fig. 10.

The application of PU nanofiber materials (1 wt% TPP) was demonstrated in a clinical study on healing chronic leg ulcers using these material as a medical covering [109] (Fig. 11). The illuminated wound dressing was applied to 89 patients with leg ulcers and resulted in a 35% reduction in wound size. The wound-related pain was reduced by 71%. Nanofiber materials meet the requirements for wound-dressings, such as high gas permeation and protection of the wound from infection and dehydration. In particular, PU nanofiber materials have many favorable properties, including their ability to exudate fluid from the wound, failure to induce

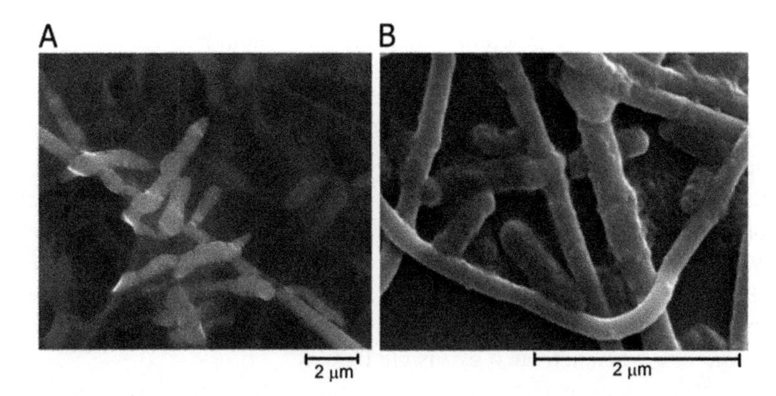

Fig. 10 WetSTEM (**A**) and SEM (**B**) of *Escherichia coli* detained on the surface of the *upper layer* of nanofibers. Adapted with permission from [124]. Copyright 2013, American Chemical Society

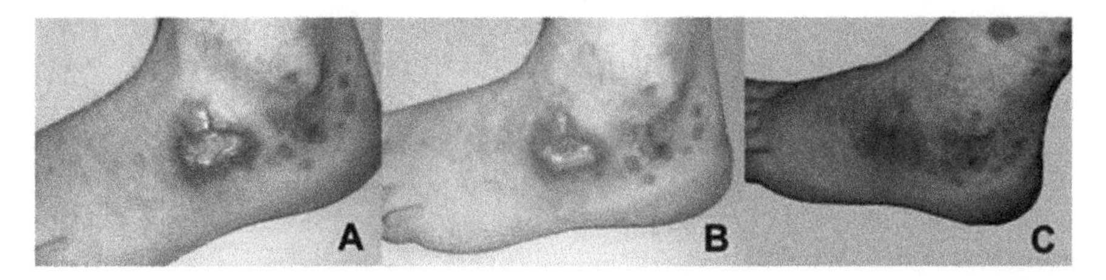

Fig. 11 Clinical photographs of a patient with leg ulcers treated with illuminated PU (Tecophilic) nanofiber textiles with encapsulated TPP at the beginning of healing (**A**) and after 15 days (**B**) and 42 days (**C**) of treatment. Adapted from [109]. Reproduced with the permission of John Wiley & Sons A/S, Copyright 2012

wound desiccation, controlled evaporative water loss property, good oxygen permeability, and fluid drainage promotion [125].

Recently, the surfaces of electrospun hydrophilic polycaprolactone and PU (Tecophilic™) nanofibers with encapsulated TPP were observed to exhibit the photoantiviral effect toward two types of viruses: non-enveloped polyomaviruses and enveloped baculoviruses [118]. The photoproduction and lifetime of $O_2(^1\Delta_g)$ are sufficient to exert a photoantiviral effect, as confirmed by a considerable decrease in virus infectivity. As expected, no antiviral effect was detected in the absence of light and/or encapsulated TPP.

3.1.2 Nanofiber Materials with Bound Photosensitizers

In addition to encapsulated PSs, covalently or ionically bound PSs inside or on the surface of polymer nanofibers have been investigated [124, 126, 127]. Photoactive nanofibers were prepared using amino end-capped polyethers with covalently attached fullerene C_{60} as a PS with a high $O_2(^1\Delta_g)$ quantum yield [126]. The use of these materials was suggested for the treatment of multidrug-resistant pathogens and the photodynamic therapy of cancer because of their photocytotoxicity toward human promonocytic THP-1 cells.

Polystyrene cation-exchange nanofiber materials with high surface areas and adsorption capacities were prepared by electrospinning followed by in situ sulfonation. The photoactivity of the materials was introduced by the ion-exchange reaction with tetra-cationic 5,10,15,20-(1-methylpyridinium-4-yl)-porphyrin (TMPyP) [124]. The photophysical properties depend on the amount of adsorbed TMPyP and its organization on the nanofiber surface. As a result, these polar nanofibers photogenerate $O_2(^1\Delta_g)$ and exhibit strong antibacterial activity towards *Escherichia coli* when activated by visible light.

Similarly, anion-exchange polystyrene nanofiber materials were prepared by electrospinning followed by two-step consecutive functionalization of the nanofiber surface using chlorosulfonic acid and ethylenediamine [127]. The photoactive character was introduced via the adsorption of tetra-anionic TPPS on the nanofiber surface. These materials have a high ion-exchange capacity, photogenerate $O_2(^1\Delta_g)$, and, as a result, induce an antibacterial effect upon irradiation by simulated daylight. The adsorption of TPPS and I^- on the surface simultaneously led to even more efficient materials with respect to phototoxicity. This behavior is a result of the oxidation of I^- by photoproduced $O_2(^1\Delta_g)$ to another cytotoxic species (I_3^-) which imparts increased antibacterial properties toward *Escherichia coli* and prolonged antibacterial effect in the dark. All processes occurring on the surfaces are schematically shown in Fig. 12.

A negatively charged anthraquinone-2,6-disulfonic acid was adsorbed onto the positively charged surfaces of cross-linked gelatin nanofiber membranes in the form of a thin coating [128]. The surfaces of the gelatin membranes efficiently

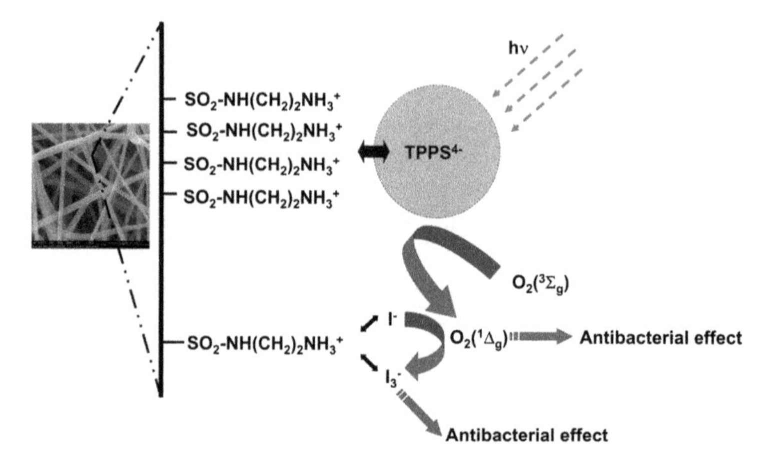

Fig. 12 Two antibacterial species, $O_2(^1\Delta_g)$ and I_3^-, are simultaneously photogenerated on the surface of anion-exchange nanofiber materials with adsorbed I^- and TPPS and efficiently destroy bacteria. Adapted from [127]. Reproduced with permission from The Royal of Chemistry and Owner Societies, Copyright 2014

photogenerate H_2O_2 after exposure to UVA and exhibit a post-irradiation antibacterial effect on *Escherichia coli*. Testing with HepG2 cells revealed that these membranes have the same biocompatibility as the original gelatin nanofibers.

3.2 Nanofiber Materials as Carriers for Other Biomedical Applications

Drug delivery via electrospun scaffolds provides flexibility in the creation of an optimal delivery vehicle for therapeutic treatments. Nanofibers made from natural polymers with low toxicities, immunogenicities, and high biocompatibilities are especially well suited to being drug release carriers. Several nanofiber-based delivery systems are under investigation for the delivery of antibiotic, antimicrobial, antifungal, antihypertensive and anticancer drugs [129–132].

Recently, nanofiber materials allowing light-controlled drug release have also been designed. These nanofibers were prepared via the electrospinning of block copolymers of vinyl-benzyl chloride and glycidyl methacrylate and subsequent coupling with sodium azide [133]. The following "click-reaction" with photoactive 4-propargyloxyazobenzene allowed the preparation of nanofibers with photoresponsive surfaces because of the *cis-trans* photoisomerization of the azo groups. The conjugate of the anticancer agent 5-fluoroucil with α-cyclodextrin (αCD) has been used as a prodrug for site-specific delivery of the drug. The surfaces of nanofibers with *trans*-azo isomers enable prodrug loading via host–guest interactions between the αCD cavity and *trans*-azo isomer. Upon irradiation with ultraviolet light ($\lambda = 365$ nm), isomerization leads to the *cis*-azo isomer and the release of the prodrug. This system exhibits a rapid photoresponse and controlled release characteristics.

Nanofiber materials with high diffusion coefficients are reported to be ideal matrixes for oxygen sensing because of their facile preparation, large area-to-volume ratios, and nanoscale characteristics, ensuring high signal intensities and the quick response of a shielded, embedded sensor (see also Sect. 2.3). Optical oxygen sensors based on highly luminescent Eu(III) complexes (Fig. 13, I) encapsulated in polystyrene nanofibers were prepared via electrospinning [134]. These materials showed high sensitivity and good linearity of the Stern–Volmer plots vs the O_2 concentration – see (3) – and have good operational stability, reproducibility, a quick response, and fast recovery times.

One of the limiting factors in regenerative medicine is the efficient and monitored oxygenation of tissue scaffolds. Many established methods for oxygen detection (such as electrodes) require mechanical disturbance of the tissue structure. To address the need for scaffold-based oxygen monitoring, a self-referenced oxygen sensor was embedded into nanofibers. The oxygen sensor is based on the dual-emissive difluoroboron iodo-dibenzylmethane dye (Fig. 13, II) encapsulated in electrospun polylactide-*co*-glycolide nanofibers and has been used to study oxygen

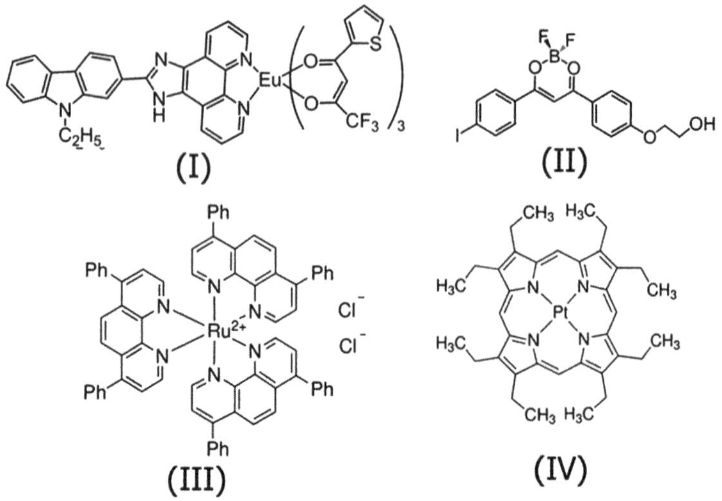

Fig. 13 Molecular structures of sensors (I–IV)

gradients in tissue constructs [135]. The boron dye emits phosphorescence and fluorescence signals whose ratio depends on the oxygen concentration.

Another optical oxygen sensor was prepared via the core-shell electrospinning of highly oxygen permeable polydimethylsiloxane (PDMS) within an envelope of biocompatible polycaprolactone. The fluorescent oxygen sensitive probes (Fig. 13, III and IV) were incorporated into the PDMS [136]. Both sensors exhibited rapid responses because of the porous nature of the nanofibers. Experiments with glioma cell viability revealed negligible effects of the sensing probes on the cells. These nanofiber-based sensors could be integrated into standard cell culture plates or bioreactors to provide information about the oxygen concentrations, which is important for cell research and tissue engineering.

Not only oxygen sensors but also other biosensors can benefit from the unique properties of nanofiber scaffolds. Composite nanofiber materials acting as light-harvesting polymers (PDMP/PLA) were used to construct an oligonucleotide (ODN) sensor based on the FRET principle [137]. Capture probe ODNs were covalently grafted onto the nanofibers, and the reporting ODN carried a reporting chromophore dye (ODN-C). Hybridization can be monitored using FRET from the polymer to the dye-labeled signaling probe (Fig. 14).

One of the most promising approaches in regenerative medicine, especially for skin and nerve regeneration, is photocurrent therapy, which includes light and electrical stimulations (the current is generated by the light stimulation) [138]. The photovoltaic polymer poly(3-hexylthiophene), which exhibits a photocurrent effect, and epidermal growth factor were encapsulated in core–shell-structured gelatin/poly(L-lactic acid)-*co*-poly(ε-caprolactone) nanofibers via coaxial spinning. The potential applications of the nanofibrous scaffold as a novel skin graft were studied. The proliferation of fibroblasts on the nanofibers was

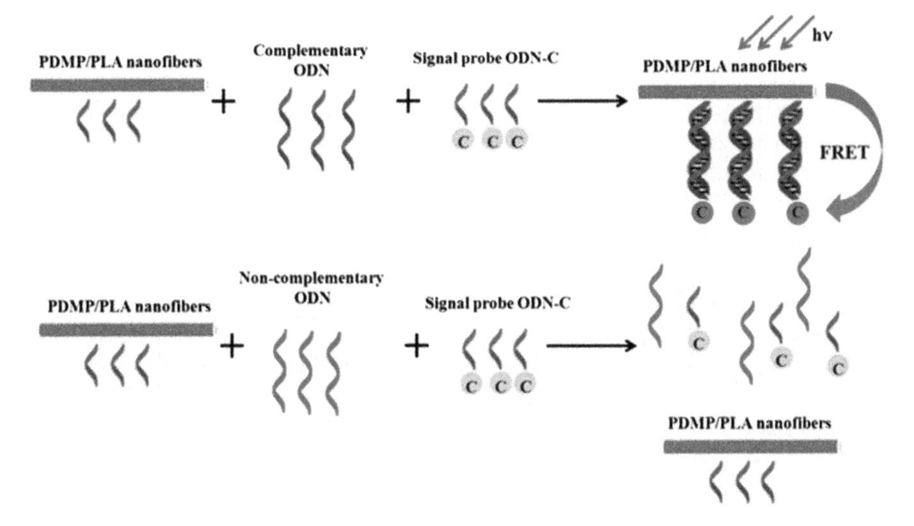

Fig. 14 Illustration of the working mechanism of oligonucleotide sensing using PDMP/PLA nanofibers. Adapted with permission from [137]. Copyright 2014, Elsevier

significantly improved under light stimulation compared to fibroblasts on the same scaffold under dark conditions. In vitro studies demonstrated the excellent wound healing ability of the nanofibers under light stimulation. These biomimetic and photosensitive nanofibers were proposed as novel scaffolds for photocurrent therapy for wound healing and skin construction [139].

4 Conclusions and Prospects

The potential applications of nanostructured surfaces for the treatment and diagnosis of diseases as well as for other biomedical purposes are currently very broad. Light irradiation has been shown to affect the surface properties of nanostructured materials, e.g., their hydrophilicity, which can improve biomaterial-cell interactions. Although examples of such materials with the associated biological response have been demonstrated, they are often special cases of a specific treatment rather than engineered solutions. Surface engineering based on nanostructured materials together with light activation can offer useful features that inhibit bacterial adhesion and biofilm growth. This approach also represents a valid alternative to classic antibiotic therapies. Sensing using luminescent sensors is an additional tool for understanding living systems and may find applications in medicine, drug discovery, process control, and food safety.

Nanostructured films and nanofibers producing $O_2(^1\Delta_g)$ have been identified as promising materials for antibacterial and antiviral treatments. The possibility of industrial-scale nanofiber production, an almost unlimited amount of oxygen from atmospheric air, the fact that bacteria and most other pathogens cannot pass through the pores of the nanostructured materials, and the engineered photosensitizing properties make these materials promising prospective sterile and photodisinfecting materials for broad applications in medicine. The progress in electrospinning which enables the preparation of polymer nanofibers with small diameters of as small as 1 nm [140] and the recent discovery that small but important pathogens, such as the HIV virus, are efficiently captured on the surfaces of polystyrene nanofibers [141] combined with the reactivity of $O_2(^1\Delta_g)$ opens new areas for the applications of $O_2(^1\Delta_g)$-producing materials. Because of their antibacterial and antiviral activities, these nanofiber materials can be used in a range of areas in which sterility needs to be maintained or enforced, including (1) coverings (e.g., wound dressings and plasters), (2) surgery supplements (e.g., surgery masks), and (3) the storage and disinfection of aqueous media. It has to be emphasized that no bacterial resistance was found towards $O_2(^1\Delta_g)$. Therefore, $O_2(^1\Delta_g)$-producing materials may be suitable alternatives to antibiotics and other antimicrobial agents because of the increasing resistance of microorganisms, which renders conventional treatments less effective.

A common feature of the materials producing $O_2(^1\Delta_g)$ is a short diffusion length of $O_2(^1\Delta_g)$, which limits efficient antibacterial action to the surface or the nearby area, especially in aqueous media, where $O_2(^1\Delta_g)$ has a short lifetime. This limitation can be overcome by grafting longer-lived photogenerated antibacterial agents (e.g., NO radicals) with longer diffusion lengths. Future studies of nanomaterials that generate $O_2(^1\Delta_g)$ and/or other antibacterial agents are foreseen, as are studies dedicated to the grafting of PS and/or other photoactivatable compounds to the surface of materials exhibiting dark antimicrobial properties.

The materials with light-triggered nanostructured surfaces with high surface area to volume ratios have great potential for being powerful materials to address various medical-, biological-, and environmental-related issues. Despite the rapid progress in this field, the possible applications have yet to be exploited fully. The intense and rational investigation may open new and facile approaches for fabrication and utilization of these materials, which we believe are far beyond what has been reported in this chapter.

Acknowledgements This work was supported by the Czech Science Foundation (No. 13-12496S).

References

1. Kim TG, Shin H, Lim DW (2012) Biomimetic scaffolds for tissue engineering. Adv Funct Mater 22:2446–2468. doi:10.1002/adfm.201103083
2. Skorb EV, Andreeva DV (2013) Surface nanoarchitecture for bio-applications: self-regulating intelligent interfaces. Adv Funct Mater 23:4483–4506. doi:10.1002/adfm.201203884
3. Tang Z, Wang Y, Podsiadlo P, Kotov NA (2006) Biomedical applications of layer-by-layer assembly: from biomimetics to tissue engineering. Adv Mater 18:3203–3224. doi:10.1002/adma.200600113

4. Chiono V, Carmagnola I, Gentile P, Boccafoschi F, Tonda-Turo C, Ballarini M, Georgieva V, Georgiev G, Ciardelli G (2012) Layer-by-layer coating of photoactive polymers for biomedical applications. Surf Coat Technol 206:2446–2453. doi:10.1016/j.surfcoat.2011.10.048
5. Volodkin D, Klitzing R, Moehwald H (2014) Polyelectrolyte multilayers: towards single cell studies. Polymers 6:1502–1527. doi:10.3390/polym6051502
6. McCullough DH III, Regen SL (2004) Don't forget Langmuir-Blodgett films. Chem Commun 2787–2791. doi:10.1039/B410027C
7. Gooding JJ, Mearns F, Yang W, Liu J (2003) Self-assembled monolayers into the 21st century: recent advances and applications. Electroanalysis 15:81–96. doi:10.1002/elan.200390017
8. Costerton JW, Montanaro L, Arciola CR (2005) Biofilm in implant infections: its production and regulation. Int J Artif Organs 28:1062–1068
9. Campoccia D, Montanaro L, Arciola CR (2013) A review of the biomaterials technologies for infection-resistant surfaces. Biomaterials 34:8533–8554. doi:10.1016/j.biomaterials.2013.07.089
10. Noimark S, Dunnill CW, Parkin IP (2013) Shining light on materials — a self-sterilising revolution. Adv Drug Deliv Rev 65:570–580. doi:10.1016/j.addr.2012.07.003
11. Vatansever F, de Melo WCMA, Avci P, Vecchio D, Sadasivam M, Gupta A, Chandran R, Karimi M, Parizotto NA, Yin R, Tegos GP, Hamblin MR (2013) Antimicrobial strategies centered around reactive oxygen species – bactericidal antibiotics, photodynamic therapy, and beyond. FEMS Microbiol Rev 37:955–989. doi:10.1111/1574-6976.12026
12. Liu K, Cao M, Fujishima A, Jiang L (2014) Bio-inspired titanium dioxide materials with special wettability and their applications. Chem Rev 114:10044–10094. doi:10.1021/cr4006796
13. Wang R, Tan H, Zhao Z, Zhang G, Song L, Dong W, Sun Z (2014) Stable ZnO@TiO_2 core/shell nanorod arrays with exposed high energy facets for self-cleaning coatings with anti-reflective properties. J Mater Chem A 2:7313–7318. doi:10.1039/C4TA00455H
14. Skorb EV, Antonouskaya LI, Belyasova NA, Shchukin DG, Möhwald H, Sviridov DV (2008) Antibacterial activity of thin-film photocatalysts based on metal-modified TiO_2 and TiO_2:In_2O_3 nanocomposite. Appl Catal B 84:94–99. doi:10.1016/j.apcatb.2008.03.007
15. Parkin IP, Palgrave RG (2005) Self-cleaning coatings. J Mater Chem 15:1689–1695. doi:10.1039/B412803F
16. Fateh R, Dillert R, Bahnemann D (2014) Self-cleaning properties, mechanical stability, and adhesion strength of transparent photocatalytic TiO_2–ZnO coatings on polycarbonate. ACS Appl Mater Interfaces 6:2270–2278. doi:10.1021/am4051876
17. Xu QF, Liu Y, Lin F-J, Mondal B, Lyons AM (2013) Superhydrophobic TiO_2–polymer nanocomposite surface with UV-induced reversible wettability and self-cleaning properties. ACS Appl Mater Interfaces 5:8915–8924. doi:10.1021/am401668y
18. Asahi R, Morikawa T, Ohwaki T, Aoki K, Taga Y (2001) Visible-light photocatalysis in nitrogen-doped titanium oxides. Science 293:269–271. doi:10.1126/science.1061051
19. Chen J, Kubota J, Wada A, Kondo JN, Domen K (2009) Time-resolved sum frequency generation reveals adsorbate migration between different surface active sites on titanium oxide/Pt(111). J Am Chem Soc 131:4580–4581. doi:10.1021/ja900052w
20. Fujishima A, Zhang X (2006) Titanium dioxide photocatalysis: present situation and future approaches. C R Chim 9:750–760. doi:10.1016/j.crci.2005.02.055
21. Li Y, Chen C, Li J, Susan Sun X (2014) Photoactivity of poly(lactic acid) nanocomposites modulated by TiO_2 nanofillers. J Appl Polym Sci 131:40241. doi:10.1002/app.40241
22. Ogilby PR (2010) Singlet oxygen: there is indeed something new under the sun. Chem Soc Rev 39:3181–3209. doi:10.1039/B926014P
23. Alves E, Faustino MAF, Neves MGPMS, Cunha Â, Nadais H, Almeida A (2014) Potential applications of porphyrins in photodynamic inactivation beyond the medical scope. J Photochem Photobiol C 22:34–57. doi:10.1016/j.jphotochemrev.2014.09.003

24. Kingshott P, Wei J, Bagge-Ravn D, Gadegaard N, Gram L (2003) Covalent attachment of poly(ethylene glycol) to surfaces, critical for reducing bacterial adhesion. Langmuir 19:6912–6921. doi:10.1021/la034032m
25. Gonçalves ES, Ogilby PR (2008) "Inside" vs "outside" photooxygenation reactions: singlet-oxygen-mediated surface passivation of polymer films. Langmuir 24:9056–9065. doi:10.1021/la801353n
26. Jesenská S, Plíštil L, Kubát P, Lang K, Brožová L, Popelka Š, Szatmáry L, Mosinger J (2011) Antibacterial nanofiber materials activated by light. J Biomed Mater Res Part A 99A:676–683. doi:10.1002/jbm.a.33218
27. Borges J, Rodrigues LC, Reis RL, Mano JF (2014) Layer-by-layer assembly of light-responsive polymeric multilayer systems. Adv Funct Mater 24:5624–5648. doi:10.1002/adfm.201401050
28. Gabriel D, Monteiro IP, Huang D, Langer R, Kohane DS (2013) A photo-triggered layered surface coating producing reactive oxygen species. Biomaterials 34:9763–9769. doi:10.1016/j.biomaterials.2013.09.021
29. Felgentrager A, Maisch T, Spath A, Schroder JA, Baumler W (2014) Singlet oxygen generation in porphyrin-doped polymeric surface coating enables antimicrobial effects on *Staphylococcus aureus*. Phys Chem Chem Phys 16:20598–20607. doi:10.1039/C4CP02439G
30. Liu L, Chen J, Wang S (2013) Flexible antibacterial film deposited with polythiophene–porphyrin composite. Adv Healthcare Mater 2:1582–1585. doi:10.1002/adhm.201300106
31. Funes MD, Caminos DA, Alvarez MG, Fungo F, Otero LA, Durantini EN (2009) Photodynamic properties and photoantimicrobial action of electrochemically generated porphyrin polymeric films. Environ Sci Technol 43:902–908. doi:10.1021/es802450b
32. Piccirillo C, Perni S, Gil-Thomas J, Prokopovich P, Wilson M, Pratten J, Parkin IP (2009) Antimicrobial activity of methylene blue and toluidine blue O covalently bound to a modified silicone polymer surface. J Mater Chem 19:6167–6171. doi:10.1039/B905495B
33. Pineiro M, Ribeiro SM, Serra AC (2010) The influence of the support on the singlet oxygen quantum yields of porphyrin supported photosensitizers. Arkivoc 2010:51–63. doi:10.3998/ark.5550190.0011.506
34. Demel J, Lang K (2012) Layered hydroxide–porphyrin hybrid materials: synthesis, structure, and properties. Eur J Inorg Chem 2012:5154–5164. doi:10.1002/ejic.201200400
35. Lang K, Bezdička P, Bourdelande JL, Hernando J, Jirka I, Káfuňková E, Kovanda F, Kubát P, Mosinger J, Wagnerová DM (2007) Layered double hydroxides with intercalated porphyrins as photofunctional materials: subtle structural changes modify singlet oxygen production. Chem Mater 19:3822–3829. doi:10.1021/cm070351d
36. Khan AI, Ragavan A, Fong B, Markland C, O'Brien M, Dunbar TG, Williams GR, O'Hare D (2009) Recent developments in the use of layered double hydroxides as host materials for the storage and triggered release of functional anions. Ind Eng Chem Res 48:10196–10205. doi:10.1021/ie9012612
37. Evans DG, Slade RCT (2006) In: Duan X, Evans DG (eds) Layered double hydroxides, vol 119. Springer, Berlin, pp 1–87
38. Liang R, Tian R, Ma L, Zhang L, Hu Y, Wang J, Wei M, Yan D, Evans DG, Duan X (2014) A supermolecular photosensitizer with excellent anticancer performance in photodynamic therapy. Adv Funct Mater 24:3144–3151. doi:10.1002/adfm.201303811
39. Káfuňková E, Lang K, Kubát P, Klementová M, Mosinger J, Šlouf M, Troutier-Thuilliez A-L, Leroux F, Verney V, Taviot-Gueho C (2010) Porphyrin-layered double hydroxide/polymer composites as novel ecological photoactive surfaces. J Mater Chem 20:9423–9432. doi:10.1039/C0JM00746C
40. Kovanda F, Jindová E, Lang K, Kubát P, Sedláková Z (2010) Preparation of layered double hydroxides intercalated with organic anions and their application in LDH/poly(butyl methacrylate) nanocomposites. Appl Clay Sci 48:260–270. doi:10.1016/j.clay.2009.11.012
41. Merchan M, Ouk TS, Kubát P, Lang K, Coelho C, Verney V, Commereuc S, Leroux F, Sol V, Taviot-Gueho C (2013) Photostability and photobactericidal properties of porphyrin-layered

double hydroxide-polyurethane composite films. J Mater Chem B 1:2139–2146. doi:10.1039/C3TB20070A
42. Xiong Z, Xu Y (2007) Immobilization of palladium phthalocyaninesulfonate onto anionic clay for sorption and oxidation of 2,4,6-trichlorophenol under visible light irradiation. Chem Mater 19:1452–1458. doi:10.1021/cm062437x
43. Káfuňková E, Taviot-Guého C, Bezdička P, Klementová M, Kovář P, Kubát P, Mosinger J, Pospíšil M, Lang K (2010) Porphyrins intercalated in Zn/Al and Mg/Al layered double hydroxides: properties and structural arrangement. Chem Mater 22:2481–2490. doi:10.1021/cm903125v
44. Perni S, Piccirillo C, Kafizas A, Uppal M, Pratten J, Wilson M, Parkin I (2010) Antibacterial activity of light-activated silicone containing methylene blue and gold nanoparticles of different sizes. J Cluster Sci 21:427–438. doi:10.1007/s10876-010-0319-5
45. Perni S, Piccirillo C, Pratten J, Prokopovich P, Chrzanowski W, Parkin IP, Wilson M (2009) The antimicrobial properties of light-activated polymers containing methylene blue and gold nanoparticles. Biomaterials 30:89–93. doi:10.1016/j.biomaterials.2008.09.020
46. Perni S, Prokopovich P, Parkin IP, Wilson M, Pratten J (2010) Prevention of biofilm accumulation on a light-activated antimicrobial catheter material. J Mater Chem 20:8668–8673. doi:10.1039/C0JM01891K
47. Wang X-d, Wolfbeis OS (2014) Optical methods for sensing and imaging oxygen: materials, spectroscopies and applications. Chem Soc Rev 43:3666–3761. doi:10.1039/C4CS00039K
48. Lu X, Winnik MA (2001) Luminescence quenching in polymer/filler nanocomposite films used in oxygen sensors. Chem Mater 13:3449–3463. doi:10.1021/cm011029k
49. Guice KB, Caldorera ME, McShane MJ (2005) Nanoscale internally referenced oxygen sensors produced from self-assembled nanofilms on fluorescent nanoparticles. J Biomed Opt 10:064031. doi:10.1117/1.2147419
50. Salavagione HJ, Diez-Pascual AM, Lazaro E, Vera S, Gomez-Fatou MA (2014) Chemical sensors based on polymer composites with carbon nanotubes and graphene: the role of the polymer. J Mater Chem A 2:14289–14328. doi:10.1039/C4TA02159B
51. Huang F, Xu H, Tan W, Liang H (2014) Multicolor and erasable DNA photolithography. ACS Nano 8:6849–6855. doi:10.1021/nn5024472
52. Chee M, Yang R, Hubbell E, Berno A, Huang XC, Stern D, Winkler J, Lockhart DJ, Morris MS, Fodor SPA (1996) Accessing genetic information with high density DNA arrays. Science 274:610–614. doi:10.1126/science.274.5287.610
53. Liu Q, Wang L, Frutos AG, Condon AE, Corn RM, Smith LM (2000) DNA computing on surfaces. Nature 403:175–179. doi:10.1038/35003155
54. Shimizu Y, Boehm H, Yamaguchi K, Spatz JP, Nakanishi J (2014) A photoactivatable nanopatterned substrate for analyzing collective cell migration with precisely tuned cell-extracellular matrix ligand interactions. PLoS One 9, e91875. doi:10.1371/journal.pone.0091875
55. Sun T, Zhang YS, Pang B, Hyun DC, Yang M, Xia Y (2014) Engineered nanoparticles for drug delivery in cancer therapy. Angew Chem Int Ed 53:12320–12364. doi:10.1002/anie.201403036
56. Sortino S, Petralia S, Compagnini G, Conoci S, Condorelli G (2002) Light-controlled nitric oxide generation from a novel self-assembled monolayer on a gold surface. Angew Chem Int Ed 41:1914–1917. doi:10.1002/1521-3773(20020603)41:11<1914::AID-ANIE1914>3.0.CO;2-J
57. Zhu C, Ninh C, Bettinger CJ (2014) Photoreconfigurable polymers for biomedical applications: chemistry and macromolecular engineering. Biomacromolecules 15:3474–3494. doi:10.1021/bm500990z
58. Burgess DJ (2012) Nanotechnology: tissue penetration of photodynamic therapy. Nat Rev Cancer 12:737–737. doi:10.1038/nrc3393

59. Jain PK, Lee KS, El-Sayed IH, El-Sayed MA (2006) Calculated absorption and scattering properties of gold nanoparticles of different size, shape, and composition: applications in biological imaging and biomedicine. J Phys Chem B 110:7238–7248. doi:10.1021/jp057170o
60. Matteini P, Tatini F, Luconi L, Ratto F, Rossi F, Giambastiani G, Pini R (2013) Photothermally activated hybrid films for quantitative confined release of chemical species. Angew Chem Int Ed 52:5956–5960. doi:10.1002/anie.201207986
61. Bitounis D, Ali-Boucetta H, Hong BH, Min D-H, Kostarelos K (2013) Prospects and challenges of graphene in biomedical applications. Adv Mater 25:2258–2268. doi:10.1002/adma.201203700
62. Matteini P, Tatini F, Cavigli L, Ottaviano S, Ghini G, Pini R (2014) Graphene as a photothermal switch for controlled drug release. Nanoscale 6:7947–7953. doi:10.1039/C4NR01622J
63. Sada T, Fujigaya T, Niidome Y, Nakazawa K, Nakashima N (2011) Near-IR laser-triggered target cell collection using a carbon nanotube-based cell-cultured substrate. ACS Nano 5:4414–4421. doi:10.1021/nn2012767
64. Zeng Y, Lu JQ (2014) Optothermally responsive nanocomposite generating mechanical forces for cells enabled by few-walled carbon nanotubes. ACS Nano 8:11695–11706. doi:10.1021/nn505042b
65. Volodkin DV, Madaboosi N, Blacklock J, Skirtach AG, Möhwald H (2009) Surface-supported multilayers decorated with bio-active material aimed at light-triggered drug delivery. Langmuir 25:14037–14043. doi:10.1021/la9015433
66. You J, Heo JS, Kim J, Park T, Kim B, Kim H-S, Choi Y, Kim HO, Kim E (2013) Noninvasive photodetachment of stem cells on tunable conductive polymer nano thin films: selective harvesting and preserved differentiation capacity. ACS Nano 7:4119–4128. doi:10.1021/nn400405t
67. Tokarev I, Motornov M, Minko S (2009) Molecular-engineered stimuli-responsive thin polymer film: a platform for the development of integrated multifunctional intelligent materials. J Mater Chem 19:6932–6948. doi:10.1039/B906765E
68. Stuart MAC, Huck WTS, Genzer J, Muller M, Ober C, Stamm M, Sukhorukov GB, Szleifer I, Tsukruk VV, Urban M, Winnik F, Zauscher S, Luzinov I, Minko S (2010) Emerging applications of stimuli-responsive polymer materials. Nat Mater 9:101–113. doi:10.1038/nmat2614
69. Kelley EG, Albert JN, Sullivan MO, Epps TH 3rd (2013) Stimuli-responsive copolymer solution and surface assemblies for biomedical applications. Chem Soc Rev 42:7057–7071. doi:10.1039/c3cs35512h
70. Caputo G, Cortese B, Nobile C, Salerno M, Cingolani R, Gigli G, Cozzoli PD, Athanassiou A (2009) Reversibly light-switchable wettability of hybrid organic/inorganic surfaces with dual micro-/nanoscale roughness. Adv Funct Mater 19:1149–1157. doi:10.1002/adfm.200800909
71. Zhou Y-N, Li J-J, Zhang Q, Luo Z-H (2014) Light-responsive smart surface with controllable wettability and excellent stability. Langmuir 30:12236–12242. doi:10.1021/la501907w
72. Xin B, Hao J (2010) Reversibly switchable wettability. Chem Soc Rev 39:769–782. doi:10.1039/B913622C
73. Chu Z, Seeger S (2014) Superamphiphobic surfaces. Chem Soc Rev 43:2784–2798. doi:10.1039/C3CS60415B
74. Galopin E, Piret G, Szunerits S, Lequette Y, Faille C, Boukherroub R (2009) Selective adhesion of bacillus cereus spores on heterogeneously wetted silicon nanowires. Langmuir 26:3479–3484. doi:10.1021/la9030377
75. Rocha L, Păiuş C-M, Luca-Raicu A, Resmerita E, Rusu A, Moleavin I-A, Hamel M, Branza-Nichita N, Hurduc N (2014) Azobenzene based polymers as photoactive supports and micellar structures for applications in biology. J Photochem Photobiol A 291:16–25. doi:10.1016/j.jphotochem.2014.06.018

76. Barillé R, Janik R, Kucharski S, Eyer J, Letournel F (2011) Photo-responsive polymer with erasable and reconfigurable micro- and nano-patterns: an in vitro study for neuron guidance. Colloids Surf B 88:63–71. doi:10.1016/j.colsurfb.2011.06.005
77. Kim M-H, Park M, Kang K, Choi IS (2014) Neurons on nanometric topographies: insights into neuronal behaviors in vitro. Biomater Sci 2:148–155. doi:10.1039/C3BM60255A
78. Hurduc N, Macovei A, Paius C, Raicu A, Moleavin I, Branza-Nichita N, Hamel M, Rocha L (2013) Azo-polysiloxanes as new supports for cell cultures. Mater Sci Eng C 33:2440–2445. doi:10.1016/j.msec.2013.01.012
79. Kenawy E-R, Bowlin GL, Mansfield K, Layman J, Simpson DG, Sanders EH, Wnek GE (2002) Release of tetracycline hydrochloride from electrospun poly(ethylene-co-vinylacetate), poly(lactic acid), and a blend. J Control Release 81:57–64. doi:10.1016/S0168-3659(02)00041-X
80. Wang X, Kim Y-G, Drew C, Ku B-C, Kumar J, Samuelson LA (2004) Electrostatic assembly of conjugated polymer thin layers on electrospun nanofibrous membranes for biosensors. Nano Lett 4:331–334. doi:10.1021/nl034885z
81. Thomas V, Dean DR, Vohra YK (2006) Nanostructured biomaterials for regenerative medicine. Curr Nanosci 2:155–177. doi:10.1097/01.hpc.0000234809.93495.e3
82. Bergshoef MM, Vancso GJ (1999) Transparent nanocomposites with ultrathin, electrospun nylon-4,6 fiber reinforcement. Adv Mater 11:1362–1365. doi:10.1002/(SICI)1521-4095(199911)11:16<1362::AID-ADMA1362>3.0.CO;2-X
83. Schreuder-Gibson H, Gibson P, Senecal K, Sennett M, Walker J, Yeomans W, Ziegler D, Tsai PP (2002) Protective textile materials based on electrospun nanofibers. J Adv Mater 34:44–55
84. Zhang Y, Lim C, Ramakrishna S, Huang Z-M (2005) Recent development of polymer nanofibers for biomedical and biotechnological applications. J Mater Sci Mater Med 16:933–946. doi:10.1007/s10856-005-4428-x
85. Buchko CJ, Chen LC, Shen Y, Martin DC (1999) Processing and microstructural characterization of porous biocompatible protein polymer thin films. Polymer 40:7397–7407. doi:10.1016/S0032-3861(98)00866-0
86. Ondarçuhu T, Joachim C (1998) Drawing a single nanofibre over hundreds of microns. Europhys Lett 42:215
87. Whitesides GM, Grzybowski B (2002) Self-assembly at all scales. Science 295:2418–2421. doi:10.1126/science.1070821
88. Martin CR (1996) Membrane-based synthesis of nanomaterials. Chem Mater 8:1739–1746. doi:10.1021/cm960166s
89. Ma PX, Zhang R (1999) Synthetic nano-scale fibrous extracellular matrix. J Biomed Mater Res 46:60–72. doi:10.1002/(SICI)1097-4636(199907)46:1<60::AID-JBM7>3.0.CO;2-H
90. Agarwal S, Wendorff JH, Greiner A (2008) Use of electrospinning technique for biomedical applications. Polymer 49:5603–5621. doi:10.1016/j.polymer.2008.09.014
91. Reneker DH, Yarin AL, Fong H, Koombhongse S (2000) Bending instability of electrically charged liquid jets of polymer solutions in electrospinning. J Appl Phys 87:4531–4547. doi:10.1063/1.373532
92. Li D, Xia YN (2004) Electrospinning of nanofibers: reinventing the wheel? Adv Mater 16:1151–1170. doi:10.1002/adma.200400719
93. Zhang YZ, Venugopal J, Huang ZM, Lim CT, Ramakrishna S (2005) Characterization of the surface biocompatibility of the electrospun PCL-collagen nanofibers using fibroblasts. Biomacromolecules 6:2583–2589. doi:10.1021/bm050314k
94. Jirsák O, Sanetrník F, Lukáš D, Kotek V, Martinová L, Chaloupek J (2003) CZ patent 294,274
95. Jirsák O, Sanetrník F, Lukáš D, Kotek V, Martinová L, Chaloupek J (2004) PCT/CZ2004/000,056
96. Jirsák O, Sanetrník F, Lukáš D, Kotek V, Martinová L, Chaloupek J (2005) WO 02,410

97. Henke P, Kozak H, Artemenko A, Kubát P, Forstová J, Mosinger J (2014) Superhydrophilic polystyrene nanofiber materials generating $O_2(^1\Delta_g)$: postprocessing surface modifications toward efficient antibacterial effect. ACS Appl Mater Interfaces 6:13007–13014. doi:10.1021/am502917w
98. Kim FS, Ren G, Jenekhe SA (2010) One-dimensional nanostructures of π-conjugated molecular systems: assembly, properties, and applications from photovoltaics, sensors, and nanophotonics to nanoelectronics. Chem Mater 23:682–732. doi:10.1021/cm102772x
99. Deitzel JM, Kleinmeyer J, Harris D, Beck Tan NC (2001) The effect of processing variables on the morphology of electrospun nanofibers and textiles. Polymer 42:261–272. doi:10.1016/S0032-3861(00)00250-0
100. Kenawy E-R, Worley SD, Broughton R (2007) The chemistry and applications of antimicrobial polymers: a state-of-the-art review. Biomacromolecules 8:1359–1384. doi:10.1021/bm061150q
101. Timofeeva L, Kleshcheva N (2011) Antimicrobial polymers: mechanism of action, factors of activity, and applications. Appl Microbiol Biotechnol 89:475–492. doi:10.1007/s00253-010-2920-9
102. Jaeger W, Bohrisch J, Laschewsky A (2010) Synthetic polymers with quaternary nitrogen atoms—synthesis and structure of the most used type of cationic polyelectrolytes. Prog Polym Sci 35:511–577. doi:10.1016/j.progpolymsci.2010.01.002
103. Bshena O, Heunis TDJ, Dicks LMT, Klumperman B (2011) Antimicrobial fibers: therapeutic possibilities and recent advances. Future Med Chem 3:1821–1847. doi:10.4155/fmc.11.131
104. Yao C, Li X, Neoh KG, Shi Z, Kang ET (2008) Surface modification and antibacterial activity of electrospun polyurethane fibrous membranes with quaternary ammonium moieties. J Membr Sci 320:259–267. doi:10.1016/j.memsci.2008.04.012
105. Kong H, Jang J (2008) Antibacterial properties of novel poly(methyl methacrylate) nanofiber containing silver nanoparticles. Langmuir 24:2051–2056. doi:10.1021/la703085e
106. Jeon HJ, Kim JS, Kim TG, Kim JH, Yu W-R, Youk JH (2008) Preparation of poly(-ε-caprolactone)-based polyurethane nanofibers containing silver nanoparticles. Appl Surf Sci 254:5886–5890. doi:10.1016/j.apsusc.2008.03.141
107. Mosinger J, Lang K, Kubát P, Sýkora J, Hof M, Plíštil L, Mosinger B (2009) Photofunctional polyurethane nanofabrics doped by zinc tetraphenylporphyrin and zinc phthalocyanine photosensitizers. J Fluoresc 19:705–713. doi:10.1007/s10895-009-0464-0
108. Mosinger J, Lang K, Plíštil L, Jesenská S, Hostomský J, Zelinger Z, Kubát P (2010) Fluorescent polyurethane nanofabrics: a source of singlet oxygen and oxygen sensing. Langmuir 26:10050–10056. doi:10.1021/la1001607
109. Arenbergerová M, Arenberger P, Bednář M, Kubát P, Mosinger J (2012) Light-activated nanofibre textiles exert antibacterial effects in the setting of chronic wound healing. Exp Dermatol 21:619–624. doi:10.1111/j.1600-0625.2012.01536.x
110. Gmurek M, Mosinger J, Miller JS (2012) 2-Chlorophenol photooxidation using immobilized meso-tetraphenylporphyrin in polyurethane nanofabrics. Photochem Photobiol Sci 11:1422–1427. doi:10.1039/C2PP25010A
111. Gmurek M, Bizukojć M, Mosinger J, Ledakowicz S (2015) Application of photoactive electrospun nanofiber materials with immobilized meso-tetraphenylporphyrin for parabens photodegradation. Catal Today 240(Part A):160–167. doi:10.1016/j.cattod.2014.06.015
112. Mosinger J, Jirsák O, Kubát P, Lang K, Mosinger B (2007) Bactericidal nanofabrics based on photoproduction of singlet oxygen. J Mater Chem 17:164–166. doi:10.1039/B614617A
113. Mosinger J, Lang K, Hostomský J, Franc J, Sýkora J, Hof M, Kubát P (2010) Singlet oxygen imaging in polymeric nanofibers by delayed fluorescence. J Phys Chem B 114:15773–15779. doi:10.1021/jp105789p
114. Suchánek J, Henke P, Mosinger J, Zelinger Z, Kubát P (2014) Effect of temperature on photophysical properties of polymeric nanofiber materials with porphyrin photosensitizers. J Phys Chem B 118:6167–6174. doi:10.1021/jp5029917

115. Zugle R, Nyokong T (2012) Physico-chemical properties of lutetium phthalocyanine complexes in solution and in solid polystyrene polymer fibers and their application in photoconversion of 4-nitrophenol. J Mol Catal A 358:49–57. doi:10.1016/j.molcata.2012.02.010
116. Masilela N, Kleyi P, Tshentu Z, Priniotakis G, Westbroek P, Nyokong T (2013) Photodynamic inactivation of *Staphylococcus aureus* using low symmetrically substituted phthalocyanines supported on a polystyrene polymer fiber. Dyes Pigm 96:500–508. doi:10.1016/j.dyepig.2012.10.001
117. Osifeko OL, Nyokong T (2014) Applications of lead phthalocyanines embedded in electrospun fibers for the photoinactivation of *Escherichia coli* in water. Dyes Pigm 111:8–15. doi:10.1016/j.dyepig.2014.05.010
118. Lhotáková Y, Plíštil L, Morávková A, Kubát P, Lang K, Forstová J, Mosinger J (2012) Virucidal nanofiber textiles based on photosensitized production of singlet oxygen. PLoS One 7, e49226. doi:10.1371/journal.pone.0049226
119. Modisha P, Nyokong T (2014) Fabrication of phthalocyanine-magnetic nanoparticles hybrid nanofibers for degradation of Orange-G. J Mol Catal A 381:132–137. doi:10.1016/j.molcata.2013.10.012
120. Goethals A, Mugadza T, Arslanoglu Y, Zugle R, Antunes E, Van Hulle SWH, Nyokong T, De Clerck K (2014) Polyamide nanofiber membranes functionalized with zinc phthalocyanines. J Appl Polym Sci 131:. doi:10.1002/app.40486
121. Gabriel D, Cohen-Karni T, Huang D, Chiang HH, Kohane DS (2014) Photoactive electrospun fibers for inducing cell death. Adv Healthcare Mater 3:494–499. doi:10.1002/adhm.201300318
122. Kirakci K, Kubát P, Dušek M, Fejfarová K, Šícha V, Mosinger J, Lang K (2012) A highly luminescent hexanuclear molybdenum cluster – a promising candidate toward photoactive materials. Eur J Inorg Chem 2012:3107–3111. doi:10.1002/ejic.201200402
123. Ma H, Hsiao BS, Chu B (2014) Functionalized electrospun nanofibrous microfiltration membranes for removal of bacteria and viruses. J Membr Sci 452:446–452. doi:10.1016/j.memsci.2013.10.047
124. Henke P, Lang K, Kubát P, Sýkora J, Šlouf M, Mosinger J (2013) Polystyrene nanofiber materials modified with an externally bound porphyrin photosensitizer. ACS Appl Mater Interfaces 5:3776–3783. doi:10.1021/am4004057
125. Khil M-S, Cha D-I, Kim H-Y, Kim I-S, Bhattarai N (2003) Electrospun nanofibrous polyurethane membrane as wound dressing. J Biomed Mater Res B 67B:675–679. doi:10.1002/jbm.b.10058
126. Stoilova O, Jérôme C, Detrembleur C, Mouithys-Mickalad A, Manolova N, Rashkov I, Jérôme R (2007) C60-containing nanostructured polymeric materials with potential biomedical applications. Polymer 48:1835–1843. doi:10.1016/j.polymer.2007.02.026
127. Plíštil L, Henke P, Kubát P, Mosinger J (2014) Anion exchange nanofiber materials activated by daylight with a dual antibacterial effect. Photochem Photobiol Sci 13:1321–1329. doi:10.1039/C4PP00157E
128. Lu A, Ma Z, Zhuo J, Sun G, Zhang G (2013) Layer-by-layer structured gelatin nanofiber membranes with photoinduced antibacterial functions. J Appl Polym Sci 128:970–975. doi:10.1002/app.38131
129. Kim K, Luu YK, Chang C, Fang D, Hsiao BS, Chu B, Hadjiargyrou M (2004) Incorporation and controlled release of a hydrophilic antibiotic using poly(lactide-co-glycolide)-based electrospun nanofibrous scaffolds. J Control Release 98:47–56. doi:10.1016/j.jconrel.2004.04.009
130. Hadjiargyrou M, Chiu JB (2008) Enhanced composite electrospun nanofiber scaffolds for use in drug delivery. Expert Opin Drug Deliv 5:1093–1106. doi:10.1517/17425247.5.10.1093
131. Liang D, Hsiao BS, Chu B (2007) Functional electrospun nanofibrous scaffolds for biomedical applications. Adv Drug Deliv Rev 59:1392–1412. doi:10.1016/j.addr.2007.04.021

132. Kumbar SG, Nair LS, Bhattacharyya S, Laurencin CT (2006) Polymeric nanofibers as novel carriers for the delivery of therapeutic molecules. J Nanosci Nanotechnol 6:2591–2607. doi:10.1166/jnn.2006.462
133. Fu G-D, Xu L-Q, Yao F, Li G-L, Kang E-T (2009) Smart nanofibers with a photoresponsive surface for controlled release. ACS Appl Mater Interfaces 1:2424–2427. doi:10.1021/am900526u
134. Wang YH, Li B, Zhang LM, Zuo QH, Li P, Zhang J, Su ZM (2011) High-performance oxygen sensors based on EuIII complex/polystyrene composite nanofibrous membranes prepared by electrospinning. ChemPhysChem 12:349–355. doi:10.1002/cphc.201000884
135. Bowers DT, Tanes ML, Das A, Lin Y, Keane NA, Neal RA, Ogle ME, Brayman KL, Fraser CL, Botchwey EA (2014) Spatiotemporal oxygen sensing using dual emissive boron dye–polylactide nanofibers. ACS Nano 8:12080–12091. doi:10.1021/nn504332j
136. Xue R, Behera P, Xu J, Viapiano MS, Lannutti JJ (2014) Polydimethylsiloxane core–polycaprolactone shell nanofibers as biocompatible, real-time oxygen sensors. Sens Actuators B 192:697–707. doi:10.1016/j.snb.2013.10.084
137. Abdul Rahman N, Gulur Srinivas AR, Travas-Sejdic J (2014) Spontaneous stacking of electrospun conjugated polymer composite nanofibers producing highly porous fiber mats. Synth Met 191:151–160. doi:10.1016/j.synthmet.2014.03.006
138. Jin G, Prabhakaran MP, Liao S, Ramakrishna S (2011) Photosensitive materials and potential of photocurrent mediated tissue regeneration. J Photochem Photobiol B 102:93–101. doi:10.1016/j.jphotobiol.2010.09.010
139. Jin G, Prabhakaran MP, Ramakrishna S (2014) Photosensitive and biomimetic core–shell nanofibrous scaffolds as wound dressing. Photochem Photobiol 90:673–681. doi:10.1111/php.12238
140. Huang C, Chen S, Lai C, Reneker Darrell H, Qiu H, Ye Y, Hou H (2006) Electrospun polymer nanofibres with small diameters. Nanotechnology 17:1558
141. Huang C, Soenen SJ, van Gulck E, Rejman J, Vanham G, Lucas B, Geers B, Braeckmans K, Shahin V, Spanoghe P, Demeester J, De Smedt SC (2014) Electrospun polystyrene fibers for HIV entrapment. Polym Adv Technol 25:827–834. doi:10.1002/pat.3310

Top Curr Chem (2016) 370: 169–202
DOI: 10.1007/978-3-319-22942-3_6
© Springer International Publishing Switzerland 2016

Gold-Based Nanomaterials for Applications in Nanomedicine

Sumaira Ashraf, Beatriz Pelaz, Pablo del Pino, Mónica Carril, Alberto Escudero, Wolfgang J. Parak, Mahmoud G. Soliman, Qian Zhang, and Carolina Carrillo-Carrion

Abstract In this review, an overview of the current state-of-the-art of gold-based nanomaterials (Au NPs) in medical applications is given. The unique properties of Au NPs, such as their tunable size, shape, and surface characteristics, optical properties, biocompatibility, low cytotoxicity, high stability, and multifunctionality potential, among others, make them highly attractive in many aspects of medicine. First, the preparation methods for various Au NPs including functionalization strategies for selective targeting are summarized. Second, recent progresses on their applications, ranging from the diagnostics to therapeutics are highlighted. Finally, the rapidly growing and promising field of gold-based theranostic nanoplatforms is discussed. Considering the great body of existing information and the high speed of its renewal, we chose in this review to generalize the data that have been accumulated during the past few years for the most promising directions in the use of Au NPs in current medical research.

S. Ashraf (✉), B. Pelaz, M.G. Soliman, Q. Zhang, and C. Carrillo-Carrion (✉)
Fachbereich Physik, Philipps Universität Marburg, 35037 Marburg, Germany
e-mail: ashraf@staff.uni-marburg.de; carolina.carrillocarrion@physik.uni-marburg.de

P. del Pino
CIC biomaGUNE, San Sebastian, Spain

M. Carril
CIC biomaGUNE, San Sebastian, Spain

Ikerbasque, Basque Foundation for Science, 48011 Bilbao, Spain

A. Escudero
Fachbereich Physik, Philipps Universität Marburg, 35037 Marburg, Germany

Instituto de Ciencia de Materiales de Sevilla, CSIC-University of Seville, 41092 Seville, Spain

W.J. Parak
Fachbereich Physik, Philipps Universität Marburg, 35037 Marburg, Germany

CIC biomaGUNE, San Sebastian, Spain

Keywords Diagnostics • Gold nanoparticles • Nanomedicine • Theranostics • Therapeutics

Contents

1 Introduction

In recent years, there has been an unprecedented expansion in the field of nanomedicine, which involves the development of novel nanoparticles (NPs) envisaged for the diagnosis and treatment of several diseases, especially cancer. NPs possess extraordinary capabilities to detect, image, and potentially treat diseases at the cellular and molecular levels [1–7]. Although micelle-based NPs (such as formulations loaded with doxorubicin, paclitaxel, or cisplatin) [8] are most advanced towards use in clinical practice, inorganic NPs also offer great potential. Among various inorganic NPs, Au NPs are important examples in the field of nanomedicine, thanks to their chemical, physical, and optical properties [9–13]. Their unique physical and chemical properties, such as inertia, biocompatibility, low cytotoxicity, stability against oxidation and degradation in vivo, and ease of conjugation to biomolecules, provide significant benefits in comparison with other NPs from a medico-biological point of view. The optical properties of Au NPs are determined by the so-called localized surface plasmon resonance band (LSPR) [14], which is associated with a collective excitation of conduction electrons. Depending on the size, shape, structure, and the NPs environment, the LSPR can be localized in a wide region from the visible to the infrared. Implementation of different surface chemistries enables them to have high stability, high carrier capacity, ability to incorporate both hydrophilic and hydrophobic substances, and compatibility with different administration routes. Because of the nano-size, Au NPs have good tumor retention capabilities as they can penetrate the leaky tumor vasculature. These properties make Au NPs interesting materials for sensing,

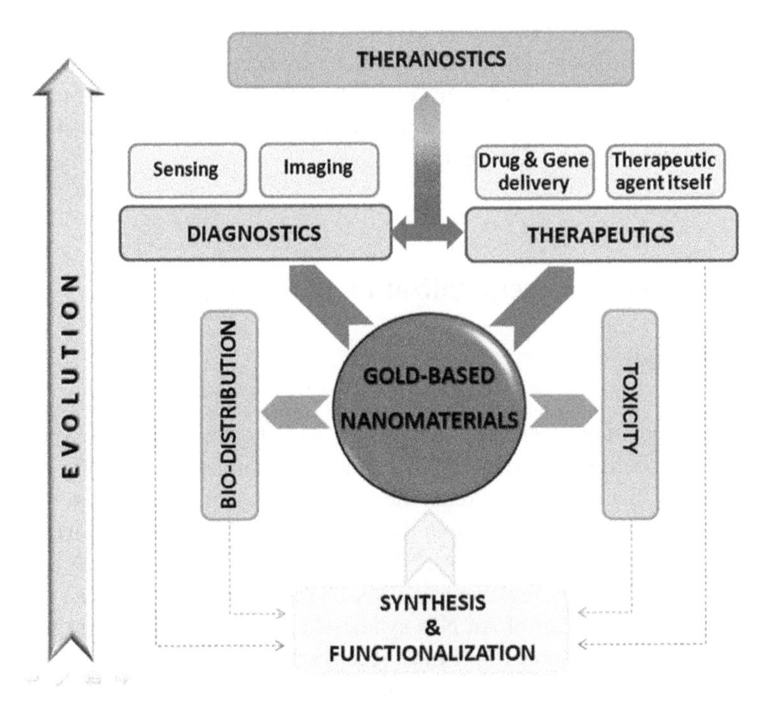

Fig. 1 Scheme showing different areas of research of Au NPs, involved in the development of their applications in nanomedicine

detection, imaging, targeted delivery of drugs and genes, photo-induced therapies, enhanced radiotherapy, and so on. Furthermore, the multifunctionality potential of Au NPs provides an ideal platform for developing the theranostic modalities combining therapeutic, targeting, and imaging functions, demonstrating synergistic effects of multi-therapies. Figure 1 shows a simplified scheme of different areas of research where Au NPs are involved in the development towards nanomedicine. The evolution of research has progressed from the synthesis and functionalization towards applications. The evolution started with simple and unimodal applications towards more complex multimodal applications (e.g., multimodal imaging, dual-mode therapies, etc.). The latest development is focused on theranostics nanoplatforms, which can diagnose, deliver targeted therapy, and monitor response to therapy. Although the natural evolution is from the bottom to the top (as drawn in the scheme), in parallel (i.e., in transversal mode), studies of toxicity and bio-distribution are key aspects to guarantee the success of their applications. These studies and results of the application performances have a direct feedback in synthesis and functionalization for improvements. These inputs are the reason why new strategies of synthesis are an area of continuous active research.

The following sections describe some recent advances in the different areas of research using Au NPs, mainly focusing on their potential for medical applications and the hurdles to be overcome to translate them into clinical trials. Bio-distribution and toxicity have been extensively reviewed and discussed in a number of recent publications [15–18], and thus we discuss it in the conclusions section. As the topic of this review involves huge amounts of information with a high speed of renewal, we focus on major ideas and some of the most recent and promising studies performed during the past few years.

2 Synthesis and Functionalization of Au NPs

2.1 Synthesis

Before the advent of nanoscience, Au NPs were already attracting interest because of their optical properties. There are many historical examples in which Au NPs were applied even without knowing it (e.g., the Lycurgus Cup or stained-glass windows). In the last few decades, the control in the synthesis of Au NPs has evolved greatly. Nowadays it is possible to produce Au NPs with different sizes and shapes in a highly controlled manner. Many reviews and book chapters have already been published about the state- of- the- art of Au NPs synthesis [13, 19–22]. Thus, here we aim to provide an overview of some of the most recent achievements in the synthesis of Au NPs using wet- chemistry (though also other routes such as laser ablation exist [23]).

Most wet-chemistry-based Au NPs are synthesized in aqueous media, but there are some important examples to produce hydrophobically-capped Au NPs (e.g., the Brust–Schriffin method) [24]. These NPs are typically spherical with a size less than 10 nm. It is well known that most of the applications of Au NPs in nanomedicine are based on their optical properties, i.e., LSPR [25]. The desire for tuning of the LSPR has been a driving force to develop synthesis strategies allowing for Au NPs of different sizes and shapes. Although controlled synthesis of Au NPs has been known since the days of Michael Faraday, most synthetic strategies to produce water soluble spherical Au NPs are based on the Turkevich method [26]. This method has been continuously improved to produce better samples (e.g., with narrow size distribution and more homogeneous NPs). This optimized methodology also allows the growing of Au NPs of sizes up to 200 nm [27]. Yet arguably the most interesting Au NPs for bio-applications are anisotropic Au NPs that exhibit their LSPR in the biological window. This biological window comprises the spectral region of 700–1100 nm, in which the body tissue components absorb less light. This range is therefore the desired region to locate the LSPR of Au NPs intended to be used for bio-applications (e.g., photothermal therapy (PTT) [28, 29] or optoacoustic imaging (OAI) [30]). Recent controls over shape during the synthesis of Au NPs make it possible to tune the position of the LSPR by changing parameters such as the shape (rods, prisms, etc.) [28, 31] or the structure of the NPs (e.g., hollow vs homogeneous) [32]. To induce the growth of anisotropic NPs it is necessary to provoke either a kinetically controlled growth of the NPs or to induce the blocking of some growing facets [33]. The sphere is the most stable shape in terms of energy. If the NPs synthesis is performed under thermodynamically controlled conditions, the NPs obtained are spherical. In general, to obtain non-spherical NPs the synthetic conditions have to be tuned to induce kinetically controlled NPs growth. This can be performed using surfactants that block some growing facets (e.g., cetyltrimethylammonium bromide (CTAB) or polymers) [34], using halides (e.g., Br^-, I^-) [35] or weak and mild reductants (e.g., $Na_2S_2O_3$) [28].

To date, Au nanorods remain the most broadly used anisotropic NPs. These rod-shaped NPs were described almost at the same time by the groups of El-Sayed [36, 37] and Murphy [28]. Since then, the synthesis of these NPs has been deeply explored. In general, the synthesis of Au nanorods is performed by using the growth seeding process. First, small spherical Au NPs are synthesized and then added to the growing solution rich in CTAB to induce rod-shaped growth. Because of the cytotoxicity of CTAB [38], Murray et al. have recently developed a modified Au nanorods synthesis method in which the required CTAB concentration is reduced by half (from 0.1 M from El-Sayed and Murphy to 0.05 M) [31, 39]. This synthesis is based on the use of aromatic additives.

Another important type of anisotropic NPs is the Au nanoshells developed by Halas et al. [40]. These structures are built using a silica core in which gold is grown. Their LSPR can be modulated by controlling the relationship between the core size and the thickness of the gold shell. By changing the shape of the core, other similar structures have also been described such as "nanorice" [41]. Halas et al. also described a synthesis for "nanomatryushkas," which are multilayered spheres. The simplest "nanomatryushka" contains a core of a gold sphere of ca. 40 nm coated with a SiO_2 shell and a second shell of Au [42]. Several bilayers of SiO_2 and Au can be deposited to obtain more complex "nanomatryushkas."

Au nanoprisms also have been described. The synthesis of triangular nanoprisms can often only be achieved with a low yield [43] and by using toxic surfactants (e.g., CTAB or CTAC (cetyltrimethylammonium chloride)) [44]. A synthesis route eliminating the use of toxic surfactants has recently been reported, which allows for tuning the LSPR position by controlling the amount of reductant. This synthesis is based on the reduction of a gold salt by thiosulfate (Fig. 2a) [28].

Au nanostars also exhibit their LSPR in the biological window. Many different synthetic procedures to produce Au nanostars have been reported. For generating Au nanostars, typically the seed-growing method is used. For instance, Liz-Marzán et al. published a method in which gold salt was reduced to metallic gold on top of the 15-nm Au NPs stabilized with poly(vinylpyrrolidone) (PVP) in the presence of dimethylformamide and PVP [45].

Finally, Au nanocages, originally developed by the group of Y. Xia, have been applied extensively with different purposes in nanomedicine [46, 47]. These cages are prepared by using a sacrificial silver nanocube, which then is oxidized to promote the reduction of gold through a galvanic replacement process. Similar approaches have been described using silver nanospheres [48] and silver nanoprisms [49].

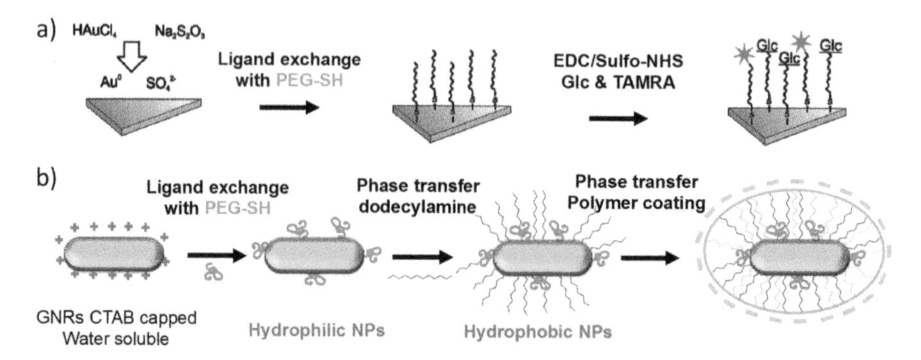

Fig. 2 Schemes for the synthesis and functionalization of Au NPs. (**a**) Synthesis of Au nanoprisms, PEGylation, and linkage of glucose (Glc) and tetramethylrhodamine (TAMRA) using carbodiimide (EDC) chemistry [28]. (**b**) Strategy of phase transfer to coat Au nanorods (GNRs) with amphiphilic polymers [75]

Aiming to use green chemistry and less toxic reagents, the production of Au NPs using natural extracts from microorganisms (e.g., micro alga [50] or fungi [51]) or plants has been also explored [50, 52]. Nevertheless, the yield and quality of these Au NPs is still far below the quality of the previously described approaches.

Au-based hybrid materials have been developed too, trying to combine the great optical properties of gold with the properties of another material. Several examples of core-shell structures of Au containing iron oxide [53–55] or semiconductors cores [54] or Au cores coated with silver (bimetallic NPs [56]) can be found in the literature. More recently, Au has been combined with more novel materials such as nanodiamonds [57] or graphene [58, 59].

2.2 *Functionalization*

After synthesis and before their use in bio-applications, NPs must be provided with stable coatings which should warrant high colloidal stability. Thus, robust organic coatings ensure that the NPs' properties remain intact in biological media [60]. Indeed, the NPs surface determine their biological fate [61]. In addition, any kind of by-product related to the synthesis, including excess of reagents or cytotoxic surfactants, should ideally be washed off to remove potential toxic effects caused by these impurities. To achieve coatings qualifying for these requirements, the NPs' surfaces need to be engineered. One of the biggest advantages of the use of Au is its high reactivity with thiol groups [62]. This reactivity permits stabilizing the NPs with ligands containing a thiol reactive group. Ligand exchange by which original surfactants are replaced by new ones is the most common stabilizing procedure for Au NPs. Ligand exchange can be used to water transfer hydrophobically-capped NPs (e.g., NPs capped with alkanethiol chains) [63] and to replace toxic surfactants (e.g., CTAB) used to produce anisotropic NPs such as Au nanorods [64, 65].

Typically, polyethylene glycol (PEG) chains which provide the NPs with a colloidal high stability in biological media and long in vivo retention times [19, 66] are the most widely used stabilizers for Au NPs. Nowadays, there are many companies which offer an endless number of hetero-functional PEG chains. Using bi-functional PEG allows for future chemical modifications for the attachment of molecules (e.g., dyes, carbohydrates, antibodies, peptides, etc.) and/or to provide charge to the NPs (Fig. 2a) [28, 67, 68]. Not only PEG is used to stabilize Au NPs – other ligands, such as dihydrolipoic acid [69], proteins (e.g., bovine serum albumin) [70], or polymers are used regularly to enhance the NPs' stability in complexes media. Polymers used for this purpose include, for example, polyelectrolytes [71], PVP [72], or amphiphilic polymers [73].

The use of amphiphilic polymers to stabilize NPs and to promote their transfer from organic solvents to aqueous solutions is based on polymer coating of the NPs. This approach can be used for virtually any kind of NPs containing aliphatic chains on the surface (e.g., oleic acid, oleylamine, etc.) [73, 74]. The advantages of this technique are many: (1) coated NPs exhibit a high colloidal stability against media with high salt concentrations and/or proteins; (2) NPs coated with the same polymer have the same surface chemistry; (3) these polymers can be made with reactive groups in their hydrophilic domain (e.g., carboxylic acids), which then can be further modified with biologically relevant molecules. The main limitation of this technique is that it can be only used with NPs soluble in organic solvents. Yet most of the anisotropic Au NPs are synthesized in water. An extension of this method based on phase transfer of the water-soluble Au NPs to organic solvents has recently been reported. This method has been demonstrated for spherical Au NPs (with size up to 15 nm) and Au nanorods (Fig. 2b) [75].

Once Au NPs are sufficiently colloidally stable in biological media, as a function of their surface chemistry, different chemical modifications can be performed. Bioconjugate chemistry protocols developed for modifying proteins, peptides, and/or surfaces can be adapted to NPs [76, 77]. Concerning bio-conjugation, we refer to some recent reviews [11, 78–80]. In summary, currently the synthesis and functionalization of Au NPs has become very versatile. This allows scientists to develop the best customized systems for each application.

3 Use of Au NPs Towards Diagnostics

3.1 Detection and Sensing

Different analytical assays involving Au NPs are widely used as sensors, ranging from the detection of ions and elements to more complex molecules, including those of biomedical interest, such as oligonucleotides, proteins, antibodies, and even bacteria and other microorganisms. The methods of designing sensing biomarkers that could be associated with the early stage diagnosis of different diseases are nowadays attracting special interest. The current challenges consist of designing sensing devices that are able to recognize more specifically different types of analytes, discriminating molecules with similar characteristics, including the use of sensor arrays, which often combine several analytical approaches. Advances in enhancing the sensitivity and reducing the time of analysis are also currently required. The physical and analytical basis of sensing with Au NPs can be summarized in different main areas [81]. They include measurements based on colorimetry and plasmon resonance, fluorescence, electrochemistry, and more recently, surface enhanced Raman spectroscopy (SERS) [82].

Colorimetric assays are based on a visible change of color of functionalized Au NPs suspensions when interacting with the appropriate analyte [83–86]. Colorimetric analyses are normally fast and can often be evaluated with the naked eye. As explained before, Au NPs exhibit plasmonic properties. The position and intensity of this LSPR band depend not only on the metal type, NPs size, shape, structure, composition, and dielectric constant of the surrounding medium [87], but also on the aggregation of the NPs [88]. Colorimetric assays are based on this effect, because analytes that produce a change in the aggregation state of Au NPs give rise to a change in the LSPR absorption band of the NPs dispersion [89]. Such an effect is not only used to sense cations and anions [89–92], but has also been applied to sense molecules of biomedical interest. For example, DNA has been detected by Au NPs wrapped with long genomic single- and double-stranded DNA (ssDNA and dsDNA) molecules [93]. Proteins such as melamine and human carbonic anhydrase II have also been sensed by cyanuric acid derivative grafted Au NPs [94] and polypeptide-functionalized Au NPs [95], respectively. More complex molecules such as folate receptors (FRs), consisting of cysteine-rich cell-surface glycoproteins that can bind folate (FA), can be sensed by FA-modified ssDNA functionalized Au NPs. In the presence of FRs, ssDNA terminally tethered to FA is protected from degradation by exonucleases, and an aggregation of the Au NPs takes place through the formation of cross-linked NPs networks, resulting in a color change of the solution from red to blue [96]. Polyethyleneimine (PEI)-stabilized Au NPs have been used for highly selective and sensitive colorimetric sensing of heparin [97]. Abnormal concentration values of human chorionic gonadotropin (hCG) can be associated with ectopic pregnancy. The concentration of this biomarker can be determined using Au NPs in the presence of positively charged hCG-specific peptides. In this case, hCG inhibits the peptide-induced aggregation of the Au NPs, giving rise to a simple, rapid, and sensitive colorimetric assay [98]. A rapid and low-cost colorimetric analysis of bacteria in drinking water has been designed by using β-galactosidase conjugated Au NPs with a colorimetric substrate (chlorophenol red-β-D-galactopyranoside (CPRG)) deposited on a paper-based test strip [99]. The aggregations of antibody-conjugated oval-shaped Au NPs that selectively target specific sites on the surface of pathogens have been used to sense Salmonella [100]. Although colorimetric assays based on Au NPs involve the large shift of the LSPR band depending on NPs aggregation, a small LSPR peak shift can also be produced when an appropriate analyte binds to the surface-bound receptors of plasmonic NPs, because of a change in the refractive index [101]. In contrast to agglomeration-based protocols, this shift in the LSPR frequency is not enough to be detected by the naked eye, but can be observed by absorption measurements. Au NPs deposited on several substrates have been used to detect analytes such as DNA [102], human IgG (Immunoglobulin G) [103], and insulin [104] in this way. Recent advances in the detection of microRNAs (miRNAs) by Au nanoprisms without the need for labels [105] and in the sensing of trace oligonucleotides biomarkers [106] have also been reported. Au NPs deposited on the metal sensing surface increase the sensitivity of planar surface plasmon resonance sensors, provided by the high dielectric constants of Au NPs and the electromagnetic coupling with the metal film [107]. This approach has been used for the sensing of different proteins [108] and oligonucleotides [108, 109].

Fluorescence assessments involving Au NPs are widely centered on fluorescence quenching-based methods. Au NPs show an important quenching effect on fluorophores close to their surface caused by their extraordinary high molar extinction coefficients and broad energy bandwidths [83, 110]. Specific interactions with the sensing molecules have been used to detect many different molecules of biomedical interest. Some assays are based on the appearance of fluorescence when the target molecules interact with the Au NP-based-sensors. For example, the quenching of a fluorophore attached to an Au NP through an oligonucleotide chain disappears in the presence of DNA [111], when the fluorophore gets detached from the NPs because of displacement by the DNA strand. Similar strategies have been used to sense proteins [112] and bacteria [113] using Au NPs conjugated with fluorescent polymers. Au NPs functionalized with enzymes have also been used to sense proteins, with an enhanced sensitivity through enzymatic catalysis [114].

Fluorescence quenching assays involving Au NPs are not only restricted to the detection of single analytes. More complicated sensing techniques, focused on the study of the interaction of different analytes, have also been reported. For example, dsDNA-conjugated Au NPs (dsDNA-Au NPs) and water-soluble conjugated polyelectrolytes are used as complementary sensing elements to construct hybrid sensors for detecting protein–DNA interactions [115]. The use of sets of sensors showing different patterns of responses in an array can provide fingerprints that allow for classification and identification of different target molecules [116]. Such an approach is used with DNA–Au NPs conjugates, in which a combination of colorimetric and fluorescence assessments enables better selectivity to distinguish different proteins [117]. Similar combination of colorimetric and fluorometric approaches has been reported for a sensor array consisting of two types of novel blue-emitting collagen-protected Au nanoclusters and macerozyme R-10-protected Au nanoclusters with lower synthetic demands, which has been recently used to sense eight different proteins [118].

The modulation of quenching of fluorescent semiconductor quantum dots (QDs) close to Au NPs in the presence of molecules which inhibit the interaction between QDs- and Au NPs-conjugated biomolecules has been used to sense molecules of biomedical interest such as avidin [119]. Normal, cancerous, and metastatic human breast cells have been distinguished by comparing the fluorescence of different cationic Au NPs functionalized with poly(*p*-phenyleneethynylene) (PPE), which show different affinities for normal and tumor cells [120].

The conductivity, roughening of the conductive sensing interface, and the catalytic properties of Au NPs have been harnessed for the huge amount of analytical assays based on electrochemical measurements that involve Au and other metal-based NPs [121]. Different immunosensors based on Au NPs have recently been reported to detect cancer biomarkers [122, 123]. Au NPs deposited on electrode surfaces are known to enhance the electrochemical detection of different analytes because of their ability to decrease the overpotentials of many electroanalytical reactions, maintaining the reversibility of redox reactions [124, 125]. This approach has been used to detect several drugs such as isoniazid [126] and hCG [127]. Au nanorods have also been used as sensing interface in pencil graphite electrodes for the electrochemical sensing of deferiprone, an anti-HIV drug, resulting in an amplification of the electrochemical sensing signal [128].

Antibody-functionalized Au NPs, showing target specificity and affinity towards different biomarkers [129], have been used to sense Salmonella by differential pulse voltammetry (DPV) [130]. Cancer circulating cells have been sensed by combining the specific labeling through antibody-modified Au NPs and the sensitivity of the Au NPs-electro-catalyzed hydrogen evolution reaction (HER) detection technique [131]. The reaction of cell surface proteins with specific antibodies conjugated to Au NPs and the catalytic properties of the Au NPs on hydrogen formation from hydrogen ions can be used to quantify the NPs internalized by cancer cells [132, 133].

Raman scattering permits the detection and analysis of many molecules, by giving a unique spectroscopic signature which potentially identifies the species [134]. The Raman scattering signal can be substantially enhanced by the presence of plasmonic NPs, resulting in SERS [135, 136]. This effect is highly influenced by the size, shape, orientation, and aggregation of the NPs [108]. In fact, Au NPs with different morphologies have been used for SERS-based detection [137–139]. Label-free and Raman-dye labeled assays are two different existing SERS-based detection methods. Label-free assays follow vibrational information about the analytes themselves, whereas the dye-labeled methods detect analytes indirectly by monitoring the SERS signal of a Raman label attached to the metallic SERS substrate [140]. Different SERS assays for sensing DNA [141, 142] and proteins [143, 144] using Au NPs can be found in the literature. An extended bi-dimensional array of Au concave nanocubes supported on a polydimethylsiloxane (PDMS) film has recently been proposed for the SERS sensing of proteins that show low intensity Raman signals [145]. The assembly of spherical Au NPs on a highly anisotropic silica-coated substrate has recently been reported for the detection of prostate specific antigen by SERS [146] and the selectively quenching of the SERS signals from the dye molecules adsorbed onto star-shaped Au NPs not internalized by cells has been used to identify intracellular distributions of Au NPs [147].

3.2 Imaging

For the treatment of many diseases and non-invasive evaluation/detection of intra-cellular and/or intra-subcellular compartments, molecular imaging based on functional nanomaterials is of paramount importance [148, 149]. For molecular imaging, different types of NPs are currently in use. Examples include polymer-based NPs [150–152], dendrimer-based NPs [153, 154], lipid-based NPs [155, 156], magnetic NPs [157–160], QDs [161–163], carbon nanotubes (CNTs) [164, 165], silica NPs [166–168], and Au NPs [169–172]. Among all the above-mentioned NPs, Au NPs possess extraordinary potential for imaging at the cellular and even molecular level. Various Au NPs are currently in use in molecular imaging, based on their different size, shape, and physical properties. Examples include spherical Au NPs [171, 172], nanorods [173–175], nanobipyramids [19], nanoshells [176], nanocages [177–179], core/shell NPs [171, 180], nanostars [181–183], and nanocubes [149], etc.

Au NPs have unique characteristics which enable their use as contrast agents in bio-medical imaging [184, 185]. In this field, they are being used as probes in dark field confocal imaging (DFCI), one- and two- photon fluorescence imaging (OTPFI), optoacoustic imaging (OI), computed tomography (CT), photothermal optical coherence tomography (POCT), positron emission tomography (PET), and imaging based on surface enhanced Raman scattering (SERS) [175, 184]. Different imaging modalities of the Au NPs can be combined, which can provide complementary information. In the following, a description of using Au NPs as contrast agents for the different imaging techniques is given.

DFCI provides contrast enhancement in unstained biological samples, but its main limitation is that it provides low light levels in images. Thus, for better visualization the biological samples should be strongly illuminated, which can, however, damage the samples. The imaging contrast of dark field microscopy can be enhanced by utilizing the high scattering properties of Au NPs [186]. For cellular detection, mostly the light scattering properties of Au NPs are utilized for straightforward image analysis. Light scattered from Au NPs is detected by using high resolution objective lenses of dark field confocal (DFC) microscopes in the form of bright spots, though the size of Au NPs is generally smaller than the diffraction limit of DFC. Using Au nanoshells it was recently possible to observe the binding and antibody mediated specific targeting of cancer cells in in vitro experiments using the dark field scattering properties of the NPs [187]. Similarly, for cancer cells localization, targeting, and real time tracking of Au nanorods-induced DNA damage in cancer cells was visualized using DFCI [188]. Scattering properties of Au NPs are also being utilized for better imaging of breast cancer cells [189]. However, despite the high scattering cross sections of Au NPs for enhancing the contrast in DFCI, their use is limited to in vitro experiments [184].

Photoluminescent properties of sub-nanometer Au nanoclusters made them attractive candidates in OTPFI based on their brightness, non-blinking behavior and stable emission [190, 191]. The luminescence of Au nanoclusters in the near infrared (NIR) window is used for fluorescence imaging and they have greater photostability than QDs [191]. Not only nanoclusters but also other Au NPs, such as nanoshells and nanostars, can be used in one photon fluorescence imaging (OPFI), after conjugation of the Au NPs with NIR active fluorophores such as indocyanine green or Cy5 [192]. After conjugation with these fluorophores, these structures help in emission enhancement of these dyes for better fluorescence imaging. Presence of a metal surface close to fluorophores does not always quench fluorescence, but can also provide fluorescence enhancement, in particular for very close distances. In OPFI, Au NPs functionalized with fluorophores, offer a suitable platform for in vitro and in vivo cancer imaging and diagnostics [193, 194]. When Au NPs are excited with femtosecond pulsed lasers whose resonance frequency matches with the LSPR band of the Au NPs, two photon absorption occurs which results in two photon luminescence from the Au NPs [182, 193, 195]. For monitoring in vivo biological events, two photon luminescence imaging (TPLI) provides sufficient penetration depth and high three-dimensional (3D) spatial resolution. The signal intensity of TPL (two photon luminescence) can be enhanced three times in magnitude by utilizing the high luminescent properties of Au nanorods and nanocages without the photo-bleaching or blinking that is observed in many fluorophores used in this technique [196]. The contrast of Au nanostars conjugated with wheat germ agglutinin in TPL-based imaging can be utilized for imaging their uptake [182]. Similarly, other Au NP structures, such as nanorods, nanocages, and nanoshells, are also being used as contrast agents in TPL with a resolution at the single NP level inside blood vessels. In this way, in vivo tracking of Au NPs and fluorescence lifetime imaging for visualizing dynamical processes in cell media is possible [197, 198]. After one and two photon luminescence-based imaging, Au nanocages are now also being utilized in three photon luminescence imaging, based on their strong multi-photon absorption capabilities, leading to in vivo detection with diminished background signals and reduced photothermal toxicity [199]. Further studies are still required for using Au NPs in multi-photon luminescence for a better understanding of their role in this imaging technique.

The penetration depth of OI-based imaging, which is typically carried out with NIR pulsed sources, is similar to ultrasound-based imaging, i.e., several centimeters in biological tissues (typically less than 5 cm). This is better compared to simple optical imaging, in which depth resolution is only on the millimeter scale. The photothermal properties of Au NPs provide high contrast in OI [30]. Upon photoexcitation, the non-radiative decay of Au NPs converts light energy into heat, which causes a sharp rise of temperature in the local environment of the NPs, resulting in thermal and acoustic response enhancement in those tissues which contain the photoexcited NPs. The increased thermal response of Au NPs enhances the pressure waves propagating through the surrounding tissues and results in improving the temporal and spatial resolution of tomographic images [200].

In clinical detection of several diseases, CT has received increasing attention because of the high spatial and density resolution. For imaging biological systems using CT, contrast agents are usually required (which can enhance the density of the imaging area) for improving the accuracy in diagnosis. Iodine-based small molecules such as "Omnipaque" are normally used in clinics, but are associated with certain drawbacks such as short imaging time, non-specificity, and renal toxicity. For overcoming these drawbacks nowadays, Au NP-based suitable contrast agents are being developed. In CT, because of the high atomic number of Au, Au NPs are providing higher spatial and density resolution compared to iodine-based contrast agents. In CT-based imaging, Au NPs attenuate X-rays much more efficiently compared to "Omnipaque," resulting in contrast enhancement by several orders of magnitude [201, 202]. Moreover, by suitable tuning the size and functionalization of the Au NPs, besides improving CT imaging it is also possible to achieve target specificity, long circulation time, and reduced renal toxicity [203].

Optical coherence tomography (OCT) can image cellular and sub-cellular structures 100 times better than CT and magnetic resonance imaging (MRI) and provides 10–25 times better spatial resolution compared to ultrasound-based images [204]. OCT is a non-invasive technique, resembling ultrasound-based imaging but, in this technique, instead of sound, reflections of NIR light are used for imaging. NIR active contrast agents, e.g., Au nanorods, can significantly improve OCT-based imaging because of their large differences in absorption-scattering profiles. Au nanoshells and nanocages, because of their strong scattering properties, can also provide enhanced optical contrast and brightness in OCT for improving the imaging of cancerous cells [205]. In this technique, tissues are illuminated with low coherent light and matching the coherence between incident and reflected beams of light helps in the detection of back-reflected light. This backscattered light thus helps in imaging. Because OCT is much more sensitive towards detection of scattering from the tissues than absorption, the scattered light helps in studying the morphology of tissues [205]. In OCT, Au NPs are being used as exogenous contrast agents based on their ability to produce distinctive backscattered light which is detectable in highly scattering tissues, thereby helping in studying the morphology of tissues [206]. Though OCT is a powerful 3D diagnostic tool in real time imaging, its resolution is low because of intense scattering from some optically dense tissues under investigation. To overcome this limitation nowadays, POCT (photothermal-OCT) imaging techniques using the photothermal properties of Au NPs are being developed [207]. In POCT, when light resonant with the plasmon energy of Au NPs strikes, Au NPs are excited and light is converted into heat and the surface temperature of the tissues is enhanced. The increase in surface temperature results in changes in the local refractive index of the medium which is then optically detected by POCT. Because of active detection of photothermal heating, POCT can identify and separate absorbing targets from scattering background, thereby helping in high resolution imaging compared to OCT [208].

In early stage diagnosis of cancer, PET – with its highly sensitive nuclear imaging modality – is extensively utilized in clinical studies using small doses of radioactive materials. However, these radioactive materials, especially small radioactive molecules, usually have short circulatory life- times in in vivo studies. After radiolabeling, Au NPs (e.g., nanocages, nanoshells, and spherical NPs) can remain inside the bloodstream for longer periods of time. Hence they facilitate long-term bio-imaging [209]. In PET, radioisotopes undergo positron emission decay or positive beta decay and positrons are emitted. These emitted positrons traverse a short distance inside tissues, lose kinetic energy, and interact with electrons. This union with electrons results in their annihilation and production of gamma photons in the form of light which is used for making images [209]. Sometimes PET is coupled with Cerenkov luminescence (CL)-based imaging for better visualization and cross-checking the imaging results. CL-based imaging is a molecular imaging technique based on Cerenkov radiation, which can originate from the decay of alpha-, beta-, or positron-emitting radionuclides [179]. Recently, a radiolabeled precursor of a gold salt ($H^{198}AuCl_4$) was used for the synthesis of radioactive Au nanocages which gave CL. The CL originating from the decay of radionuclides helps in real-time CL-based imaging and monitoring of tumors over extended periods of time. CL of radionuclides can be increased by using high refractive index materials such as gold in conjugation with higher energy radionuclides. Thus CL imaging based on Au NPs can effectively bridge the gap between nuclear and optical imaging [179]. CL-based imaging can use radionuclides for diagnosis of diseases, which are routinely used for PET based imaging. CL based imaging improves PET-based imaging in terms of resolution. CL imaging signal can be modulated by using smart imaging agents such as NPs, and hence better insight in tumor biology can be obtained [210]. Over the last decade there have been numerous studies for enhancing the efficacy of SERS-based molecular imaging using Au NPs conjugated with Raman active moieties. Au NPs enhance the Raman scattering of vicinal molecules by means of chemical and electromagnetic enhancements. Au NPs enable identification of single molecules spectroscopically at room temperature by amplifying (ca. 10^{15}-fold) the Raman scattering signals of adsorbed species. The LSPR of Au NPs enhances Raman signals in SERS-based imaging, which helps in better detection of tumor margins during the surgical removal of tumors [211]. The use of Au NPs provides photostability, improved contrast, and higher spectral specificity in SERS-based imaging [184].

Among all enlisted imaging techniques, no single modality can be considered as ideal and sufficient for getting all requisite information for a particular question. Nowadays, multimodal imaging probes based on Au NPs are being designed which possess integrated or complementary functions. For example, SERS-based imaging is highly sensitive and multiplexed imaging is possible with this technique, but it has poor penetration depth. On the other hand, OI has better penetration depth and high spatial resolution, but its sensitivity is limited. For multiplexed imaging, Au nanorod-based SERS/OI can be used for early stage detection of cancer. Similarly, Au NPs having triple modality are being used for MRI/SERS/OI-based imaging [212]. For disease (particularly cancer) intervention, Au NP-based multifunctional/ multimodal imaging platforms have enough potential for the development of future contrast agents useful for nanomedicine. One can envision the potential applications of Au NPs for multiplexed detection and imaging of cancer and other such types of diseases by proper tailoring of their functionalization, size, shape, composition, and hybridization with other materials [184].

4 Use of Au NPs in Therapeutics

The properties of Au NPs can be exploited in therapy in two ways – as passive carriers for delivery, in which the therapeutic effect arise from active molecules bound to the carrier, or as active therapeutic agents, where the therapeutic effect directly originates from the Au NPs. The first use of Au NPs for therapeutic applications was as delivery vehicles for drugs and genes, because NPs in the size range 2–100 nm can interact with biological systems at the molecular level, and can allow for targeted delivery and passage through biological barriers. Later on, also investigations showing that Au NPs can be intrinsically therapeutic became available. This is because Au NPs can actively mediate molecular processes to regulate cell functions. In this section we summarize the potential of Au NPs in therapy, providing examples of currently investigated strategies.

4.1 Au NPs as Passive Carriers in Delivery Systems

Au NPs are widespread in the field of delivery of different kinds of therapeutic molecules. They are very attractive as nanocarriers because of their colloidal stability, ease of preparation, and their on-demand tunable size and surface modification possibilities. In addition, they are essentially bio-inert, non-toxic and lack immunogenicity. All these features in combination with their ability to enter cells by endocytic pathways naturally make them good vehicles for a plethora of biomolecules and drugs to be delivered inside cells [13, 213]. Normally, the cargoes are loaded onto the Au NPs' surface either by non-covalent binding (via ionic or hydrophobic interactions) or by direct binding on the Au surface by thiolated linkers. Once inside the cells, the cargo may slowly detach from the surface of the Au NPs or its release may be triggered by internal stimuli such as pH [214] or cytosolic glutathione [215], which is able to reduce disulfide bonds, releasing molecules linked to the NPs surface via that kind of linkage. Furthermore, Au NPs are well known for their optical properties which are not only useful in the field of imaging and biosensing, but also serve to detach molecules selectively from the NPs surface when irradiated with light. Upon irradiation of plasmonic Au NPs at their LSPR frequency with a continuous wave laser, they absorb energy and reduce the attraction between the Au surface and non-covalently linked molecules, which eventually produces desorption of the cargo. In contrast, high energy pulse irradiation provokes the reshaping of the nanocarriers and the rupture of Au–S bonds, releasing more strongly linked cargoes. Interestingly, light irradiation of plasmonic nanomaterials allows for spatio-temporal controlled release of drugs and biomolecules because the delivery only takes place during irradiation of the NPs and stops when the laser is off [216–218]. The most commonly used Au NPs for delivery applications are nanospheres, and nanorods.

Au NPs have been used for decades for the purpose of gene delivery, traditionally developed for the transfection of plants using gene guns [219]. Nowadays, Au NPs are also used in gene therapy as non-viral carriers for delivering nucleic acids inside the cells [220]. Gene therapy may include the delivery of DNA inside cells to induce certain protein expression, or the introduction of miRNA able to interfere with the correct translation of messenger RNA inside the cells, avoiding protein production and silencing a particular gene responsible for a cellular malfunction. Gene therapy is increasingly important, particularly in the field of cancer treatment. However, delivery of naked nucleic acids is hampered by their fast degradability inside the body and their polyanionic nature that inhibits the cellular uptake. There are many reports in the literature that have demonstrated that adsorption of nucleic acids onto positively charged Au NPs drastically reduced their degradability and helps their internalization into cells. This could be achieved, for example, by coating Au NPs with positively charged lysine amino acids [221] or with positively charged polymers [222–225]. However, to ensure that the genetic material reaches the nucleus, more sophisticated constructions may be required. Such constructions may involve the use of PEI, a poly-cationic polymer able to escape endosomes because of the "proton sponge" effect causing membrane disruption [226], although it has been shown to be cytotoxic after a certain dose threshold. Hence, several authors have coated Au NPs with PEI to trap DNA or RNA, taking advantage of the endosomal escaping capacity of PEI, and substantially reducing its cytotoxicity [227–232]. Chen et al. have recently described a smart three-layered nanocarrier, based on the layer-by-layer (LBL) deposition of PEI/chitosan-aconitic anhydride (CS-Aco)/PEI/ shRNA (short/small hairpin RNA) onto Au NPs. CS-Aco was introduced as a pH-triggered charge-reversible compound which hydrolyzed into positively charged CS once inside the lysosomes, causing the disassembly of the nanocarrier layers. The as-released Au-PEI NPs facilitated lysosomal membrane disruption and hence the successful delivery of shRNA-PEI into the cytoplasm [214]. Although PEI is particularly attractive for nucleic acids delivery, similar results have been achieved by other poly-cationic polymers adsorbed onto Au NPs [220, 233]. For instance, Lee and co-workers fabricated siRNA-loaded Au NPs using the LBL approach with alternating positively charged poly-L-lysine (PLL) and negatively charged siRNA. They successfully coated the Au NPs with four layers of PLL and three layers of siRNA, which were slowly released inside the cells by protease degradation of PLL and displayed gene silencing capability [234]. As already pointed out, another way of delivering DNA or RNA avoiding normal cell internalization pathways is adsorption onto naked Au NPs or projectiles and direct bombardment inside the cells using gene guns. This strategy is frequently used in plants, but nowadays it has also been explored for mammalian tissues [235, 236].

Drug delivery is also an important field in which Au NPs are utilized. This field is of particular interest in the case of cancer treatment to avoid systemic toxicity during chemotherapy and to facilitate the delivery of hydrophobic drugs [237]. Indeed, the group of Rotello showed how to entrap two different hydrophobic drugs (tamoxifen and β-lapachone) in hydrophobic pockets created within alkanethiol monolayers surrounding Au NPs and their effective delivery

[238]. Some other strategies involve the chemical modifications of drugs to link them covalently onto NPs, which could, in some cases, compromise the drug performance. Gibson et al. covalently functionalized Au NPs with approximately 70 molecules of anticancer drug paclitaxel per NP, but they did not report about the delivery and biological activity of the drug [239]. Nonetheless, most studies are based on systems to deliver doxorubicin (DOX) [240, 241], which is known for its properties for treating cancer, but also for its toxicity and side effects. The group of Li has recently reported high loading of PEGylated hollow Au nanospheres with DOX (up to 63% in weight) and their delivery after NIR light irradiation [242]. Apart from drugs, Au NPs have also been used as carriers for vaccines by decorating their surfaces with appropriate ligands (selected antigens and T-helper peptides) which were able to elicit an immunogenic response [243].

Interestingly, Au NPs have been used not only as carriers but also as smart container openers. Plasmonic Au NPs were embedded in between polyelectrolyte layers in LBL constructed capsules carrying a therapeutic cargo within the capsule cavity. In a similar fashion, as explained before, irradiation of light onto those NPs led to the spatio-temporal controlled disassembly of the polymeric capsules producing the immediate cargo release [244–248].

Nowadays, all these delivery systems are evolving into more sophisticated constructions, which take advantage of the optical properties of Au NPs for combined drug therapy with photothermal ablation (PTA) and imaging [249] and in combination with other materials such as carbon [250, 251]. For instance, some recent reports along this dimension describe the wrapping of Au NPs or nanorods with hydrophilic graphene oxide nanosheets as carriers for gene therapy and improved PTA therapy [252–254].

4.2 Au NPs as Active Therapeutic Agents

When Au NPs act as therapeutic agents per se, several therapies can be distinguished depending on the NPs properties exploited. In the following we illustrate the wide range of potential therapies using Au NPs with some selected recent examples.

One of the most promising groups of therapeutic strategies using Au NPs are light-based therapies. These utilize the application of light to irradiate photosensible materials, whereby this light-activation is directly responsible for the desired therapeutic effects (i.e., destroying tumor cells). This group of light-based therapies includes photothermal therapy (PTT), photodynamic therapy (PDT), and photoimmunotherapy (PIT) [255, 256]. Although PDT and PIT use a photosensitizer (i.e., light-activated drugs) for the release of reactive oxygen species (ROS) and the activation of immune responses, PTT is based on the use of photosensible materials (e.g., Au NPs) to generate local heat after being irradiated with electromagnetic radiation. The main difference between PIT and PDT is that in PIT monoclonal antibodies are associated with photosensitizers to improve the selective binding to the target tissues [255]. PTT has lately received more interest because it does not require oxygen to interact with the target cells or tissues and is able to use longer wavelength light, which is less energetic and therefore less harmful to other cells and tissues. PTT using Au NPs, also called plasmonic photothermal therapy (PPTT), exploits the unique LSPR properties of Au NPs. When an energy source such as electromagnetic radiation is applied, conversion to heat energy efficiently occurs in Au NPs because of electron excitation and subsequent non-radiative relaxation through electron–phonon and phonon–phonon coupling. This generated thermal energy can induce temperature increases of more than 20 °C (i.e., hyperthermia), which can thereby induce tumor tissue ablation [255, 256].

There are several advantages of using Au NPs for PTT: (1) Au NPs have high absorption cross sections, and thus only minimal irradiation energy is required, (2) the conversion of light into heat is very fast (about 1 ps), (3) Au NPs are biocompatible, and (4) the ability of tuning the LSPR absorption (changing the size and shape of Au NPs) to absorb light in the visible up to the NIR region. Although visible light is successful in destroying cells labeled with spherical Au NPs, the NIR region is especially crucial to penetrate deep into tissues, with minimal attenuation by water and hemoglobin. The light can penetrate up to 10 cm in soft tissues in the "biological window" (650–900 nm), a region ideal for the LSPR absorption of Au nanoshells, nanorods, nanoprisms, and nanocages [28, 257]. When comparing the different NPs structures in terms of their applications in PTT, Au nanorods exhibit the best efficient NIR photothermal heat conversion. Although nanoshells have a larger absorption cross-section because of their larger size, and as a result they produce more heat, the nanorod shape has been shown to be twice as efficient in converting light radiation into thermal energy (photothermal efficiency) [258]. El-Sayed et al. determined the most effective Au nanorods size for PTT heat generation [259]. In this context, 28×8 nm Au nanorods were found to be the most effective, both in theoretical calculations and in in vitro experiments with human oral squamous cell carcinoma. Au nanorods in this dimension were the best compromise between the total light absorbed and the fraction that is converted into heat. Additionally, nanorods in this size led to an intense electromagnetic field that extends far enough from the NPs surface to allow for field coupling between NPs aggregates, resulting in enhanced experimental photothermal heating in solution. For example, Lin et al. [260] synthesized PEG-coated Au nanorods that showed enhanced PTT when used in the soft tissues of a genetically engineered mouse model (GEMM) of sarcoma. This model recapitulates the human disease more accurately in terms of structure and biology than subcutaneous xenograft models. This study represented a nice demonstration of a therapeutic, NPs-mediated thermal ablation protocol in a GEMM. Untargeted PEG-Au nanorods accumulated in the sarcomas at levels comparable to those in subcutaneous xenografts, providing evidence that passive targeting is indeed sufficient for PEG-Au nanorods to accumulate in a physiologic tumor microenvironment. Significant delays in tumor growth with no progression in some instances demonstrated the success of this method. A similar approach was used by Chen et al. [261], where PEG- Au nanocages could be passively delivered and accumulated into animal tumors, causing irreversible damage to tumor cells after exposure to NIR laser. Interestingly, PEG-Au nanocages were found not only on the surface but also in the core of the tumor.

There are increasing efforts to enhance therapeutic treatments by combining therapy methods that show synergistic effects, as in the case of PTT and PDT. For example, Choi et al. have reported a method which combines both phototherapies using Au nanorods-photosensitizer complexes and two different light sources to excite the photosensitizers and photothermal NPs separately because of their absorption mismatch [262]. In this work, the negatively charged photosensitizer Al(III) phthalocyanine chloride tetrasulfonic acid (AlPcS4) was attached onto the positively charged surface of Au nanorods by electrostatic interaction, and the photodynamic effect of the AlPcS4 photosensitizer was temporarily suppressed after complex formation with Au nanorods. In the intracellular environment the photosensitizer was released and it could finally be optically activated for phototherapeutic effect. Two different light sources were used to excite Au nanorods (810 nm laser) and AlPcS4 photosensitizer (675 nm laser) separately. Tumor growth was suppressed by 95% with PTT/PDT dual therapy, whereas the suppression was only 79% with PDT alone.

These examples successfully demonstrate the potential of NIR-active Au NPs for use in light-based therapies. The current challenge in these phototherapies is to increase the level of selectivity to act on tumor tissues with minimum damage to the surrounding healthy tissue. Furthermore, better control over bio-distribution and clearance are critical issues to be addressed.

Au NPs are also used for enhanced X-ray radiotherapy. A challenge of X-ray radiation therapy in general is that high-dose X-rays under therapeutic conditions damage normal cells. Au NPs, upon X-ray irradiation, can act as dose enhancers and/or generate radicals that damage cancer cells and induce cell apoptosis. There are two main features of Au NPs which make them very good candidates for acting as X-ray radiosensitizers. First, Au has high number of protons ($Z=79$) and neutrons, compared with the previous elements evaluated for dose enhancing such as iodine ($Z=53$) and gadolinium ($Z=64$). This translates into an increased photoelectric cross-section. Second, the size of Au NPs is critical for escaping the tumor vasculature using the enhanced permeability and retention (EPR) phenomena. Thus, Au NPs have been proposed as potential radiosensitizers for X-rays mediated cancer therapy, allowing for a reduction in X-ray dose with improved therapeutic results [263–266]. Yang et al. have recently demonstrated the potential effects of radiation-induced killing of melanoma cells as mediated by amphiphilic Au NPs embedded within the walls of lipid nanocapsules. Interestingly, the membrane-penetrating properties of these amphiphilic Au NPs allowed for significant enhancement of the radiotherapy efficiency, which opens a path for improving the efficacy of frontline radiotherapy treatments [264]. An additional way to improve the radiotherapeutic enhancement effects has been reported by using Au NPs with glucose (Glc) and PEG as ligands (PEG-Glc-Au NPs) [265]. The enormous reduction in tumor size after 47 days of treatment was also because of the role of PEG and Glc in improving uptake and bio-distribution, which led to a concentration of PEG-Glc-Au NPs in tumor tissue 20 times higher than in healthy cells 48 h after injection. Alternatively, the potential of Au NPs to aggregate within tumors can be exploited in this direction [266]. In addition, 15 nm Au NPs have been designed to aggregate and remain largely in the tumor, after direct intratumoral infusion, thus changing from NIR-transparent to NIR-absorbent, enabling tumor-specific heating upon NIR illumination. Aggregation within tumors seems to be induced by the lower pH of the tumor milieu and endosomes/lysosomes or other mechanisms, such as by labile ligand bonds and lysosomal enzymes. This aggregation effect, and subsequent heating by NIR followed by X-ray treatment, was able to reduce the X-ray dose needed for tumor control by a factor >3. Because of the limited penetration of NIR, certain superficial or accessible tumors (e.g., a subset of head, neck, and melanoma) would be immediate candidates to evaluate the potential of this strategy. These examples illustrate the huge potential of Au NPs to enhance radiotherapy treatments, providing useful insights for further clinical studies.

It is important to note that the mechanism by which Au NPs can lead to dose enhancements in radiation therapy differs when comparing photon and proton radiations for NPs excitation. The dose enhancement using protons can be up to 14% and is independent of proton energy, in contrast to photon excitation where the dose enhancement is highly dependent on the photon energy used. A theoretical Monte Carlo simulation study [267] concluded that the potential of Au NPs to enhance radiation therapy depends on the type of radiation source, and proton therapy can be enhanced significantly only if the Au NPs are in close proximity to target tissues.

Radioactive Au NPs are being used to make radiation therapy more effective. The radioactive properties of ^{198}Au ($\beta_{max} = 0.96$ MeV; half-life 2.7 days) make it an ideal candidate for use in radiotherapeutic applications [268]. A major challenge in cancer therapy has been delivery and retention, as it is necessary to increase the therapeutic payload to get an effective tumor treatment. In this regard, NPs containing radioactive isotopes can concentrate within the tumor and provide an opportunity to tune the radioactive therapeutic dose delivered to tumor cells. Furthermore, ^{198}Au NPs have extraordinary tumor retention capabilities because of their natural affinity to leaky tumor vasculature. In this area, relevant advances have been achieved [268–270]. Khan et al. developed a method for the encapsulation of radioactive Au within a dendrimeric composite and demonstrated that radioactive Au NPs could deliver therapeutic doses to tumors [269]. More recently, gum arabic glycoprotein (GA)-functionalized Au NPs, consisting of beta-emitting ^{198}Au, were used for reducing the sizes of inoperable prostate tumors [268, 270]. Interestingly, the optimum hydrophobicity of the GA matrix allowed for effective penetration across tumor membranes. The therapeutic efficacy of GA-^{198}Au NPs was demonstrated in prostate tumor-bearing severely compromised immunodeficient (SCID) mice models, reaching an unprecedented 82%, 3 weeks after single-dose intratumoral administration of GA-^{198}Au NPs (408 μCi). The findings of significant therapeutic efficacy, good in vivo tolerance, and non-toxic features make these NPs potentially ideal candidates for future human applications.

Radiofrequency (RF) fields can be used to induce Au NP-mediated thermal ablation in a similar manner to that of photothermal and radio-sensitization therapies. The efficiency of RF-based therapy is significantly enhanced by using Au NPs, which are accumulated in the tumor area and then absorb main RF radiation power to heat cancer cells and thus cause their selective destruction. In particular, an intense source of RF radiation with frequency of 13.6 MHz and the power of 600 W induced the heating of suspensions of Au NPs with a heating rate of ~20 K/s, which resulted in considerable cell necrosis [271]. The NPs heating mechanism in an RF field is a very complex phenomenon. Glazer et al. have demonstrated that Au NPs heat primarily via Joule heating [271, 272]. Briefly, the Au NPs are hypothesized to function as tiny resistors, where free electrons on the surface have restricted movements. Therefore the friction created at the individual NP level releases heat into the surrounding aqueous solution [273].

The potential use of Au NPs coupled to RF waves was evaluated for the treatment of human hepatocellular and pancreatic cancer cells [274]. Direct injection of citrate-Au NPs into the tumor allowed focusing of the RF waves (13.56 MHz RF field) for selective heating of cancer cells. The resulting induced heat was lethal to these cancer cells bearing Au NPs in vitro. It was also demonstrated that the Au NPs had no intrinsic cytotoxicity or antiproliferative effects in the two human cancer cell lines studied. In another example, Curley et al. designed a method using Au NPs functionalized with the epidermal growth factor receptor (EGFR) inhibitor cetuximab in Panc-1 (pancreatic adenocarcinoma) and Difi (colorectal adenocarcinoma) cells which express high levels of EGFR [275]. This method proved to be cytotoxic to nearly 100% of the pancreatic and colorectal cells, but hardly any of the cells from the control group were damaged. The advantages of this therapy are that shortwave (megahertz range) RF energy is non-ionizing, penetrates deeply into biological tissues with no adverse side effects, and heats Au NPs efficiently. Thus, this technique may represent an effective treatment for numerous human malignant diseases using non-invasive RF hyperthermia.

The finding that Au NPs are able to inhibit angiogenesis (i.e., the formation of new vessels in organs or tissues) has also opened a new path to control the growth and spread of cancerous tissues via angiogenesis therapy. One method to inhibit angiogenesis in vivo is to block the function of pro-angiogenic heparin-binding growth factors (HB-GFs) such as vascular endothelial growth factor 165 (VEGF165) and basic fibroblast growth factor (bFGF). Mukherjee et al. demonstrated that Au NPs inhibit VEGF165-induced proliferation of endothelial cells in a dose-dependent manner [276]. This inhibition effect was tested in vivo using a nude mouse ear model, showing that after a week of daily intraperitoneal injections, the ascites volume had reduced in the NPs treated mice compared to the non-treated tumor-bearing mice. More recently, detailed studies of the antiangiogenic properties of Au NPs concluded that Au NPs not only inhibit VEGF165-induced HUVEC (Human Umbilical Vein Endothelial Cells) proliferation but also repress endothelial cell migration and tube formation [277]. Using Au NPs of different sizes and surface charges, it was demonstrated that a naked Au NPs surface is required and that the core size plays an important role to inhibit the function of heparin-binding growth factors (HB-GFs) and subsequent intracellular signaling events. Furthermore, the inhibitory effect of Au NPs was produced by the change in HB-GFs conformation/configuration (denaturation) by the NPs, whereas the conformations of non-HB-GFs remained unaffected [278]. The antiangiogenic properties of Au NPs have also been exploited for the treatment of chronic inflammatory diseases such as rheumatoid arthritis. Intra-articular delivery of Au NPs has been demonstrated to be an effective treatment strategy for collagen-induced arthritis [279].

5 Use of Au NPs Towards Theranostics

Recent research has paved the way for multimodal 'theranostic' (i.e., a combination of therapy and diagnosis) nanocarriers designed for carrying out simultaneous detection/diagnosis and treatment of the disease following administration [184, 280–282]. Au NPs are suitable for developing theranostic NPs thanks to their unique characteristics that enable their use as contrast agents, as therapeutic entities, and as scaffolds to adhere functional molecules, therapeutic cargoes (e.g., drugs/genes), and targeting ligands [184, 281]. Several examples of Au-based theranostic NPs are illustrated in Fig. 3, which are explained below.

Au-based theranostic NPs that utilize light-based techniques for monitoring and treating diseases are of special interest as they allow for spatially and temporally controllable drug release, localized therapy, and minimally invasive treatment modalities that reduce patients discomfort [282]. An interesting photo-triggered theranostic system has been developed by Khlebtsov et al. [283], consisting of a silver/gold (Ag/Au) nanocage core surrounded by a silica shell containing the NIR photosensitizer Yb–2,4-dimethoxyhematoporphyrin (Yb–HP) for monitoring tumors and simultaneous dual therapy, i.e., PTT/PDT (Fig. 3a). A significant higher death rate of HeLa cervical cancer cells was observed in vitro when they were incubated with the composite NPs and irradiated by 630-nm light because of PTT by the Ag/Au NPs as well as PDT using the presence of Yb-HP. Furthermore, the IR luminescence of Yb-HP (900–1060 nm, originating from Yb^{3+} ions, and located in the tissue transparency window) could be used for diagnostic purposes and for controlling the accumulation and bio-distribution of the composite NPs in tumors. Another example of theranostic NPs for simultaneous X-rays/ CT dual-imaging and dual-mode enhanced radiation therapy (RT) and PTT was reported by Huan et al. [284]. Folic acid-conjugated and silica-modified Au nanorods were synthesized and showed highly selective targeting, excellent X-ray/CT imaging ability, and enhanced RT and PTT effects (Fig. 3b). These multifunctional NPs could specifically bind to folate receptors on the surface of MGC803 gastric cancer cells and were imaged in vivo using both X-ray and CT imaging followed by treatment via RT or PTT. Alternatively, activatable theranostic NPs were developed by using

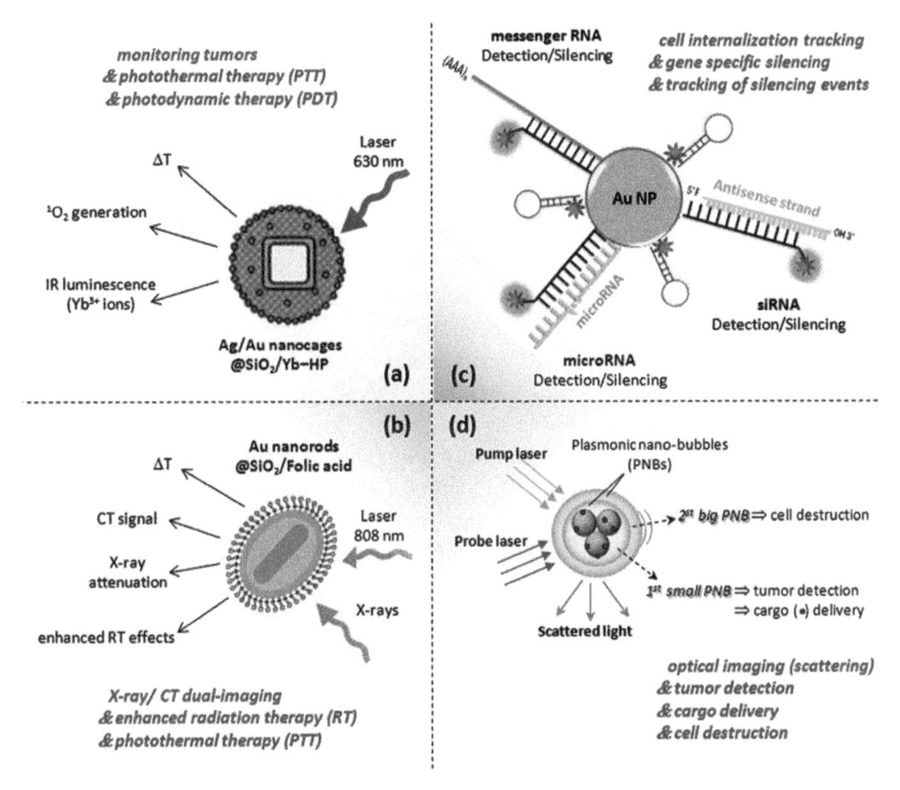

Fig. 3 Simplified examples of diverse Au-based theranostic NPs. (**a**) Silver/gold (Ag/Au) nanocages surrounded by a silica shell containing the NIR photosensitizer Yb–2,4-dimethoxyhematoporphyrin (Yb–HP) for monitoring tumors *via* IR luminescence and simultaneous dual-therapy PTT/PDT [283]. (**b**) Au nanorods conjugated with folic acid for selective targeting, cancer cells X-ray/CT dual-imaging and treatment *via* enhanced-RT or PTT [284]. (**c**) Au NPs functionalized with a fluorophore labeled hairpin-DNA for simultaneous gene specific silencing and intracellular tracking of the silencing events [285]. (**d**) Use of Au NPs, either alone or with linked cargo molecules, for generating plasmonic nano-bubbles (PNBs) which allow tumor detection *via* light scattering, cargo delivery *via* creation of transient holes on the cell membrane and finally cell destruction *via* mechanical impact [286]

Au@Ag/Au NPs assembled with activatable aptamer probes, which provided high-contrast image-guided site-specific PTT therapy [287]. The Au@Ag/Au NPs simultaneously serve as an optical heater and a fluorescence quencher. The activatable aptamer probes comprised a thiolated aptamer and a fluorophore-labeled complementary DNA. Thus, the activatable theranostic NPs with quenched fluorescence in the free state could undergo signal activation through target binding-induced conformational change of the activatable aptamer probes in specific tumor tissues, and then achieve on-demand treatment under image-guided irradiation. By using S6 aptamer as a model, in vitro and in vivo studies of A549 lung cancer cells verified that these NPs greatly improved imaging contrast and specific destruction. This strategy might be explored as a versatile platform for simultaneous detection and treatment of multiple kinds of cancer cells with the use of specific aptamers for varying cancer targets.

Conde et al. [285] recently developed an interesting Au-based theranostic system capable of intersecting all RNA pathways: from gene specific downregulation to silencing the silencers, i.e., siRNA and miRNA pathways (Fig. 3c). The system consists of Au NPs functionalized with a fluorophore labeled hairpin-DNA, which allows one to downregulate a specific gene directly and also to silence single gene expression, exogenous siRNA, and endogenous miRNAs, simultaneously tracking cell internalization and identifying the cells where silence is occurring (i.e., the fluorescence signal is directly proportional to the level of silencing). The usefulness of this approach was demonstrated for silencing an endogenous miRNA (miR-21) commonly upregulated in cancer, such as in colorectal carcinoma cells (HCT-116). The photothermal properties of Au NPs can also be used to generate transient vapor nano-bubbles to produce a tunable nanoscale theranostic agent, described as PNBs [286]. These PNBs are generated when Au NPs are locally overheated with short laser pulses because of the evaporation of a very thin volume of the surrounding medium, which in turn creates a vapor nanobubble that expands and collapses within nanoseconds. The bubble scatters the light, thus acting as an optical probe which allows for tumor detection, and the fast expansion of the PNB produces a localized mechanical impact which damages cell membranes, resulting in cell death, and therefore acting as a therapeutic agent. This novel theranostic system has been successfully applied as an in vivo tunable theranostic cellular agent in zebrafish hosting prostate cancer xenografts, presenting higher therapeutic selectivity when compared with Au NPs alone [288]. Au NPs conjugated with anti-EGFR antibody C225 could actively target EGFR-positive A549 lung carcinoma cells. Following cellular uptake, single human prostate cancer cells could be detected and ablated under optical guidance in vivo by tunable PNBs in a single theranostic procedure. By varying the energy of the laser pulse, the PNBs size could be dynamically tuned in a theranostic sequence of two PNBs: an initial small PNB detected a cancer cell through optical scattering, followed by a second bigger PNB, which mechanically ablated this cell without damaging the surrounding tissues, and its optical scattering confirmed the destruction of the cells. This innovative and promising theranostic strategy concept of a 'cell theranostics' approach that unites diagnosis, therapy, and confirmation (guidance) of the results of therapy in a single process at cellular level principally can help to improve both the rapidity and the precision of treatment [288]. Recently, the same group has used this concept for both, localized delivery of molecular cargo as well as mechanical destruction of cells by generation of a transient PNB around the Au NPs with a single incident laser pulse. Small PNBs can create a transient hole on the cell membrane to 'inject' molecular cargo without damage to the cells. Large PNBs, on the other hand, can cause mechanical destruction of the cells of interest [289] (Fig. 3d).

Based on these examples, it is apparent that theranostic Au NPs have opened the door to novel and advanced treatment strategies that combine therapeutics with diagnostics, aiming to monitor the response to treatment and increase drug efficacy and safety, which would be a key part of personalized medicine.

6 Applications of Au NPs in Clinical Trials

Some of the above-mentioned medical applications using Au NPs are already in the stage of pre-clinical or clinical trials.

The diagnostics company "Nanosphere" has developed the so-called Nanosphere's Verigene® System, which utilizes advanced automation and Au NPs to enable rapid direct detection of nucleic acids and high-sensitivity protein detection on the same platform. This technology has already received food and drug administration (FDA) approval in the United States. It is based on Au NPs of 13–20 nm diameter functionalized with either a defined number of oligonucleotides (i.e., short pieces of DNA or RNA) or a defined number of antibodies specific to a particular protein of interest.

One therapy using Au NPs which has reached clinical trials is CYT-6091, 27 nm citrate-coated Au NPs conjugated with thiolated-PEG and tumor necrosis factor-α (TNF-α) (Aurimmune; CytImmune Sciences). The NPs have the dual effect of increasing tumor targeting and tumor toxicity in comparison with the use of TNF-α alone [290]. In this trial the side effects and best dose of CYT-6091 in treating patients with advanced solid tumors by intravenous administration have been studied [291]. Future clinical studies should focus on combining CYT-6091 with approved chemotherapies for the systemic treatment of non-resectable cancers.

Using the same CYT-6091 NPs, another clinical trial has been carried out to evaluate the tissue distribution and the selective tumor trafficking of CYT-6091 in patients with primary and metastatic cancers [292]. Patients, stratified according to cancer type, received CYT-6091 and then underwent standard-care surgery. Tumor and normal tissues were removed during surgery for analysis of antitumor effects and tissue distribution of CYT-6091 by electron microscopy.

Au nanoshells (AuroShells®, Nanospectra Biosciences), which consist of a silica core of 120 nm diameter with a 15-nm gold shell, were used in clinical trials to treat head and neck cancers using PPTT. This therapy, called AuroLase® Therapy, consisted of an injection of Au nanoshell NPs into the patient's bloodstream. After 12–24 h (enough time for the NPs to accumulate inside the tumor), an 808 nm IR laser was used to heat the NPs and destroy tumor cells [293]. These NPs are currently under clinical trials in patients with primary and/or metastatic lung cancer where there is airway obstruction. In this study, patients are given a systemic infusion of NPs and a subsequently escalating dose of laser radiation delivered by an optical fiber via bronchoscopy.

In the treatment of atherosclerotic lesions, two delivery techniques for NPs and PPTT are under clinical trials (NANOM FIM) [294]. Patients underwent nano-intervention either with the delivery of silica-Au NPs in a mini-surgery implanted bioengineered on-artery patch, or with the delivery of silica-Au iron-bearing NPs with targeted micro-bubbles or stem cells by means of magnetic navigation system vs stent implantation. The primary results showed a similar degree of regression of total atheroma volume after 12 months for both approaches of delivery.

Another technique being tested in pre-clinical trials works is on validating polyvalent Au NPs functionalized with RNAi (RNA interference) as anti-glioma therapeutics [295]. This nano-RNAi platform can be used to target signature lesions of glioblastoma, which play an important role in driving glioma pathogenesis, mediating therapeutic resistance, and instigating neurologically debilitating necrogenesis. RNAi-Au NPs are being validated on multiple levels, using glioma stem cell cultures, derived xenografts, and genetically engineered glioma mouse models.

Despite these examples, the full clinical impact of Au NP-based therapies is not yet known. There is clearly a need to translate already developed applications to clinical trials in a timely but safe manner.

7 Concluding Remarks and Future Outlook

We have discussed novel strategies for the synthesis and functionalization of Au NPs to evaluate their potential use in nanomedicine. In addition, their detection and sensing properties have been explored for diagnosing some diseases. Au NPs either alone or in hybrid form can also improve the performance of practically used imaging techniques. Moreover, Au NP-based therapies are generally superior in terms of specificity, selectivity, efficiency, and cytotoxicity compared to the same methods without Au NPs. Additionally, Au composite NPs have recently been evaluated for their theranostic potential both in vitro and in vivo. Nowadays, the main focus is the transition of Au NPs from laboratories to the clinics. Though the initial theranostic efficacy of Au NPs shows promising results, there are still many challenges which need to be addressed before their use in clinical practice. The first challenge involves the long-term retention, cytotoxicity, and ultimate renal clearance of the NPs. Though the biodistribution and toxicity of Au NPs have been extensively studied, reliable predictions based on these results are rare. Therefore, more studies need to be performed to ensure their safety before use in humans. The biodistribution of Au NPs is dependent on their size, geometry and surface chemistry. Dissimilarities of reported results dealing with Au NPs of the same size and shape have been attributed to the type of coating or stabilizing agents used. In order to overcome this problem, strategies to improve comparability and standardization of nanotoxicological studies are needed. Moreover, there should be a shift of the focus of toxicological experiments from 'live–dead' assays to the assessment of cell function, allowing observation of bioresponses at lower doses, which are more relevant for in vivo scenarios. Second, detection and sensing of analytes in complex biological fluids (such as urine, blood, etc.) are still complicated to achieve. Third, non-invasive clinical trials at the molecular level need to be better explored. Fourth, the development of personalized medicines for the treatment of individual patients according to their genetic profiles is so far merely a vision described in scientific papers. Last, but not least, vaccinations based on Au NPs for humans and/or animals against biologically active factors or diseases still remain a dream to be fulfilled. Addressing these and other such types of challenges may help in the future to shift Au NP-based nanomedicines further into clinics.

Acknowledgments This work was supported by the European Commission (project Futurenanoneeds to WJP). SA, BP, and CCC acknowledge the Alexander von Humboldt Foundation for a PostDoc fellowship. QZ acknowledges CSC for funding. MGS acknowledges the Youssef Jameel Foundation for a PhD fellowship. AE acknowledges Junta de Andalucia for a Talentia Postdoc Fellowship, co-financed by the European Union's Seventh Framework Programme, grant agreement no 267226. MC acknowledges a Research Fellow Grant from Ikerbasque, Basque Foundation for Science.

References

1. Patra CR, Bhattacharya R, Mukhopadhyay D, Mukherjee P (2010) Adv Drug Deliv Rev 62:346
2. Jain KK (2008) Nanomedicine 17:89
3. Chakraborty M, Jain S, Rani V (2011) Appl Biochem Biotechnol 165:1178
4. Shi J, Votruba AR, Farokhzad OC, Langer R (2010) Nano Lett 10:3223
5. Parveen S, Misra R, Sahoo SK (2012) Nanomed Nanotechnol Biol Med 8:147
6. Hauert S, Bhatia SN (2014) Trends Biotechnol 32:448
7. Steichen SD, Caldorera-Moore M, Peppas NA (2013) Eur J Pharm Sci 48:416
8. Nishiyama N, Okazaki S, Cabral H, Miyamoto M, Kato Y, Sugiyama Y, Nishio K, Matsumura Y, Kataoka K (2003) Cancer Res 63:8977
9. Cai W, Gao T, Hong H, Sun J (2008) Nanotechnol Sci Appl 1:17
10. Arvizo R, Bhattacharya R, Mukherjee P (2010) Expert Opin Drug Deliv 7:753
11. Giljohann DA, Seferos DS, Daniel WL, Massich MD, Patel PC, Mirkin CA (2010) Angew Chem Int Ed 49:3280
12. Dykman L, Khlebtsov N (2012) Chem Soc Rev 41:2256
13. Sperling RA, Rivera-Gil P, Zhang F, Zanella M, Parak WJ (2008) Chem Soc Rev 37:1896
14. Myroshnychenko V, Rodriguez-Fernandez J, Pastoriza-Santos I, Funston AM, Novo C, Mulvaney P, Liz-Marzan LM, Abajo FJG (2008) Chem Soc Rev 37:1792
15. Khlebtsov N, Dykman L (2011) Chem Soc Rev 40:1647
16. Fraga S, Brandão A, Soares ME, Morais T, Duarte JA, Pereira L, Soares L, Neves C, Pereira E, de Lourdes BM (2014) Nanomed Nanotechnol Biol Med 10:1757
17. Simpson CA, Salleng KJ, Cliffel DE, Feldheim DL (2013) Nanomed Nanotechnol Biol Med 9:257
18. Bednarski M, Dudek M, Knutelska J, Nowiński L, Sapa J, Zygmunt M, Nowak G, Luty-Błocho M, Wojnicki M, Fitzner K (2014) Pharmacol Rep 67:405
19. Dreaden EC, Alkilany AM, Huang X, Murphy CJ, El-Sayed MA (2012) Chem Soc Rev 41:2740
20. Pelaz B, del Pino P (2012) Front Nanosci 4:3
21. Bao C, Conde J, Polo E, del Pino P, Moros M, Baptista PV, Grazu V, Cui D, de la Fuente JM (2014) Nanomedicine 9:2353
22. Li N, Zhao P, Astruc D (2014) Angew Chem Int Ed 53:1756
23. Petersen S, Barcikowski S (2009) J Phys Chem C 113:19830

24. Brust M, Walker M, Bethell D, Schiffrin DJ, Whyman R (1994) J Chem Soc Chem Commun 1994:801
25. Liz-Marzán LM (2004) Mater Today 7:26
26. Turkevich J, Stevenson PC, Hillier J (1951) Discuss Faraday Soc 11:55
27. Bastús NG, Comenge J, Puntes V (2011) Langmuir 27:11098
28. Pelaz B, Grazu V, Ibarra A, Magen C, del Pino P, de la Fuente JM (2012) Langmuir 28:8965
29. Perez-Hernandez M, del Pino P, Mitchell SG, Moros M, Stepien G, Pelaz B, Parak WJ, Galvez EM, Pardo J, De La Fuente JM (2014) ACS Nano 9:52
30. Bao C, Beziere N, del Pino P, Pelaz B, Estrada G, Tian F, Ntziachristos V, de la Fuente JM, Cui D (2013) Small 9:68
31. Ye X, Zheng C, Chen J, Gao Y, Murray CB (2013) Nano Lett 13:765
32. Skrabalak SE, Chen J, Sun Y, Lu X, Au L, Cobley CM, Xia Y (2008) Acc Chem Res 41:1587
33. Personick ML, Mirkin CA (2013) J Am Chem Soc 135:18238
34. Perez-Juste J, Pastoriza-Santos I, Liz-Marzan LM, Mulvaney P (2005) Coord Chem Rev 249:1870
35. DuChene JS, Niu W, Abendroth JM, Sun Q, Zhao W, Huo F, Wei WD (2012) Chem Mater 25:1392
36. Nikoobakht B, El-Sayed MA (2003) Chem Mater 15:1957
37. Jana NR, Gearheart L, Murphy CJ (2001) J Phys Chem B 105:4065
38. Alkilany AM, Nagaria PK, Hexel CR, Shaw TJ, Murphy CJ, Wyatt MD (2009) Small 5:701
39. Ye X, Jin L, Caglayan H, Chen J, Xing G, Zheng C, Doan-Nguyen V, Kang Y, Engheta N, Kagan CR (2012) ACS Nano 6:2804
40. Pham T, Jackson JB, Halas NJ, Lee TR (2002) Langmuir 18:4915
41. Wang H, Brandl DW, Le F, Nordlander P, Halas NJ (2006) Nano Lett 6:827
42. Bardhan R, Mukherjee S, Mirin NA, Levit SD, Nordlander P, Halas NJ (2009) J Phys Chem C 114:7378
43. Ha TH, Koo H-J, Chung BH (2007) J Phys Chem C 111:1123
44. Chen L, Ji F, Xu Y, He L, Mi Y, Bao F, Sun B, Zhang X, Zhang Q (2014) Nano Lett 14:7201
45. Kumar PS, Pastoriza-Santos I, Rodriguez-Gonzalez B, de Abajo FJG, Liz-Marzan LM (2008) Nanotechnology 19:015606
46. Chen J, Wiley B, Li ZY, Campbell D, Saeki F, Cang H, Au L, Lee J, Li X, Xia Y (2005) Adv Mater 17:2255
47. Cobley CM, Chen J, Cho EC, Wang LV, Xia Y (2011) Chem Soc Rev 40:44
48. Goodman AM, Cao Y, Urban C, Neumann O, Ayala-Orozco C, Knight MW, Joshi A, Nordlander P, Halas NJ (2014) ACS Nano 8:3222
49. Aherne D, Charles DE, Brennan-Fournet ME, Kelly JM, Gun'ko YK (2009) Langmuir 25:10165
50. Luangpipat T, Beattie IR, Chisti Y, Haverkamp RG (2011) J Nanopart Res 13:6439
51. Gericke M, Pinches A (2006) Hydrometallurgy 83:132
52. Mittal AK, Chisti Y, Banerjee UC (2013) Biotechnol Adv 31:346
53. Levin CS, Hofmann C, Ali TA, Kelly AT, Morosan E, Nordlander P, Whitmire KH, Halas NJ (2009) ACS Nano 3:1379
54. Bardhan R, Grady NK, Ali T, Halas NJ (2010) ACS Nano 4:6169
55. Salado J, Insausti M, Lezama L, de Muro IG, Moros M, Pelaz B, Grazu V, de la Fuente J, Rojo T (2012) Nanotechnology 23:315102
56. Cardinal MF, Rodriguez-Gonzalez B, Alvarez-Puebla RA, Perez-Juste J, Liz-Marzan LM (2010) J Phys Chem C 114:10417
57. Cheng L-C, Chen HM, Lai T-C, Chan Y-C, Liu R-S, Sung JC, Hsiao M, Chen C-H, Her L-J, Tsai DP (2013) Nanoscale 5:3931
58. Tan C, Huang X, Zhang H (2013) Mater Today 16:29
59. Moon H, Kumar D, Kim H, Sim C, Chang J-H, Kim J-M, Kim H, Lim D-K (2015) ACS Nano 9:2711
60. Rivera-Gil P, Jimenez De Aberasturi D, Wulf V, Pelaz B, Del Pino P, Zhao Y, De La Fuente JM, Ruiz De Larramendi I, Rojo T, Liang X-J (2012) Acc Chem Res 46:743

61. Pelaz B, Charron G, Pfeiffer C, Zhao Y, de la Fuente JM, Liang X-J, Parak WJ, del Pino P (2013) Small 9:1573
62. Love JC, Estroff LA, Kriebel JK, Nuzzo RG, Whitesides GM (2005) Chem Rev 105:1103
63. Yang J, Lee JY, Ying JY (2011) Chem Soc Rev 40:1672
64. Indrasekara A, Wadams RC, Fabris L (2014) Part Part Syst Charact 31:819
65. Kinnear C, Dietsch H, Clift MJ, Endes C, Rothen-Rutishauser B, Petri-Fink A (2013) Angew Chem 125:1988
66. Flynn NT, Tran TNT, Cima MJ, Langer R (2003) Langmuir 19:10909
67. Pino P, de la Fuente JM (2013) Chem Commun 49:3676
68. Puertas S, Batalla P, Moros M, Polo E, del Pino P, Guisan JM, Grazu V, de la Fuente JM (2011) ACS Nano 5:4521
69. Roux S, Garcia B, Bridot J-L, Salomé M, Marquette C, Lemelle L, Gillet P, Blum L, Perriat P, Tillement O (2005) Langmuir 21:2526
70. Brewer SH, Glomm WR, Johnson MC, Knag MK, Franzen S (2005) Langmuir 21:9303
71. Mayya KS, Schoeler B, Caruso F (2003) Adv Funct Mater 13:183
72. Mahl D, Greulich C, Meyer-Zaika W, Köller M, Epple M (2010) J Mater Chem 20:6176
73. Lin C-AJ, Sperling RA, Li JK, Yang T-Y, Li P-Y, Zanella M, Chang WH, Parak WJ (2008) Small 4:334
74. Pellegrino T, Kudera S, Liedl T, Muñoz Javier A, Manna L, Parak WJ (2005) Small 1:48
75. Soliman MG, Pelaz B, Parak WJ, del Pino P (2015) Chem Mater 27:990
76. Fuentes M, Mateo C, Guisán J, Fernández-Lafuente R (2005) Biosens Bioelectron 20:1380
77. Tang W, Becker ML (2014) Chem Soc Rev 43:7013
78. Sperling RA, Parak WJ (2010) Phil Trans R Soc A 368:1333
79. Montenegro J-M, Grazu V, Sukhanova A, Agarwal S, Jesus M, Nabiev I, Greiner A, Parak WJ (2013) Adv Drug Deliv Rev 65:677
80. Avvakumova S, Colombo M, Tortora P, Prosperi D (2014) Trends Biotechnol 32:11
81. Jimenez de Aberasturi D, Montenegro J-M, Ruiz de Larramendi I, Rojo T, Klar TA, Alvarez-Puebla R, Liz-Marzán LM, Parak WJ (2012) Chem Mater 24:738
82. Saha K, Agasti SS, Kim C, Li X, Rotello VM (2012) Chem Rev 112:2739
83. Sapsford KE, Berti L, Medintz IL (2006) Angew Chem Int Ed 45:4562
84. Leuvering J, Thal P, Waart M, Schuurs A (1980) Fresenius J Anal Chem 301:132
85. Leuvering JH, Thal PJ, Van der Waart M, Schuurs AH (1981) J Immunol Methods 45:183
86. Elghanian R, Storhoff JJ, Mucic RC, Letsinger RL, Mirkin CA (1997) Science 277:1078
87. Mie G (1976) Ann Phys 25:377
88. Srivastava S, Frankamp BL, Rotello VM (2005) Chem Mater 17:487
89. Lin S-Y, Liu S-W, Lin C-M, Chen C-H (2002) Anal Chem 74:330
90. Reynolds AJ, Haines AH, Russell DA (2006) Langmuir 22:1156
91. Watanabe S, Seguchi H, Yoshida K, Kifune K, Tadaki T, Shiozaki H (2005) Tetrahedron Lett 46:8827
92. Daniel WL, Han MS, Lee J-S, Mirkin CA (2009) J Am Chem Soc 131:6362
93. Deng H, Zhang X, Kumar A, Zou G, Zhang X, Liang X-J (2012) Chem Commun 49:51
94. Ai K, Liu Y, Lu L (2009) J Am Chem Soc 131:9496
95. Aili D, Selegård R, Baltzer L, Enander K, Liedberg B (2009) Small 5:2445
96. Zhu Y, Wang G, Sha L, Qiu Y, Jiang H, Zhang X (2015) Analyst 140:1260
97. Wen S, Zheng F, Shen M, Shi X (2013) Colloids Surf A 419:80
98. Chang C-C, Chen C-Y, Chen C-P, Lin C-W (2015) Anal Methods 7:29
99. Creran B, Li X, Duncan B, Kim CS, Moyano DF, Rotello VM (2014) ACS Appl Mater Interfaces 6:19525
100. Wang S, Singh AK, Senapati D, Neely A, Yu H, Ray PC (2010) Chem Eur J 16:5600
101. Malinsky MD, Kelly KL, Schatz GC, Duyne RPV (2001) J Am Chem Soc 123:1471
102. Moon S, Kim DJ, Kim K, Kim D, Lee H, Lee K, Haam S (2010) Appl Opt 49:484
103. Wang J, Wang L, Sun Y, Zhu X, Cao Y, Wang X, Zhang H, Song D (2010) Colloids Surf B 75:520
104. Frasconi M, Tortolini C, Botre F, Mazzei F (2010) Anal Chem 82:7335
105. Joshi GK, Deitz-McElyea S, Johnson M, Mali S, Korc M, Sardar R (2014) Nano Lett 14:6955

106. Hu Y, Zhang L, Ying Z, Wang B, Wang Y, Fan Q, Huang W, Wang L (2015) ACS Appl Mater Interfaces 7:2459
107. Wang J (2005) Small 1:1036
108. Mitchell JS, Lowe TE (2009) Biosens Bioelectron 24:2177
109. He L, Musick MD, Nicewarner SR, Salinas FG, Benkovic SJ, Natan MJ, Keating CD (2000) J Am Chem Soc 122:9071
110. Jain PK, El-Sayed IH, El-Sayed MA (2007) Nano Today 2:18
111. Dubertret B, Calame M, Libchaber AJ (2001) Nat Biotechnol 19:365
112. You C-C, Miranda OR, Gider B, Ghosh PS, Kim I-B, Erdogan B, Krovi SA, Bunz UH, Rotello VM (2007) Nat Nanotechnol 2:318
113. Phillips RL, Miranda OR, You CC, Rotello VM, Bunz UH (2008) Angew Chem Int Ed 47:2590
114. Miranda OR, Chen H-T, You C-C, Mortenson DE, Yang X-C, Bunz UH, Rotello VM (2010) J Am Chem Soc 132:5285
115. Lukman S, Aung KMM, Liu J, Liu B, Su X (2013) ACS Appl Mater Interfaces 5:12725
116. Albert KJ, Lewis NS, Schauer CL, Sotzing GA, Stitzel SE, Vaid TP, Walt DR (2000) Chem Rev 100:2595
117. Sun W, Lu Y, Mao J, Chang N, Yang J, Liu Y (2015) Anal Chem 87:3354
118. Xu S, Lu X, Yao C, Huang F, Jiang H, Hua W, Na N, Liu H, Ouyang J (2014) Anal Chem 86:11634
119. Oh E, Hong M-Y, Lee D, Nam S-H, Yoon HC, Kim H-S (2005) J Am Chem Soc 127:3270
120. Bajaj A, Miranda OR, Kim I-B, Phillips RL, Jerry DJ, Bunz UH, Rotello VM (2009) Proc Natl Acad Sci 106:10912
121. Katz E, Willner I, Wang J (2004) Electroanal 16:19
122. Munge BS, Coffey AL, Doucette JM, Somba BK, Malhotra R, Patel V, Gutkind JS, Rusling JF (2011) Angew Chem 123:8061
123. Jensen GC, Krause CE, Sotzing GA, Rusling JF (2011) Phys Chem Chem Phys 13:4888
124. Li Y, Schluesener HJ, Xu S (2010) Gold Bull 43:29
125. Chen A, Chatterjee S (2013) Chem Soc Rev 42:5425
126. Jena BK, Raj CR (2010) Talanta 80:1653
127. Chai R, Yuan R, Chai Y, Ou C, Cao S, Li X (2008) Talanta 74:1330
128. Narang J, Malhotra N, Singh G, Pundir C (2015) Biosens Bioelectron 66:332
129. Omidfar K, Khorsand F, Azizi MD (2013) Biosens Bioelectron 43:336
130. Afonso AS, Pérez-López B, Faria RC, Mattoso LH, Hernández-Herrero M, Roig-Sagués AX, Maltez-da Costa M, Merkoçi A (2013) Biosens Bioelectron 40:121
131. Maltez-da Costa M, de la Escosura-Muñiz A, Nogués C, Barrios L, Ibáñez E, Merkoçi A (2012) Nano Lett 12:4164
132. de la Escosura-Muñiz A, Sánchez-Espinel C, Díaz-Freitas B, González-Fernández A, Maltez-da Costa M, Merkoçi A (2009) Anal Chem 81:10268
133. Maltez-da Costa M, de la Escosura-Muñiz A, Nogués C, Barrios L, Ibáñez E, Merkoçi A (2012) Small 8:3605
134. Kneipp K, Kneipp H, Itzkan I, Dasari RR, Feld MS (1999) Chem Rev 99:2957
135. Willets KA (2009) Anal Bioanal Chem 394:85
136. Cialla D, März A, Böhme R, Theil F, Weber K, Schmitt M, Popp J (2012) Anal Bioanal Chem 403:27
137. La Porta A, Grzelczak M, Liz-Marzán LM (2014) ChemistryOpen 3:146
138. Scarabelli L, Coronado-Puchau M, Giner-Casares JJ, Langer J, Liz-Marzán LM (2014) ACS Nano 8:5833
139. Quaresma P, Osório I, Dória G, Carvalho PA, Pereira A, Langer J, Araújo JP, Pastoriza-Santos I, Liz-Marzán LM, Franco R (2014) RSC Adv 4:3659
140. Hughes J, Izake EL, Lott WB, Ayoko GA, Sillence M (2014) Talanta 130:20
141. Thuy NT, Yokogawa R, Yoshimura Y, Fujimoto K, Koyano M, Maenosono S (2010) Analyst 135:595

142. Harpster MH, Zhang H, Sankara-Warrier AK, Ray BH, Ward TR, Kollmar JP, Carron KT, Mecham JO, Corcoran RC, Wilson WC (2009) Biosens Bioelectron 25:674
143. Wang Y, Lee K, Irudayaraj J (2010) Chem Commun 46:613
144. Da P, Li W, Lin X, Wang Y, Tang J, Zheng G (2014) Anal Chem 86:6633
145. Matteini P, de Angelis M, Ulivi L, Centi S, Pini R (2015) Nanoscale 7:3474
146. Panikkanvalappil SR, El-Sayed MA (2014) J Phys Chem B 118:14085
147. Xie H-N, Lin Y, Mazo M, Chiappini C, Sánchez-Iglesias A, Liz-Marzán LM, Stevens MM (2014) Nanoscale 6:12403
148. Liu G, Swierczewska M, Lee S, Chen X (2010) Nano Today 5:524
149. Doane TL, Burda C (2012) Chem Soc Rev 41:2885
150. Sun X, Zhang N (2010) Mini Rev Med Chem 10:108
151. Koo H, Jin G-W, Kang H, Lee Y, Nam K, Bai CZ, Park J-S (2010) Biomaterials 31:988
152. Danhier F, Ansorena E, Silva JM, Coco R, Le Breton A, Préat V (2012) J Control Release 161:505
153. Menjoge AR, Kannan RM, Tomalia DA (2010) Drug Discov Today 15:171
154. Zhang L, Gu FX, Chan JM, Wang AZ, Langer RS, Farokhzad OC (2008) Clin Pharmacol Ther 83:761
155. Mulder WJ, Strijkers GJ, van Tilborg GA, Griffioen AW, Nicolay K (2006) NMR Biomed 19:142
156. Puri A, Loomis K, Smith B, Lee J-H, Yavlovich A, Heldman E, Blumenthal R (2009) Crit Rev Ther Drug Carrier Syst 26:523
157. Talelli M, Rijcken CJ, Lammers T, Seevinck PR, Storm G, van Nostrum CF, Hennink WE (2009) Langmuir 25:2060
158. Gupta AK, Naregalkar RR, Vaidya VD, Gupta M (2007) Future Med Chem 2:23
159. Figuerola A, Di Corato R, Manna L, Pellegrino T (2010) Pharmacol Res 62:126
160. Singh SP (2011) J Biomed Nanotechnol 7:95
161. Wang Y, Chen L (2011) Nanomed Nanotechnol Biol Med 7:385
162. Smith AM, Duan H, Mohs AM, Nie S (2008) Adv Drug Deliv Rev 60:1226
163. Bentolila LA, Ebenstein Y, Weiss S (2009) J Nucl Med 50:493
164. Kostarelos K, Bianco A, Prato M (2009) Nat Nanotechnol 4:627
165. Son SJ, Bai X, Lee S (2007) Drug Discov Today 12:657
166. Jokerst JV, Lobovkina T, Zare RN, Gambhir SS (2011) Nanomedicine 6:715
167. Lee D-E, Koo H, Sun I-C, Ryu JH, Kim K, Kwon IC (2012) Chem Soc Rev 41:2656
168. Tasciotti E, Liu XW, Bhavane R, Plant K, Leonard AD, Price BK, Cheng MMC, Decuzzi P, Tour JM, Robertson F, Ferrari M (2008) Nat Nanotechnol 3:151
169. Huang X, Jain PK, El-Sayed IH, El-Sayed MA (2007) Nanomedicine 2:681
170. Qian X, Peng X-H, Ansari DO, Yin-Goen Q, Chen GZ, Shin DM, Yang L, Young AN, Wang MD, Nie S (2008) Nat Biotechnol 26:83
171. Boisselier E, Astruc D (2009) Chem Soc Rev 38:1759
172. Murphy CJ, Gole AM, Stone JW, Sisco PN, Alkilany AM, Goldsmith EC, Baxter SC (2008) Acc Chem Res 41:1721
173. Huff TB, Tong L, Zhao Y, Hansen MN, Cheng JX, Wei A (2007) Nanomedicine 2:125
174. Tong L, Wei Q, Wei A, Cheng JX (2009) Photochem Photobiol 85:21
175. Ye S, Marston G, McLaughlan JR, Sigle DO, Ingram N, Freear S, Baumberg JJ, Bushby RJ, Markham AF, Critchley K (2015) Adv Funct Mater. doi:10.1002/adfm.201404358:1-11
176. Bardhan R, Lal S, Joshi A, Halas NJ (2011) Acc Chem Res 44:936
177. Xia X, Xia Y (2014) Front Phys 9:378
178. Moon GD, Choi S-W, Cai X, Li W, Cho EC, Jeong U, Wang LV, Xia Y (2011) J Am Chem Soc 133:4762
179. Wang Y, Liu Y, Luehmann H, Xia X, Wan D, Cutler C, Xia Y (2013) Nano Lett 13:581
180. Salgueiriño-Maceira V, Correa-Duarte MA (2007) Adv Mater 19:4131
181. Yuan H, Fales AM, Vo-Dinh T (2012) J Am Chem Soc 134:11358
182. Yuan H, Khoury CG, Hwang H, Wilson CM, Grant GA, Vo-Dinh T (2012) Nanotechnology 23:075102

183. Yuan H, Khoury CG, Wilson CM, Grant GA, Bennett AJ, Vo-Dinh T (2012) Nanomed Nanotechnol Biol Med 8:1355
184. Webb JA, Bardhan R (2014) Nanoscale 6:2502
185. Latterini L, Tarpani L (2014) Bio Bioinspir Nanomater 7:173
186. Wang Z (2013) Sci China Phys Mech Astron 56:506
187. Melancon MP, Lu W, Yang Z, Zhang R, Cheng Z, Elliot AM, Stafford J, Olson T, Zhang JZ, Li C (2008) Mol Cancer Ther 7:1730
188. Kang B, Mackey MA, El-Sayed MA (2010) J Am Chem Soc 132:1517
189. Rosman C, Pierrat S, Henkel A, Tarantola M, Schneider D, Sunnick E, Janshoff A, Sönnichsen C (2012) Small 8:3683
190. Wu Z, Jin R (2010) Nano Lett 10:2568
191. Wu X, He X, Wang K, Xie C, Zhou B, Qing Z (2010) Nanoscale 2:2244
192. Dam DHM, Lee JH, Sisco PN, Co DT, Zhang M, Wasielewski MR, Odom TW (2012) ACS Nano 6:3318
193. Zhu J, Yong K-T, Roy I, Hu R, Ding H, Zhao L, Swihart MT, He GS, Cui Y, Prasad PN (2010) Nanotechnology 21:285106
194. Hong G, Tabakman SM, Welsher K, Chen Z, Robinson JT, Wang H, Zhang B, Dai H (2011) Angew Chem Int Ed 50:4644
195. Li J-L, Gu M (2010) Biomaterials 31:9492
196. Durr NJ, Larson T, Smith DK, Korgel BA, Sokolov K, Ben-Yakar A (2007) Nano Lett 7:941
197. Gao L, Vadakkan TJ, Nammalvar V (2011) Nanotechnology 22:365102
198. Au L, Zhang Q, Cobley CM, Gidding M, Schwartz AG, Chen J, Xia Y (2009) ACS Nano 4:35
199. Tong L, Cobley CM, Chen J, Xia Y, Cheng J-X (2010) Angew Chem Int Ed 49:3485
200. Wang LV, Hu S (2012) Science 335:1458
201. Hainfeld JF, Slatkin DN, Focella TM, Smilowitz HM (2006) Br J Radiol 79:248
202. Ahn S, Jung SY, Lee SJ (2013) Molecules 18:5858
203. Peng C, Qin J, Zhou B, Chen Q, Shen M, Zhu M, Lu X, Shi X (2013) Polym Chem 4:4412
204. Fujimoto JG (2003) Nat Biotechnol 21:1361
205. Zagaynova E, Shirmanova M, Kirillin MY, Khlebtsov B, Orlova A, Balalaeva I, Sirotkina M, Bugrova M, Agrba P, Kamensky V (2008) Phys Med Biol 53:4995
206. Ntziachristos V (2010) Nat Methods 7:603
207. Sebastián V, Lee S-K, Zhou C, Kraus MF, Fujimoto JG, Jensen KF (2012) Chem Commun 48:6654
208. Boyer D, Tamarat P, Maali A, Lounis B, Orrit M (2002) Science 297:1160
209. Karmani L, Labar D, Valembois V, Bouchat V, Nagaswaran PG, Bol A, Gillart J, Levêque P, Bouzin C, Bonifazi D (2013) Contrast Media Mol Imaging 8:402
210. Das S, Grimm J, Thorek DL (2014) Emerg Appl Mol Imaging Oncol 124:213
211. Singhal S, Nie S, Wang MD (2010) Annu Rev Med 61:359
212. Li J, Gupta S, Li C (2013) Quant Imaging Med Surg 3:284
213. Ghosh P, Han G, De M, Kim CK, Rotello VM (2008) Adv Drug Deliv Rev 60:1307
214. Chen Z, Zhang L, He Y, Shen Y, Li Y (2014) Small 11:952
215. Kong WH, Bae KH, Hong CA, Lee Y, Hahn SK, Park TG (2011) Bioconjug Chem 22:1962
216. Bansal A, Zhang Y (2014) Acc Chem Res 47:3052
217. Braun GB, Pallaoro A, Wu G, Missirlis D, Zasadzinski JA, Tirrell M, Reich NO (2009) ACS Nano 3:2007
218. Huschka R, Zuloaga J, Knight MW, Brown LV, Nordlander P, Halas NJ (2011) J Am Chem Soc 133:12247
219. Yang N, Burkholder J, Roberts B, Martinell B, McCabe D (1990) Proc Natl Acad Sci 87:9568
220. Park J, Kim WJ (2012) J Drug Target 20:648
221. Ghosh PS, Kim C-K, Han G, Forbes NS, Rotello VM (2008) ACS Nano 2:2213
222. YoungáChoi S, YeongáLee S (2011) J Mater Chem 21:13853
223. Yan X, Blacklock J, Li J, Möhwald H (2011) ACS Nano 6:111
224. Ramos J, Rege K (2013) Mol Pharm 10:4107

225. Ramos J, Rege K (2012) Biotechnol Bioeng 109:1336
226. Wagner E (2011) Acc Chem Res 45:1005
227. Hu C, Peng Q, Chen F, Zhong Z, Zhuo R (2010) Bioconjug Chem 21:836
228. Elbakry A, Zaky A, Liebl R, Rachel R, Goepferich A, Breunig M (2009) Nano Lett 9:2059
229. Song W-J, Du J-Z, Sun T-M, Zhang P-Z, Wang J (2010) Small 6:239
230. Guo S, Huang Y, Jiang Q, Sun Y, Deng L, Liang Z, Du Q, Xing J, Zhao Y, Wang PC (2010) ACS Nano 4:5505
231. Wang F, Shen Y, Zhang W, Li M, Wang Y, Zhou D, Guo S (2014) J Control Release 196:37
232. Lee M-Y, Park S-J, Park K, Kim KS, Lee H, Hahn SK (2011) ACS Nano 5:6138
233. Bonoiu AC, Bergey EJ, Ding H, Hu R, Kumar R, Yong K-T, Prasad PN, Mahajan S, Picchione KE, Bhattacharjee A (2011) Nanomedicine 6:617
234. Lee SK, Han MS, Asokan S, Tung CH (2011) Small 7:364
235. Uchida M, Li XW, Mertens P, Alpar HO (2009) Biochim Biophys Acta 1790:754
236. Martin-Ortigosa S, Wang K (2014) Transgenic Res 23:743
237. Zhu Y, Liao L (2015) J Nanosci Nanotechnol 15:4753
238. Kim CK, Ghosh P, Pagliuca C, Zhu ZJ, Menichetti S, Rotello VM (2009) J Am Chem Soc 131:1360
239. Gibson JD, Khanal BP, Zubarev ER (2007) J Am Chem Soc 129:11653
240. Mohammad F, Yusof NA (2014) J Colloid Interface Sci 434:89
241. You J, Zhang R, Zhang G, Zhong M, Liu Y, Van Pelt CS, Liang D, Wei W, Sood AK, Li C (2012) J Control Release 158:319
242. You J, Zhang G, Li C (2010) ACS Nano 4:1033
243. Hartwell BL, Antunez L, Sullivan BP, Thati S, Sestak JO, Berkland C (2015) J Pharm Sci 104:346
244. Skirtach AG, Javier AM, Kreft O, Köhler K, Alberola AP, Möhwald H, Parak WJ, Sukhorukov GB (2006) Angew Chem Int Ed 45:4612
245. del Mercato LL, Rivera-Gil P, Abbasi AZ, Ochs M, Ganas C, Zins I, Sönnichsen C, Parak WJ (2010) Nanoscale 2:458
246. Carregal-Romero S, Ochs M, Parak WJ (2012) Nanophotonics 1:171
247. Muñoz Javier A, Pino P, Bedard M, Skirtach AG, Ho D, Sukhorukov G, Plank C, Parak WJ (2009) Langmuir 24:12517
248. De Geest BG, De Koker S, Sukhorukov GB, Kreft O, Parak WJ, Skirtach AG, Demeester J, De Smedt SC, Hennink WE (2009) Soft Matter 5:282
249. Choi KY, Liu G, Lee S, Chen X (2012) Nanoscale 4:330
250. Wang C, Li J, Amatore C, Chen Y, Jiang H, Wang X-M (2011) Angew Chem Int Ed 50:11644
251. Modugno G, Ménard-Moyon C, Prato M, Bianco A (2015) Br J Pharmacol 172:975
252. Xu C, Yang D, Mei L, Lu B, Chen L, Li Q, Zhu H, Wang T (2013) ACS Appl Mater Interfaces 5:2715
253. Xu C, Yang D, Mei L, Li Q, Zhu H, Wang T (2013) ACS Appl Mater Interfaces 5:12911
254. Ma X, Qu Q, Zhao Y, Luo Z, Zhao Y, Ng KW, Zhao Y (2013) J Mater Chem B 1:6495
255. Sanchez-Barcelo EJ, Mediavilla MD (2014) Recent Pat Endocr Metab Immune Drug Discovery 8:1
256. Menon JU, Jadeja P, Tambe P, Vu K, Yuan B, Nguyen KT (2013) Theranostics 3:152
257. Huang X, Jain P, El-Sayed I, El-Sayed M (2008) Lasers Med Sci 23:217
258. Pattani VP, Tunnell JW (2012) Lasers Surg Med 44:675
259. Mackey MA, Ali MR, Austin LA, Near RD, El-Sayed MA (2014) J Phys Chem B 118:1319
260. Lin KY, Bagley AF, Zhang AY, Karl DL, Yoon SS, Bhatia SN (2010) Nano Life 1:277
261. Chen J, Glaus C, Laforest R, Zhang Q, Yang M, Gidding M, Welch MJ, Xia Y (2010) Small 6:811
262. Jang B, Park J-Y, Tung C-H, Kim I-H, Choi Y (2011) ACS Nano 5:1086
263. Kong T, Zeng J, Wang X, Yang X, Yang J, McQuarrie S, McEwan A, Roa W, Chen J, Xing JZ (2008) Small 4:1537
264. Yang Y-S, Carney RP, Stellacci F, Irvine DJ (2014) ACS Nano 8:8992

265. Geng F, Xing JZ, Chen J, Yang R, Hao Y, Song K, Kong B (2014) J Biomed Nanotechnol 10:1205
266. Hainfeld JF, Lin L, Slatkin DN, Dilmanian FA, Vadas TM, Smilowitz HM (2014) Nanomed Nanotechnol Biol Med 10:1609–1617
267. Lin Y, McMahon SJ, Scarpelli M, Paganetti H, Schuemann J (2014) Phys Med Biol 59:7675
268. Chanda N, Kan P, Watkinson LD, Shukla R, Zambre A, Carmack TL, Engelbrecht H, Lever JR, Katti K, Fent GM (2010) Nanomed Nanotechnol Biol Med 6:201
269. Khan MK, Minc LD, Nigavekar SS, Kariapper MS, Nair BM, Schipper M, Cook AC, Lesniak WG, Balogh LP (2008) Nanomed Nanotechnol Biol Med 4:57
270. Kannan R, Zambre A, Chanda N, Kulkarni R, Shukla R, Katti K, Upendran A, Cutler C, Boote E, Katti KV (2012) Wiley Interdiscip Rev Nanomed Nanobiotechnol 4:42
271. Moran CH, Wainerdi SM, Cherukuri TK, Kittrell C, Wiley BJ, Nicholas NW, Curley SA, Kanzius JS, Cherukuri P (2009) Nano Res 2:400
272. Glazer ES, Curley SA (2010) Cancer 116:3285
273. Glazer ES, Curley SA (2011) Ther Deliv 2:1325
274. Gannon CJ, Patra CR, Bhattacharya R, Mukherjee P, Curley SA (2008) J Nanobiotechnol 6:1
275. Curley SA, Cherukuri P, Briggs K, Patra CR, Upton M, Dolson E, Mukherjee P (2007) J Exp Ther Oncol 7:313
276. Mukherjee P, Bhattacharya R, Wang P, Wang L, Basu S, Nagy JA, Atala A, Mukhopadhyay D, Soker S (2005) Clin Cancer Res 11:3530
277. Pan Y, Wu Q, Qin L, Cai J, Du B (2014) BioMed Res Int 2014:1
278. Arvizo RR, Rana S, Miranda OR, Bhattacharya R, Rotello VM, Mukherjee P (2011) Nanomed Nanotechnol Biol Med 7:580
279. Tsai C-Y, Shiau A-L, Chen S-Y, Chen Y-H, Cheng P-C, Chang M-Y, Chen D-H, Chou C-H, Wang C-R, Wu C-L (2007) Arthritis Rheum 56:544
280. Muthu MS, Leong DT, Mei L, Feng S-S (2014) Theranostics 4:660
281. Akhter S, Ahmad MZ, Ahmad FJ, Storm G, Kok RJ (2012) Expert Opin Drug Deliv 9:1225
282. Rai P, Mallidi S, Zheng X, Rahmanzadeh R, Mir Y, Elrington S, Khurshid A, Hasan T (2010) Adv Drug Deliv Rev 62:1094
283. Khlebtsov B, Panfilova E, Khanadeev V, Bibikova O, Terentyuk G, Ivanov A, Rumyantseva V, Shilov I, Ryabova A, Loshchenov V (2011) ACS Nano 5:7077
284. Huang P, Bao L, Zhang C, Lin J, Luo T, Yang D, He M, Li Z, Gao G, Gao B (2011) Biomaterials 32:9796
285. Conde J, Rosa J, Jesús M, Baptista PV (2013) Biomaterials 34:2516
286. Lukianova-Hleb EY, Oginsky AO, Samaniego AP, Shenefelt DL, Wagner DS, Hafner JH, Farach-Carson MC, Lapotko DO (2011) Theranostics 1:3
287. Shi H, Ye X, He X, Wang K, Cui W, He D, Li D, Jia X (2014) Nanoscale 6:8754
288. Wagner DS, Delk NA, Lukianova-Hleb EY, Hafner JH, Farach-Carson MC, Lapotko DO (2010) Biomaterials 31:7567
289. Lukianova-Hleb EY, Mutonga MB, Lapotko DO (2012) ACS Nano 6:10973
290. National Institutes of Health Clinical Center (CC), National Cancer Institute (NCI) (2012) http://www.clinicaltrials.gov; identifier: NCT00356980
291. Libutti S, Paciotti G, Myer L, Haynes R, Gannon W, Walker M, Seidel G, Byrnes A, Yuldasheva N, Tamarkin L (2009) ASCO annual meeting proceedings 27:3586
292. National Institutes of Health Clinical Center (CC), National Cancer Institute (NCI) (2012) http://www.clinicaltrials.gov; identifier: NCT00436410
293. Nanospectra Biosciences (2014) http://www.clinicaltrials.gov; identifier: NCT00848042
294. Nanospectra Biosciences (2012–to date) http://www.clinicaltrials.gov; identifier: NCT01679470
295. Stegh AH (2012–to date) http://www.nu-ccne.org/Research_project4.htm

Top Curr Chem (2016) 370: 203–224
DOI: 10.1007/978-3-319-22942-3_7
© Springer International Publishing Switzerland 2016

Diagnostic and Therapeutic Applications of Quantum Dots in Nanomedicine

Sukanta Kamila, Conor McEwan, David Costley, Jordan Atchison, Yinjie Sheng, Graham R.C. Hamilton, Colin Fowley, and John F. Callan

Abstract The interest in Quantum Dots as a class of nanomaterials has grown considerably since their discovery by Ekimov and Efros in the early 1980s. Although this early work focussed primarily on CdSe-based nanocrystals, the field has now expanded to include various classes of nanoparticles with different types of core, shell or passivation chemistry. Such differences can have a profound effect on the optical properties and potential biocompatibility of the resulting constructs. Although QDs have predominantly been used for imaging and sensing applications, more examples of their use as therapeutics are beginning to emerge. In this chapter we discuss the progress made over the past decade in developing QDs for imaging and therapeutic applications.

Keywords Cancer • Fluorescence • Microbiology • Photodynamic therapy • Quantum dots • ROS

Contents

S. Kamila, C. McEwan, D. Costley, J. Atchison, Y. Sheng, G.R.C. Hamilton, C. Fowley, and J.F. Callan (✉)
Biomedical Sciences Research Institute, University of Ulster, Coleraine, Northern Ireland BT52 1SA, UK
e-mail: j.callan@ulster.ac.uk

1 Introduction

Quantum Dots (QDs) are a unique class of fluorescent nanoparticles with superior optical properties when compared to organic dyes [1–3]. Although several different classes of QDs have emerged over the past couple of decades, by far the most commonly used QDs are those prepared from semiconductor materials [4]. These QDs typically consist of a group IIB metal, usually cadmium, combined with a chalcogenic element, e.g. CdS, CdSe or CdTe, with all three nanocrystal dimensions in the 2–10 nm range [5]. Although possessing the same crystal structure as the bulk semiconductor material, QDs consist of only a few hundred to a few thousand atoms. At such sizes (smaller than the exciton Bohr radius of the bulk material) these nanoparticles behave differently from the bulk solid because of size quantization effects [6]. The result of quantum confinement means that the electron and hole (the exciton) energy levels, and therefore the band gap, is a function of the QD diameter and composition. As a result, increasing the size of the QD decreases its band gap and, as a consequence, increases its emission wavelength. The combination of size-tuneable emission with broad absorption spectra means it is possible to excite several QDs simultaneously, a feature that enables multiplexing applications [7].

A significant volume of work has been undertaken in the past two decades to develop new classes of QDs (1) with even better optical properties, (2) that are prepared from more convenient and greener synthetic methods and (3) that provide improved biocompatibility when considering biological applications (see www.invitrogen.com). Although it is not possible to cover the whole body of literature available on these developments, this review chapter provides representative examples of how QDs can been used in biomedical imaging or therapeutic applications. Although the particular focus is on the use of these nanoparticles in cancer diagnosis and treatment, other areas such as medical microbiology are also discussed.

2 Biomedical Imaging Using QDs

2.1 *Passive Targeting of QDs to Tumours for Imaging Applications*

In biology, passive targeting refers to the ability of a molecule or particle to accumulate in a particular physiological location without the assistance of a targeting ligand or the application of a stimulus [8]. In essence, the process is an inherent characteristic of the molecule/particle itself and is not derived through pre-determined structural modification. In the context of nanoparticles, their size means they undergo passive targeting to solid tumours via a phenomenon known as the "enhanced permeability and retention" (EPR) effect [9]. This process refers to the observation that nanoparticles, such as QDs, preferentially accumulate at tumour sites because of the latter's characteristic abnormal vasculature enabling QDs to leak from the blood vessel into the tumour [10]. In combination with a compromised lymphatic system that impedes the removal of these particles, the end result is a greater concentration of nanoparticles in tumour tissue compared to healthy tissue.

Gao et al. have investigated the potential of NIR-emitting (800 nm) InAs/InP/ZnSe QDs passivated with mercaptopropionic acid (QD-MPA) to accumulate in both squamous cell carcinoma (22B) and colorectal cancer (LS174T) tumours via the EPR effect [11]. These nanoparticles, with a hydrodynamic diameter of 8.2 nm, were injected intravenously and in vivo NIR images were recorded 4 h later. A high fluorescent intensity was observed in the tumour region with good contrast compared to other tissue illustrating effective tumour uptake. When the individual organs were harvested and subsequently imaged ex vivo 4 h post treatment, the fluorescence observed in the kidneys and bladder was high, illustrating the potential of these nanoparticles to undergo effective renal clearance. However, significant fluorescence intensity was also observed in the liver. The uptake of the QD-MPA were also compared to commercially available 800 nm-emitting carboxyl functionalised QDs (QD-COOH) with a hydrodynamic diameter of 25 nm. In contrast to the QD-MPA, the QD-COOH demonstrated poor tumour accumulation with a higher intensity observed from the liver than for the QD-MPA, with QD fluorescence also observed in the bone marrow and spleen. The lower tumour accumulation and high non-specific uptake by the reticuloendothelial system for the QD-COOH compared to QD-MPA was attributed to the larger size of the QDs. In an attempt to reduce uptake by the liver, the negatively charged QD-MPA were further coated with human serum albumin (HSA) via electrostatic interactions to decrease the probability of engulfment by macrophage cells and reduce the accumulation of the nanoparticles in the mononuclear phagocytic system (MPS) organs such as the liver and spleen. Although the hydrodynamic diameter of these nanoparticles increased to 17 nm, they still showed effective tumour uptake, and their content in the liver and spleen was substantially reduced compared to the unmodified QD-MPA. These results demonstrate the ability of QDs to accumulate effectively in tumours as a result of the EPR effect but also emphasises the importance of physical and chemical properties such as size and surface chemistry in dictating their effective clearance from the body and potential long term toxicity.

2.2 *Active Targeting of QDs to Tumours for Imaging Applications*

In contrast to passive targeting, active tumour targeting involves the conjugation of a specific ligand to the surface of the QD whose function is to bind to a receptor that is either unique or significantly overexpressed in target cells compared to normal

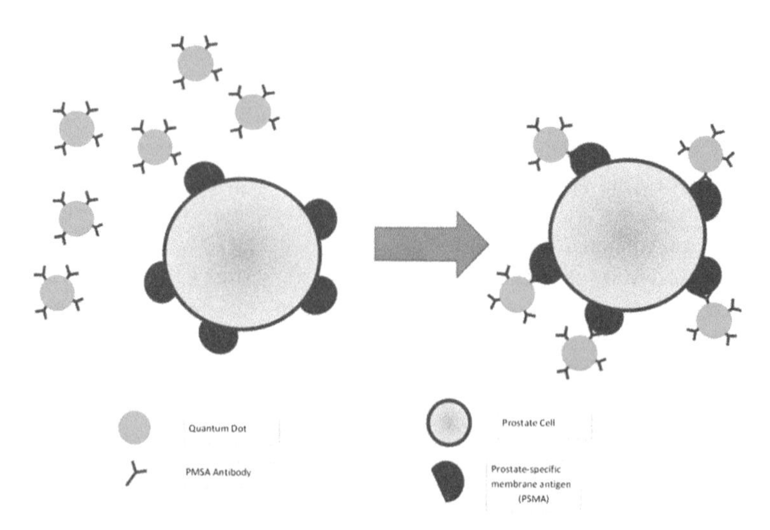

Fig. 1 Schematic representation of illustrating how QD-antibody conjugates (*green*) target prostate cells (*light blue*) via binding to the prostate-specific membrane antigen (*dark blue*) [12]

cells [12]. Many different types of targeting ligands have been combined with QDs to enhance their uptake by specific cells or cellular compartments. Off course, decorating the surface of the QD with targeting ligands does not remove the passive targeting capability of QDs but instead enhances the specificity of the resulting ensemble. In the following sections we discuss representative examples of how antibodies, peptides and oligonucleotides have been combined with QDs to improve targeting to specific cells.

2.2.1 QD-Antibody Conjugates

The specificity of antibodies for target proteins is unrivalled by low molecular mass ligands with many QD-antibody combinations having been reported in the literature and some now commercially available [13, 14]. In an early example, Gao et al. reported the preparation of PEG-coated CdSe/ZnS QDs conjugated with a prostate-specific membrane antigen (PSMA) antibody [15]. The QD-Ab conjugate demonstrated effective uptake in C4-2 cells which are known to express PSMA but displayed poor uptake in PMSA negative PC-3 cells. When administered intravenously to mice bearing C4-2 tumours, rapid accumulation of the QD-Ab conjugate was observed in the tumour with accumulation also observed in the liver and spleen. In contrast, control QDs lacking the PSMA specific antibody demonstrated poor tumour uptake. A similar study by Chen et al. utilised CdSe/ZnS QDs decorated with an alpha-fetoprotein (AFP) antibody which is a biomarker for hepatocellular carcinoma [16]. Again, the QD-Ab conjugates showed a substantially higher degree of tumour accumulation compared to the unmodified QDs (Fig. 1).

Instead of using antibodies with specificity for a target protein associated with a single cell type, antibodies with broad applicability to many different cell types are also available [17]. Epidermal growth factor receptor (EGFR) is involved in a number of biological processes, such as cell proliferation, angiogenesis and metastasis [18]. Although EGFR is ubiquitously expressed in normal tissues it is overexpressed in many cancers such as lung, breast and pancreatic cancer [19]. Diagaradjane et al. conjugated human recombinant EGF to 800 nm-emitting CdSeTe/ZnS QDs to enable selective targeting of human colon carcinoma xenografts in mice [20]. In vivo imaging demonstrated a rapid uptake of both the unmodified and EGF-labelled conjugate 60 min after administration. After this time the fluorescent intensity from tumours in animals treated with the unmodified QDs reduced significantly although the QD-EGF emission remained intense for up to 6 h.

2.2.2 QD-Peptide Conjugates

Oligopeptides with amino acid sequences complementary for binding to important biomolecules such as antibodies or enzymes can also be utilised as targeting entities for QDs [21]. One advantage of this approach is that oligopeptides are much smaller than antibodies and their attachment to QDs reduces the overall size of the resulting conjugate. Integrins are transmembrane receptors involved in communication processes between cells and their extracellular matrix [22]. The integrin $\alpha_v\beta_3$ plays an important role in angiogenesis and binds to arginine-glycine-aspartate acid (RGD) regions of the interstitial matrix such as vitronectin [23]. It is significantly upregulated in many different types of cancer and so has emerged as a popular targeting strategy in drug design [24]. Cai et al. conjugated RGD to 705 nm-emitting CdSe/ZnS QDs and investigated the QD-RGD conjugates' ability to bind to integrin positive (MDA-MB-435 and U87MG) and negative (MCF-7) cell lines [25]. Using fluorescent microscopy, both MDA-MB-435 and U87MG cells were clearly visualised when stained with the QD-RGD conjugates although the MCF-7cells were not. When the QD-RGD conjugate and unmodified QDs were administered intravenously to nude mice bearing ectopic U87MG tumours, a significant difference was observed in the intensity of the tumour fluorescence in those mice receiving the conjugate relative to non-labelled QDs. Using a similar approach, Gao et al. used 800 nm-emitting InAs/InP/ZnSe QDs with RGD and again found improved targeting to $\alpha_v\beta_3$-positive U87MG human tumours when compared to the unmodified QDs [26].

Peptide neurotoxins have emerged as high affinity ligands with both diagnostic and therapeutic potential [27]. Because of their high specificity for neuronal receptors, these peptides can greatly assist in our understanding of complex neurological processes [28, 29]. In addition, peptide neurotoxins may also play an important role in the treatment of neurological disease with compounds such as Ziconotide, commercially available as an N-type voltage-gated calcium channel blocker [30]. Orndoff et al. utilised chlorotoxin (CTX) and dendrotoxin-1 (DTX-1)

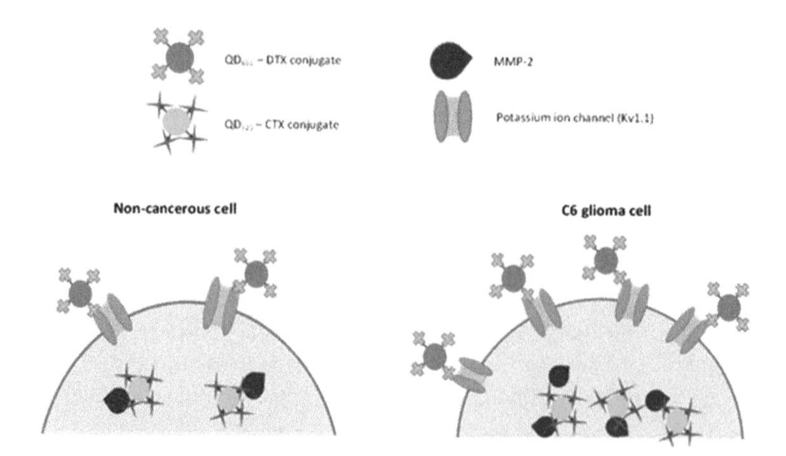

Fig. 2 Overexpression of the MMP-2 and Kv1.1 proteins in C6 glioma cells can be detected using QD-CTX and QD-DTX conjugates, respectively. More QDs bind to cells overexpressing these proteins, leading to a stronger fluorescence signal compared to non-cancerous cells [31]

as targeting ligands to image cancer cells selectively [31]. These peptides were conjugated to 525 nm- or 655 nm-emitting CdSe/ZnS QDs, respectively, and the resulting QD_{525}-CTX and QD_{655}-DTX conjugates used to detect simultaneously their respective targets in a multiplex manner. CTX is known to target MMP-2 that is involved in glioma cell proliferation whereas DTX targets potassium ion channels of the subtype Kv1.1, also present in glioma cells. By incubating living C6 glioma cells with the QD conjugates, the authors were able to label MMP-2 and Kv1.1 successfully and demonstrated that levels of these receptors were to be four times higher in cancer cells than non-cancer cells. The combination of high affinity peptide neurotoxins with QDs offers the possibility of long term detection of potential disease state markers because of the reduced photobleaching offered by QDs (Fig. 2).

2.2.3 QD-Oligonucleotide Conjugates

Molecular beacons (MBs) are commonly used in biology to probe for specific nucleotide sequences. The sequence-driven complementarity of nucleotides makes them highly selective ligands for target DNA sequences [32, 33]. Conventionally, the MB contains an oligonucleotide sequence complementary to the sequence being targeted with a fluorophore attached at the 5′ end and a quencher dye at the 3′ end. In the absence of the target sequence, Förster Resonance Energy Transfer (FRET) occurs between the fluorophore and quencher so that fluorescence remains low. However, upon hybridisation with the target sequence, the separation distance between the fluorophore and the quencher increases so that FRET is no longer possible and fluorescence is switched on. FRET involves the non-radiative transfer of excitation energy from an energy donor to the ground state of a nearby acceptor molecule through a dipole-dipole interaction. The efficiency of energy transfer between the two fluorophores is governed by two main factors: (1) the distance between the donor and acceptor molecules and (2) a good spectral overlap between the donor's emission and acceptor's absorption spectra. In addition to these criteria, it is essential that the dipoles of both donor and acceptor molecules are parallel to each other for efficient energy transfer to occur. In spectroscopic terms, an increase in FRET efficiency manifests itself in a reduction in the donor fluorophore emission with a concomitant increase in the emission from the acceptor fluorophore. As there is a sixth power dependence of the FRET efficiency on the donor-acceptor separation distance, small changes in this separation can manifest itself in a significant modulation of the acceptor emission. Thus, FRET has earned the term "spectroscopic ruler" as it can be used to probe separation distances between 10 and 100 Å in biomolecules and to provide information on their conformational arrangement in different environments or upon binding a target molecule (Fig. 3).

Although organic dye fluorophores have traditionally been the favourite choice in the assembly of MBs, QDs are increasingly being used as alternatives because of their impressive photophysical properties. One early example by Ozkan et al. attached mercaptoacetic acid-coated CdSe/ZnS QDs to the 5′ end of an MB whose 3′ end was attached to a organic quencher molecule 4-(4′-dimethylamino-phenylazo)benzoic acid (DABCYL) [35]. The FRET efficiency of the QD/DABCYL MBs was investigated and compared with all organic MBs containing a 6-FAM fluorophore with DABCYL quencher. Three-dimensional molecular modelling studies using AMBER force field calculations predicted the QD-quencher centre to centre distances to be slightly greater than the 6-FAM—quencher pair with a concomitant decrease in FRET efficiency. This translated to lower fluorescence enhancement for the QD-MB (sixfold) compared to the all organic MB (tenfold) upon hybridisation with the complementary sequence. However, a major benefit of the QD-based MBs was their high resistance to photobleaching with practically no loss in optical performance after 10 min irradiation compared to a 15% drop in intensity for the 6-FAM analogue.

One limitation of conventional molecular beacons is high background fluorescence produced by incomplete quenching as a result of a single donor–acceptor pair. In an attempt to overcome this, Wang and co-workers cleverly designed a quencher-free FRET system based on emissive QDs and the organic fluorophore Cy5 [34]. Two single-stranded DNA sequences complementary to the target oligonucleotide were used, one labelled with Cy5 at the 5′ end and the second with biotin at the 3′ end. These two derivatised sequences were incubated with streptavidin-coated CdSe/ZnS QDs. In the absence of the target DNA strand the biotin-labelled complementary sequence can bind with the streptavidin-coated QDs which cause no distortion of the QD fluorescence. However, when the target DNA sequence is present, the complementary sequences hybridise, producing a hybrid with the biotin and Cy5 fluorophore at opposite ends. This hybrid can also bind to the QDs through the biotin—streptavidin interaction, and given that the emission spectrum of the QDs at 605 nm overlaps well with the Cy5 absorption spectrum, FRET occurs

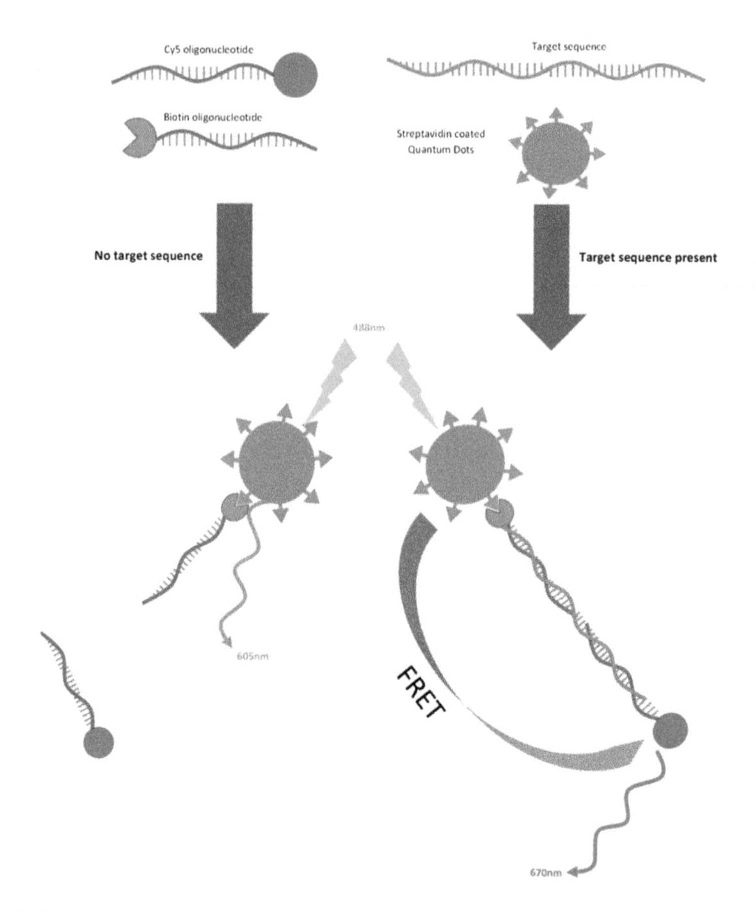

Fig. 3 Schematic representation of how a FRET system can be utilised for the detection of specific nucleotide sequences. Strands with complementary sequences to the target were conjugated to Biotin and a fluorophore (Cy5) at opposite ends. If the target sequence was not present then only the biotin conjugated 3′ end attaches to the QD and does not affect its fluorescence properties. However, when the target sequence was present, both complementary strands can hybridise, bringing the Cy5 attached to the 5′ strand into close proximity to the QD, thus allowing FRET to occur [34]

between the QD and Cy5 with a resulting decrease in QD emission and enhancement of Cy5 emission. The authors compared the effectiveness of their sensor to a commercially available molecular beacon and found that a 100-fold increase in sensing responsitivity was observed for the QD probe.

Given the similarity in size between gold nanoparticles (AuNPs) and QDs, coupled with the quenching properties of the former, it is not surprising that several groups have utilised the supramolecular association of these two distinct nanoparticles in a sensing modality. For example, Melvin and co-workers have conjugated CdSe/ZnS QDs to the 5′ end of single stranded DNA (5′-TGC AGA TAG ATA GCA G-3′) by an amide bond linkage using carbodiimide mediated coupling. The complementary strand (5′-CTG CTA TCT ATC TGC-3′) was attached to AuNP, again using an amide bond linkage [36]. When the AuNP-DNA was hybridised with the QD-DNA, an 85% quenching of the QD fluorescence intensity was observed, which could be recovered when an excess (10 equiv,) of the unlabelled strand (i.e. pure 5′-CTG CTA TCT ATC TGC-3′) was added to the mixture. Although this clearly demonstrates the reversibility of the system, the time taken (3.5 h in total) limits its use practically. However, the authors were successful in attaching the QD-DNA complex to a glass surface and observe a quenching effect similar to that observed in solution upon addition of the Au-labelled complementary sequence.

As mentioned previously, QDs are ideal for use in multiplex experiments because of their broad absorption spectra and narrow size dependent emission spectra. Krull and co-workers exploited these advantageous properties when they used two different sized mercaptoacetic acid-coated CdSe/ZnS QDs emitting in the green and red regions of the visible spectrum [37]. The green-emitting QDs were attached to a nucleotide sequence complementary to the sequence diagnostic for spinal muscular dystrophy using standard EDC coupling techniques. The larger red-emitting QDs were attached to a nucleotide sequence complementary to the sequence diagnostic for *Escherichia coli*. In a slightly different approach to those outlined above, the target sequences themselves were labelled with organic dyes whose absorption spectra overlapped with those of the respective QDs (Cy3 and Alexa 647 for the green and red QDs, respectively). The addition of the labelled target sequences resulted in a concentration-dependent increase in the acceptor emission caused by FRET from the QD. This system enabled the detection of both target sequences simultaneously in the same solution using dual wavelength detection. However, the sensitivity of the dual nucleotide detection system relative to single nucleotide detection was significantly less, a feature attributed to non-specific absorption. Another hurdle that must be overcome when designing multiplex systems containing more than one dye is optimizing the ratio of dyes per nanoparticle produced by different fluorescence quantum yields of each dye (and possibly QD). Krull and co-workers subsequently demonstrated an improved version of their work when they immobilised the two different coloured QD-nucleic acid probes on optical fibres [38].

2.3 Sentinel Lymph Node Imaging

A sentinel lymph node (SLN) is defined as the first lymph node to which cancer cells are most likely to spread from a primary tumour [39]. Therefore, SLN biopsy is an important predictor of metastasis in cancer patients [40–42]. Fluorescent imaging using near-infrared-emitting dyes such as indocyanine green (ICG) have proven useful in the real time analysis of SLN [43]. Ballou et al. were the first to investigate the possibility of using QDs to map SLN in mice [44]. Using two ectopic tumour models (M21 human melanoma and MH-15 mouse teratocarcinoma) and CdSe/ZnS QDs with three different surface chemistries (i.e. positive, negative and neutral charge) they imaged the lymphatic drainage of the QDs following intratumoral injection. The tumours were located in the right thigh of the mice so that they would drain into the inguinal node which is a large and superficial node which can easily be imaged in vivo. All three types of QDs were found to drain effectively from the tumour to the inguinal node, illustrating that QD surface charge was not an important factor in dictating effective drainage.

Pons et al. have investigated the use of NIR-emitting cadmium free $CuInS_2/ZnS$ QDs to image two regional lymph nodes in healthy mice following sub-cutaneous injection and compared the relative toxicity of these QDs to cadmium-containing CdTeSe/CdZnS QDs by measuring the inflammatory response following injection [45]. The two regional lymph nodes could easily be visualised minutes after injection with a direct correlation between the dose administered and the observed intensity 24 h post injection. When the degree of inflammation in the right axillary lymph node and right lateral thoracic lymph node were measured for the both types of QDs, the onset of inflammation for the cadmium-free QDs occurred at a ten times higher dose than for the cadmium-containing QDs, indicating the former produced significantly less acute toxicity. In another study by Kosaka et al., 545 nm- and 645 nm-emitting CdSe/ZnS QDs were used to enable multiplex mapping of lymph fluid tracking in nude mice following interstitial injection of the QDs into the chin or ear of the mouse [46]. The different injection sites resulted in accumulation of the QDs in separate lymph nodes which could be imaged in real time using fluorescence microscopy because of the different coloured QDs being excited using the same excitation wavelength.

2.4 Use of QDs in Medical Microbiology

Many detection methods for bacteria and other micro-organisms employ the use of immunofluorescent antibodies (IFA) in an ELISA type approach [47]. Not unsurprisingly, QDs have also been investigated as an alternative fluorophore to conventional organic dyes for such applications given their advantageous optical properties [48, 49]. Zhu et al. utilised the non-covalent avidin–biotin interaction to attach 565 nm- and 605 nm-emitting CdSe/ZnS QDs to antibodies specific for *Cryptosporidium parvum* and *Giardia lamblia*, respectively [50]. Protozoa were fixed onto slides and incubated with the QD-Ab conjugates for 30 min at 37 °C. After washing to remove unbound conjugates, the slide was then imaged using fluorescence microscopy. The results demonstrated the QD-Ab conjugates could effectively label their respective bacterial cells with signal to noise ratios some 1.5–9 times greater than commercially available organic dye-based test kits. Furthermore, using dual-colour imaging, it was possible to discriminate between the two different bacterial cells when present in the same sample.

Following a similar strategy, Hahn and co-workers attached a biotinylated antibody specific for the pathogenic bacterium *E. coli* serotype 0157:H and compared its performance against a commercially available fluorescein isothiocyanate

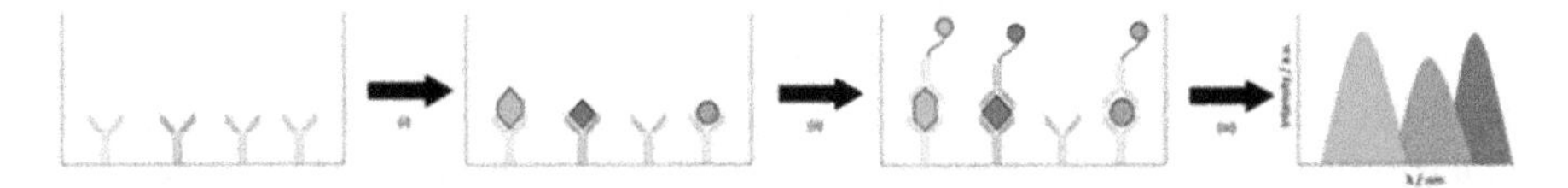

Fig. 4 Schematic illustration of a multiplexing fluorescence immunoassay using QDs with three distinct emission colours. Capture antibodies are fixed to the bottom of a well and (i) antigen mixture added followed with a wash step (ii) reporter antibody-QD conjugates added followed by wash step (iii) fluorescence intensities analysed of each specific QD

(FITC) *E. coli* detection kit [51]. The QD-based fluorescence immunoassay was found to more sensitive by two orders of magnitude compared to the FITC-based assay. In addition, the QD-Ab conjugate was shown to be highly selective for this particular serotype and showed little affinity for alternative serotypes (i.e. *E. coli* serotype DH5α). Using a combination of magnetic microparticles with three capture antibodies attached (specific for *S. typhimurium*, *S flexneri* and *E. coli* O157: H7) and three different sized CdTe QDs (with emission maxima at 520, 560 and 620 nm), each with a reporter antibody corresponding to the capture antibody present on the magnetic beads, Zhao et al. demonstrated multiplex detection of the three bacteria in a complex mixture [52]. Using a magnet to fix the microparticle-capture antibody to the bottom of a micro-centrifuge tube, captured bacteria were identified when the QD reporter conjugates were added. Indeed, the intensity of the signals at 525, 570 and 625 nm were observed to increase linearly when each of the target bacterial strains increased in number from 10^3 to 10^6 cfu. Such multiplexing capability would be difficult if not impossible to achieve using conventional dyes because of different excitation wavelengths required for three different dyes, not to mention the potential for overlap of their typically broader emission spectra (Fig. 4).

As an alternative to the use of antibodies, Edgar et al. attached CdSe/ZnS QDs to T7 bacteriophage particles, which are specific for *E. coli* [53]. The phage was engineered to express a biotinylated capsid protein that would allow the attachment of streptavidin coated QDs. The method enabled highly sensitive detection of bacteria (10 bacteria per mL) and displayed a 100-fold amplification in the signal to background ratio after 1 h, with no evidence of photobleaching. Although not undertaken in this study, the authors suggest the possibility of detecting multiple bacterial strains using their specific phage combined with different emitting QDs in a multiplex arrangement.

3 Therapeutic Applications Involving QDs

3.1 QD-Based Theranostics

We open this section with examples where QDs have been combined with cytotoxic drugs to produce ensembles with both therapeutic and diagnostic potential. The key features of QDs that make them appealing for use in such constructs are (1) their nanoparticle size means they exhibit the aforementioned EPR effect, (2) their high single- and two-photon extinction coefficient means they can be used to image to depth in mammalian tissue and (3) their broad absorption and size-tuneable emission spectra offer a wide selection of possible excitation and emission wavelengths for improved imaging.

The first example we discuss involves a report by Bagalkot et al., who covalently attached an amine terminated prostate specific RNA aptamer (A10 PSMA) to carboxylic acid functionalised 490 nm-emitting CdSe/ZnS QDs using carbodiimide based chemistry [54]. The potent anthracycline drug Doxorubicin (Dox) is known to intercalate with a single CG sequence present in this aptamer, forming a reversible 1:1 complex. Dox is also inherently fluorescent with an absorbance band overlapping the emission band of the 490 nm-emitting QDs, so incorporation of Dox within the QD bound aptamers resulted in quenched QD emission through a FRET process. However, intercalation of Dox within the aptamer also quenches Dox fluorescence. Therefore, although the excitation energy passed from the QD excites Dox, the latter's emission remains quenched when part of the QD-aptamer complex. The authors tested the complex for Dox release in both PMSA positive (LNCaP) and negative (Pc3) cell lines and demonstrated a significant difference in cytotoxicity between the LNCaP and Pc3 cell lines relative to Dox alone. Release of Dox from the complex was demonstrated by confocal microscopy 90 min after incubation, monitoring both the CQD and Dox emission, and was attributed to either physical dissociation of Dox from the conjugates or biodegradation of the PSMA aptamer by lysosomal enzymes.

Using a poly(methacrylate) comb type copolymer with both amine and PEG terminated side chains as a capping ligand, Wang et al. have developed a pH sensitive CdSe/ZnS—Dox conjugate for pH-mediated release of Dox [55]. The rationale for this approach is the knowledge that cancerous tissue typically has a lower pH than healthy tissue. Surprisingly, the authors used the lower affinity amine side chains (compared to higher affinity thiol groups) as a multidentate ligand for attachment of the polymer to the QD surface and subsequently attached Dox to the polymer-coated QDs using a pH sensitive imine bond. In contrast to Bagalkot et al.'s work described above, minimal FRET was observed between the QD and Dox produced by poor overlap between the absorbance of Dox and the longer emitting QD (λ_{max} 543 nm) used in this study. The QD-Dox system demonstrated improved Dox release (65%) after 10 h incubation in buffer at pH 5.0 whereas only 10% was released when the experiment was repeated at pH 7.4. However, the lack of cell-based and/or in vivo experiments was a major limitation of this study.

As an alternative to using CdSe/ZnS QDs as the fluorescent reporter, Zheng et al. used carbon quantum dots (CQDs) in a bid to overcome the perceived toxicity issues surrounding Cd-based nanoparticles for in vivo applications [56]. CQDs are more recent addition to the QD family and can be prepared from relatively cheap and readily available carbon precursors such as graphite, citric acid or glucose [57]. In contrast to their semiconductor counterparts, CQDs display excitation-dependent emission spectra where the emission maximum red-shifts with increasing excitation wavelength [58]. Although the explanation for such a phenomenon is not yet fully understood, these nanoparticles possess good fluorescence quantum yields, high two-photon absorption cross section and can be decorated with biocompatible ligands to make them suitable for biological applications [59]. Zheng et al. prepared CQDs from citric acid passivated with an amine terminated polymer [56]. Using carbodiimide-based coupling they attached a carboxylic acid functionalised platinum complex (oxaliplatin) to the CQD surface. Oxaliplatin is currently approved for the treatment of metastatic colorectal cancer but, as with most cancer chemotherapeutics, it suffers from significant off-target toxicity. Confocal fluorescence imaging, using both single- and two-photon excitation, demonstrated that the CQD-oxaliplatin conjugate crossed the membrane of living HeLa cells and localised in the cytosol, although the cytotoxicity of the conjugate in liver carcinoma cells (HepG2) was comparable to free oxaliplatin alone. The conjugate also displayed good efficacy in vivo when administered by intratumoural injection to ectopic H22 tumours in mice with an 82% reduction in tumour volume 3 days after treatment, although in vivo imaging confirmed clearance of the conjugates from the tumours 24 h after treatment. Although the authors suggested the CQD-oxaliplatin complex remains stable in circulation but becomes activated by the reductive environment present in tumour cells, evidence for such a claim was not provided in this study. Indeed, although convincing reasons were put forward for using an intra-tumoural injection, it would have been interesting to determine the conjugates distribution in other organs following intravenous administration. Nonetheless, this work provides a nice example of the use of CQDs in combined imaging and therapeutic applications.

As an alternative to using QDs in conjunction with chemo agents, Fakhroueian and co-workers reported the use of zinc oxide QDs as chemotherapeutic agents in their own right [60]. Incubating both cancerous and non-cancerous cell lines with oleic acid coated ZnO QDs, this study reported an elevated cytotoxic effect in cancerous cell lines compared to the non-cancerous cell line tested. Using the breast cancer cell line MCF7 and the colon cancer cell line HT29, LC_{50} values of 10.66 and 5.75 μg/mL were obtained. However, concentrations of 200 μg/mL in the non-cancerous cell line MDBK failed to show a cytotoxic effect. The authors failed to report a definitive reason for these observations, although it was suggested reactive oxygen species may play a role. In addition to showing anti-cancer effects, these QDs also demonstrated anti-fungal and anti-bacterial properties.

3.2 *QDs for Use in Photodynamic Therapy*

Photodynamic therapy (PDT) is a clinical treatment that uses a combination of light, photosensitising drugs and molecular oxygen to kill diseased cells. Typically, a light source is used to excite an electron in the photosensitiser (PS) molecule from the ground state (S_0) to the first excited singlet state (S_1), as shown in Fig. 5.

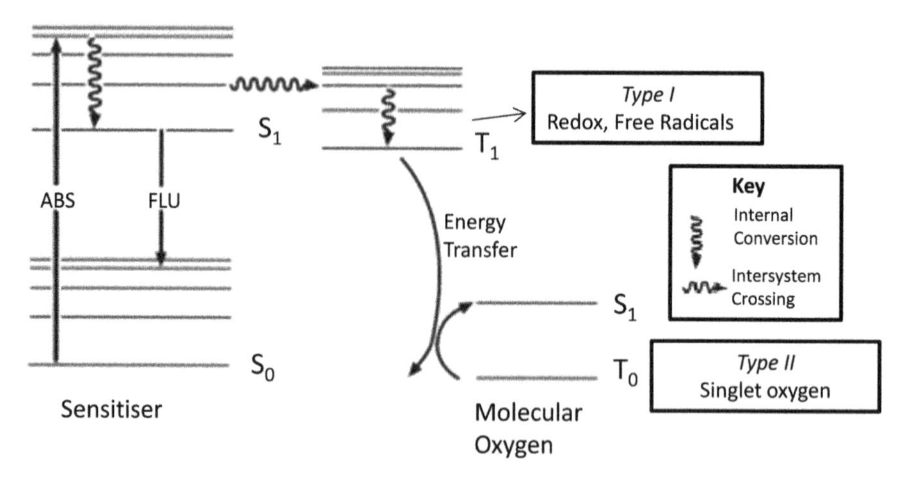

Fig. 5 Modified Jablonski diagram illustrating the key photophysical processes involved PDT

At this stage, the excited electron has several options. It can return to the ground state and emit its excited energy in the form of fluorescence (FLU), or it can engage in intersystem crossing, where the spin of the excited electron flips, resulting in the formation of a triplet excited state (T_1). Triplet states are much longer lived than singlet states and commonly lead to phosphorescent emission. However, triplet excited states may also engage in electron transfer (with biomolecules or water) or energy transfer (with molecular oxygen) to generate free radicals commonly referred to as Reactive Oxygen Species (ROS). ROS are potent intracellular cytotoxins and ultimately lead to cell death [61].

Since 1993, when the first PDT agent (Photofrin®) was approved for clinical use, there has been a dramatic increase in the use of PDT as a therapeutic treatment. Although traditionally considered as a treatment for skin cancers, PDT is now emerging as an option in the treatment of other cancers, such as head and neck, lung, oesophageal, prostate and bladder. It can also be used to treat certain non-cancerous conditions such as psoriasis [62]. However, there are several factors that have limited a more widespread use of PDT in the clinic. For instance, currently approved photosensitisers absorb in the visible region (below 700 nm), limiting light penetration to only a few millimetres. This is appropriate treatment for superficial tumours, but unsuitable for deeper seated tumours. Endogenous compounds, such as melanin or haem pigments, compete with PS for light absorption, meaning light intended for PS activation may be absorbed by natural pigments first and may not reach the target molecules in situ [63]. This interference can be reduced if PS are developed that absorb in the near-infrared (NIR) region where PS absorbance is much greater—in fact the penetration depth of light through the skin is four times greater (8 mm) at 800 nm than at 630 nm (2 mm) [64]. In addition, the PS should ideally possess no dark cytotoxicity, i.e. it should be non-toxic in ambient light. To date, following treatment with first generation photosensitisers, patients need to remain in low level lighting for several weeks to avoid excessive burning of the skin. Furthermore, the PS should be as selective as possible for targeted tissues to avoid damage of healthy cells. Otherwise, a lack of selectivity combined with poor visualisation of affected tissue means healthy tissue is also irradiated, leading to an increased prevalence of unwanted side effects [65].

In an attempt to improve the targeting of PS drugs for use in PDT, Choi et al. developed a CQD prepared from an α-cyclodextrin precursor passivated with poly (ethyleneglycol) diamine covalently attached to folic acid [66]. The overexpression of the folic acid receptor by certain cancers has been well documented. In addition to folic acid, the sensitiser zinc phthalocyanine (ZnPc) was loaded onto the folate-labelled CQDs using π–π stacking interactions between the sp^2 hybridised CQD surface and the aromatic sensitiser. The benefit of the targeting group was determined by incubating the conjugate in folate receptor positive (MDA-MB-31) and negative (FR-A549) cells and monitoring the fluorescence of both the CQD and ZnPc using confocal microscopy. Both CQD and ZnPc emission was observed from MDA-MB-31 cells with minimal emission observed from the FR-A549 cells, clearly illustrating effective folic acid mediated targeting. PDT experiments, conducted in HeLa cells which also overexpress the FR, revealed 91.8% and 39.1% reductions in viability for the conjugate and free ZnPc, respectively, following light irradiation, with the enhanced cytotoxicity of the conjugate being attributed to improved uptake resulting from the presence of the folic acid group. Intravenous administration of the conjugates in BALB/c nude mice bearing ectopic HeLa tumours showed enhanced ZnPc-mediated tumour fluorescence compared to conjugates lacking the folic acid group. Unfortunately, the authors failed to mention whether they attempted to measure the CQD fluorescence in vivo. Nonetheless, irradiation of the tumours with 660-nm light resulted in a significant inhibition of tumour growth for those mice treated with the conjugate relative to the controls. It should be mentioned that these results are for a single treatment and one would expect additional advantages following multiple treatment cycles as is common with conventional chemotherapy.

Although the previous examples utilised QDs primarily as fluorescent tracers, we, and other groups, are also investigating the potential of utilising QDs for use in two-photon excitation PDT (TPE-PDT) [67–69]. The improved tissue penetration capability and spatial resolution offered by two-photon excitation has led to it being investigated as a possible excitation source in PDT [68]. Indeed, it has previously been demonstrated in breast and lung cancer xenografts that it is possible to activate sensitisers to a depth of 2 cm using two-photon excitation [70]. However, because of their inability to absorb two-photon irradiation effectively, currently approved sensitisers, such as Photofrin, require very high two-photon excitation powers that are close to the threshold of tissue damage. In fact, Karotki et al. suggested that sensitisers would require two-photon absorption levels some two to three orders of magnitude greater than that of Photofrin to be clinically effective [71]. In contrast, QDs with two-photon absorption cross sections as high as 48,000 Göppert-Mayer units are easily excited with laser powers as low as 1 mW [72]. Unfortunately, the majority of available QDs are poor at generating ROS and so are unsuitable as potential sensitisers in PDT. However, it is possible to combine QDs with conventional sensitisers so that, following TPE of the QD, its excitation energy can be passed to the sensitiser which in turn becomes excited, generating ROS [73]. The end result is an indirect excitation of the sensitiser using two-photon wavelengths in the NIR where tissue transparency to light is greatest.

Although Samia et al. first demonstrated the indirect excitation of a ZnPc sensitiser by FRET from a CdSe QD using single photon excitation in 2003 [74], it was not until 9 years later that Fowley et al. illustrated that two-photon excitation could also be used to excite Rose Bengal indirectly following QD excitation [67]. The end-result was ROS generation which proved cytotoxic to HeLa cells. The same group later combined CQDs with protoporphyrin IX (PPIX) and again demonstrated it was possible to excite the PPIX indirectly by exciting the CQD with TPE at 800 nm [75]. HeLa cells incubated with the CQD-PPIX conjugate and treated with TPE at 800 nm reduced in viability by over 80% whereas cells maintained in the dark showed only a minor reduction. When these experiments were repeated using PPIX alone there was no difference in viability between cells maintained in the dark or those treated with TPE at 800 nm. These results revealed the importance of the CQD in absorbing the TPE and passing it to the PPIX in order for cytotoxicity to be observed. When C3H/HeN mice bearing ectopic RIF-1 tumours were treated with the CQD-PPIX conjugate and exposed to TPE at 800 nm for 3 min, they reduced by 60% from their original pre-treatment size 72 h after treatment whereas tumours treated with just the conjugate or just TPE treatment alone grew by greater than 60% over the same time period.

In a very promising development, Ge et al. have reported the use of graphene QDs (GQDs) as "stand-alone" sensitisers with extremely high $^1O_2$ quantum yields (greater than unity) and broad absorption spectra which extended into the red [76]. When HeLa cells were incubated with increasing concentrations of the GQDs (0.036–18 μM), minimal dark toxicity was observed across the entire concentration range. However, when the experiment was repeated with PPIX alone using the same concentrations, cell viability reduced to about 40% of its original value at concentrations of 1.8 μM or above. In addition, when cells treated with the GQDs were irradiated with a white light source for 10 min, cell viability reduced to 20% of the original value at a concentration of 0.18 μM, illustrating excellent light induced toxicity. Furthermore, when BALB/nu mice bearing GFP expressing MDA MB-231 breast cancer tumours were treated by intra-lesional injection of the GQDs followed by 10 min white-light treatment at 80 mW cm^{-2}, complete clearance of tumours was observed 21 days after treatment. In contrast, control mice receiving GQDs alone or light treatment alone grew significantly (>100%) over the same time period. These extremely impressive results can be attributed to the exceptionally high $^1O_2$ quantum yield of the GQDs ($\Phi^1O_2 = 1.33$), which is in contrast to other semiconducting or carbon-based QDs which are typically poor singlet oxygen generators. Although the authors put forward a multi-state mechanism (MSS) where $^1O_2$ generation was a result of energy transfer from both the T_1 and S_1 states of the GQD to the triplet ground state of molecular oxygen, a full explanation of the choice and role of the polythiophene carbon precursor was not provided. Nonetheless, the potential of such nanoparticles to the PDT field is significant.

The efficacies of the examples discussed above rely primarily on the availability of molecular oxygen as a substrate for ROS production and thus are heavily dependent on oxygen availability to enable effective PDT. However, hypoxia is a common feature associated with solid tumours and is a result of the atypical vasculature associated with growing tumours [77]. Essentially, this atypical vasculature results in the development of areas within the tumour where oxygen demand outstrips oxygen supply. Indeed, hypoxic fractions ranging from 10% to 30% are present in most solid tumours regardless of size [78]. Therefore, hypoxia also presents a significant challenge for the effective treatment of solid tumours using PDT, and several strategies are currently being investigated that (1) improve tumour oxygenation during PDT [79] or (2) utilise compounds that generate an oxygen-independent cytotoxic species upon light activation. With reference to the latter, Sortino's group have demonstrated the cytotoxic potential of nitric oxide radical (NO·) in several cancerous cell lines [80]. The production of NO radical is oxygen independent and has similar advantageous characteristics to that of singlet oxygen, i.e. it reacts with various biological substrates minimising the likelihood of resistance and its cytotoxic action is restricted to short distances from the site of its generation (<200 μm), reducing the likelihood of systemic toxicity [81]. However, the excitation wavelength required to activate such NO photodonors is in the blue region of the visible spectrum, where the penetration of light through human tissue is restricted to only a few millimetres [82]. To overcome this problem, Fowley et al. have combined an amine functionalised nitroaniline compound with blue-emitting carboxylic acid-functionalised CQDs in a FRET format similar to that described previously for the CQD-PPIX conjugate [83]. They demonstrated NO radical generation upon TPE of the CQD-nitroaniline conjugate at 800 nm in cell free solution and confirmed the cytotoxic potential of the NO radical generation in HeLa cells. To prove the effectiveness of their approach in vivo, human pancreatic BxPC-3 tumours were induced in Balb/c SCID mice, injected with conjugate and exposed to two-photon irradiation at 800 nm for 3 min. An 8% reduction in tumour volume was observed for mice treated with both the conjugate and TPE whereas mice treated with conjugate alone grew by 52% over the same time period. Although the magnitude of the tumour reduction observed may appear modest, the BxPC-3 tumours used were considerably hypoxic, exhibiting a pO_2 of 2 mmHg. Therefore, the use of such oxygen independent photo-generated radicals may complement ROS in the treatment of solid tumours using PDT.

4 Conclusions

This purpose of this review chapter was to provide the interested reader with an overview of how QDs can be used as fluorescent tags for the direct visualisation or quantification of particular cell types or biomolecules, as well as illustrating how they can serve as effective energy donors in sensing and therapeutic applications. These studies demonstrate how QDs provide many advantages in terms optical performance when compared to organic dyes. However, certain issues still remain before these nanoparticles become the first choice fluorophore when undertaking biomedical research. Batch to batch reproducibility, controlling the valence of surface functionality and eliminating blinking remain a challenge. The perceived toxicity of cadmium-based QDs has also limited the use of these nanoparticles for in vivo applications. In this context, the emergence of new classes of QDs prepared from non-cadmium sources may help to reinvigorate this area. However, more research is still required to gain a fuller understanding of the structure–function relationship of these nanoparticles, i.e. how their physical and chemical properties influence factors such as uptake and clearance from the body. Nonetheless, QDs remain an exciting and versatile class of fluorophores with a bright future in the biomedical diagnostics and therapeutics arena.

Acknowledgements JFC acknowledges support from Norbrook Laboratories Ltd for an endowed chair.

References

1. Michalet X, Pinaud F, Lacoste TD, Dahan M, Bruchez MP, Alivisatos AP, Weiss S (2001) Properties of fluorescent semiconductor nanocrystals and their application to biological labelling. Single Mol 2(4):261
2. Medintz IL, Uyeda HT, Goldman ER, Mattoussi H (2005) Quantum dot bioconjugates for imaging, labelling and sensing. Nat Mater 4(6):435
3. Alivisatos P (2004) The use of nanocrystals in biological detection. Nat Biotechnol 22(1):47
4. Anderson KE, Fong CY, Pickett WE (2002) Quantum confinement in CdSe nanocrystallites. J Non Cryst Solids 299:1105
5. Tomasulo M, Yildiz I, Raymo FM (2006) pH-Sensitive quantum dots. J Phys Chem B 110:3853
6. Schmidt KH, Medeiros-Ribeiro G, Garcia J, Petroff PM (1997) Cyclic crystalline-amorphous transformations of mechanically alloyed Co75Ti25. Appl Phys Lett 70:1679
7. Clapp AR, Medintz IL, Mauro JM, Fisher BR, Bawendi MG, Mattoussi H (2004) Fluorescence resonance energy transfer between quantum dot donors and dye-labeled protein acceptors. J Am Chem Soc 126:301
8. Lin G, Wang X, Yin F, Yong KT (2015) Passive tumor targeting and imaging by using mercaptosuccinic acid-coated near-infrared quantum dots. Int J Nanomedicine 10:335
9. Greish K (2010) Enhanced permeability and retention (EPR) effect for anticancer nanomedicine drug targeting. Methods Mol Biol 624:25
10. Iyer AK, He J, Amiji MM (2012) Image-guided nanosystems for targeted delivery in cancer therapy. Curr Med Chem 19:3230
11. Gao J, Chen K, Xie R, Lee S, Cheng Z, Peng X, Chen X (2010) Ultrasmall near-infrared non-cadmium quantum dots for in vivo tumor imaging. Small 6:256
12. Gao X, Cui Y, Levenson RM, Chung LWK, Nie S (2004) In vivo cancer targeting and imaging with semiconductor quantum dots. Nat Biotechnol 22:969–976
13. Engvall E, Perlmann P (1971) Enzyme-linked immunosorbent assay (Elisa) quantitative assay of immunoglobulin-G. Immunochemistry 8:871
14. Vanweeme BK, Schuurs AHW (1971) Immunoassay using antigen-enzyme conjugates. FEBS Lett 15:232

15. Gao X, Yang L, Petros JA, Marshall FF, Simons JW, Nie S (2005) In vivo molecular and cellular imaging with quantum dots. Curr Opin Biotechnol 16:63–68
16. Chen LD, Liu J, Yu XF, He M, Pei XF, Tang ZY, Wang QQ, Pang DW, Li Y (2008) The biocompatibility of quantum dot probes used for the targeted imaging of hepatocellular carcinoma metastasis. Biomaterials 29(31):4170–4176
17. Eichmüller S, Stevenson PA, Paus R (1996) A new method for double immunolabelling with primary antibodies from identical species. J Immunol Methods 190:255
18. Herbst RS (2004) Review of epidermal growth factor receptor biology. Int J Radiat Oncol Biol Phys 59:21
19. Zhang H, Berezov A, Wang Q, Zhang G, Drebin J, Murali R, Greene MI (2007) ErbB receptors: from oncogenes to targeted cancer therapies. J Clin Invest 117:2051
20. Diagaradjane P, Deorukhkar A, Gelovani JG, Maru DM, Krishnan S (2010) Gadolinium chloride augments tumor-specific imaging of targeted quantum dots in vivo. ACS Nano 27:4131
21. Olivo M, Bhuvaneswari R, Lucky SS, Dendukuri N, Thong PSP (2010) Targeted therapy of cancer using photodynamic therapy in combination with multi-faceted anti-tumor modalities. Pharmaceuticals 3:1507
22. Hynes R (2002) Integrin: bidirectional, allosteric signalling machines. Cell 110:673
23. Jenne D, Stanley KK (1987) Nucleotide sequence and organization of the human S-protein gene: repeating peptide motifs in the "pexin" family and a model for their evolution. Biochemistry 26(21):6735
24. Felding-Habermann B, Cheresh DA (1993) Vitronectin and its receptors. Curr Opin Cell Biol 5:864
25. Cai W, Shin D-W, Chen K, Gheysens O, Cao Q, Wang SX, Gambhir SS, Chen X (2006) Peptide-labeled near-infrared quantum dots for imaging tumor vasculature in living subjects. Nano Lett 6:669
26. Gao J, Chen K, Xie R, Xie J, Yan Y, Cheng Z, Peng X, Chen X (2010) In vivo tumor-targeted fluorescence imaging using near-infrared non-cadmium quantum dots. Bioconjug Chem 21 (4):604
27. Pringos E, Vignes M, Martinez J, Rolland V (2011) Peptide neurotoxins that affect voltage-gated calcium channels: a close-up on ω-agatoxins. Toxins (Basel) 3:17
28. Choi D (1988) Calcium-mediated neurotoxicity: relationship to specific channel types and role in ischemic damage. Trends Neurosci 11:465
29. Choi DW, Rothman SM (1990) The role of glutamate neurotoxicity in hypoxic-ischemic neuronal death. Annu Rev Neurosci 13:171
30. Skov MJ, Beck JC, de Kater AW, Shopp GM (2007) Nonclinical safety of ziconotide: an intrathecal analgesic of a new pharmaceutical class. Int J Toxicol 26:411
31. Orndorff RL, Rosenthal SJ (2009) Neurotoxin quantum dot conjugates detect endogenous targets expressed in live cancer cells. Nano Lett 9:2589
32. Didenko VV (2001) DNA probes using fluorescence resonance energy transfer (FRET): designs and applications. Biotechniques 31(5):1106–1110
33. Tyagi S, Kramer FR (1996) Molecular beacons: probes that fluoresce upon hybridization. Nat Biotechnol 14(3):303–308
34. Zhang CY, Yeh HC, Kuroki MT, Wang TH (2005) Single-quantum-dot-based DNA nanosensor. Nat Mater 4:826
35. Kim JH, Morikis D, Ozkan M (2004) Adaptation of inorganic quantum dots for stable molecular beacons. Sens Actuator B Chem 102:315
36. Dyadyusha L, Yin H, Jaiswal S, Brown T, Baumberg JJ, Booy FP, Melvin T (2005) Quenching of CdSe quantum dot emission, a new approach for biosensing. Chem Commun 25:3201
37. Algar WR, Krull UJ (2007) Towards multi-colour strategies for the detection of oligonucleotide hybridization using quantum dots as energy donors in fluorescence resonance energy transfer (FRET). Anal Chim Acta 581:193

38. Algar WR, Krull UJ (2009) Toward a multiplexed solid-phase nucleic acid hybridization assay using quantum dots as donors in fluorescence resonance energy transfer. Anal Chem 81:4113
39. Gould EA, Winship T, Philbin PH, Kerr HH (1960) Observations on a "sentinel node" in cancer of the parotid. Cancer 13:77
40. Liedberg FCG, Davidsson T, Gudjonsson S, Månsson WJ (2006) Intraoperative sentinel node detection improves nodal staging in invasive bladder cancer. J Urol 175(1):84–88
41. Ganswindt U, Paulsen F, Corvin S, Eichhorn K, Glocker S, Hundt I, Birkner M, Alber M, Anastasiadis A, Stenzl A, Bares R, Budach W, Bamberg M, Belka C (2005) Intensity modulated radiotherapy for high risk prostate cancer based on sentinel node SPECT imaging for target volume definition. BMC Cancer 28:5–9
42. Brouwer OR, Valdés Olmos RA, Vermeeren L, Hoefnagel CA, Nieweg OE, Horenblas SJ (2011) SPECT/CT and a portable gamma-camera for image-guided laparoscopic sentinel node biopsy in testicular cancer. Nucl Med 52(4):551–554
43. Korn JM, Tellez-Diaz A, Bartz-Kurycki M, Gastman B (2014) Indocyanine green SPY elite-assisted sentinel lymph node biopsy in cutaneous melanoma. Plast Reconstr Surg 133(4):914–922
44. Ballou B, Ernst LA, Andreko S, Harper T, Fitzpatrick JAJ, Waggoner AS, Bruchez MP (2007) Sentinel lymph node imaging using quantum dots in mouse tumor models. Bioconjug Chem 18:389–396
45. Pons T, Pic E, Lequeux N, Cassette E, Bezdetnaya L, Guillemin F, Marchal F, Dubertret B (2010) Cadmium-free CuInS2/ZnS quantum dots for sentinel lymph node imaging with reduced toxicity. ACS Nano 4(5):2531–2538
46. Kosaka N, McCann TE, Mitsunaga M, Choyke PL, Kobayashi H (2010) Real-time optical imaging using quantum dot and related nanocrystals. Nanomedicine (Lond) 5:765–776
47. Parija SC (2012) Textbook of microbiology and immunology, 2nd edn. Elsevier, New Delhi
48. Hahn MA, Tabb JS, Krauss TD (2005) Detection of single bacterial pathogens with semiconductor quantum dots. Anal Chem 77:4861–4869
49. Aldeek F, Mustin C, Balan L, Roques-Carmes T, Fontaine-Aupart MP, Schneider R (2011) Surface-engineered quantum dots for the labeling of hydrophobic microdomains in bacterial biofilms. Biomaterials 32:5459–5470
50. Zhu L, Ang S, Liu W (2004) Quantum dots as a novel immunofluorescent detection system for *Cryptosporidium parvum* and *Giardia lamblia.* Appl Environ Microbiol 70:597–598
51. Hahn MA, Keng PC, Krauss TD (2008) Flow cytometric analysis to detect pathogens in bacterial cell mixtures using semiconductor quantum dots. Anal Chem 80:864–872
52. Zhao Y, Ye M, Chao Q, Jia N, Ge Y, Shen H (2008) Simultaneous detection of multifood-borne pathogenic bacteria based on functionalized quantum dots coupled with immunomagnetic separation in food samples. J Agric Food Chem 57:517–524
53. Edgar R, McKinstry M, Hwang J, Oppenheim AB, Fekete RA, Giulian G (2006) High-sensitivity bacterial detection using biotin-tagged phage and quantum-dot nanocomplexes. Proc Natl Acad Sci U S A 103:4841–4845
54. Bagalkot V, Zhang L, Levy-Nissenbaum E, Jon S, Kantoff PW, Langer R, Farokhzad OC (2007) Quantum dot−aptamer conjugates for synchronous cancer imaging, therapy, and sensing of drug delivery based on bi-fluorescence resonance energy transfer. Nano Lett 7:3065–3070
55. Wang Y, Zhang X, Xu Z, Huang H, Li Y, Wang J (2014) A pH-sensitive theranostics system based on doxorubicin with comb-shaped polymer coating of quantum dots. J Mater Sci 49:7539–7546
56. Zheng M, Liu S, Li J, Qu D, Zhao H, Guan X, Hu X, Xie Z, Jing X, Sun Z (2014) Integrating oxaliplatin with highly luminescent carbon dots: an unprecedented theranostic agent for personalized medicine. Adv Mater 26:3554–3560
57. Zhang R, Liu Y, Yu L, Li Z, Sun S (2013) Preparation of high-quality biocompatible carbon dots by extraction, with new thoughts on the luminescence mechanisms. Nanotechnology 24:225601

58. Zhang W, Dai D, Chen X, Guo X, Fan J (2014) Red shift in the photoluminescence of colloidal carbon quantum dots induced by photon reabsorption. Appl Phys Lett 104:091902
59. Yang S, Wang X, Wang H, Lu F, Luo PG, Cao L, Meziani MJ, Liu J, Liu Y, Chen M, Huang Y, Sun Y (2009) Carbon dots as nontoxic and high-performance fluorescence imaging agents. J Phys Chem 113:18110–18114
60. Fakhroueian Z, Dehshiri AM, Katouzian F, Esmaeilzadeh P (2014) In vitro cytotoxic effects of modified zinc oxide quantum dots on breast cancer cell lines (MCF7), colon cancer cell lines (HT29) and various fungi. J Nanopart Res 16:2483
61. Donnelly RF, McCarron PA, Woolfson D (2009) Drug delivery systems for photodynamic therapy. Recent Pat Drug Deliv Formul 3:1–7
62. NHS Clinical trials and Medical Research Website (2015) Clinical trials for photodynamic therapy. http://www.nhs.uk/Conditions/photodynamic-therapy-NGPDT-sonodynamic-therapy/Pages/clinical-trial.aspx
63. Wainwright M (2010) Therapeutic applications of near-infrared dyes. Color Technol 126:115–126
64. Ochsner M (1996) Light scattering of human skin: a comparison between zinc (II)-phthalocyanine and photofrin II. J Photochem Photobiol B 32:3–9
65. Samia ACS, Dayal S, Burda CJ (2006) Quantum dot-based energy transfer: perspectives and potential for applications in photodynamic therapy. Photochem Photobiol 82:617–625
66. Choi Y, Kim S, Choi M, Ryoo S, Park J, Min D, Kim B (2014) Highly biocompatible carbon nanodots for simultaneous bioimaging and targeted photodynamic therapy in vitro and in vivo. Adv Funct Mater 24:5781–5789
67. Fowley C, Nomikou N, McHale AP, McCarron PA, McCaughan B, Callan JF (2012) Water soluble quantum dots as hydrophilic carriers and two-photon excited energy donors in photodynamic therapy. J Mater Chem 22:6456–6462
68. Qi Z, Li D, Jiang P, Jiang F, Li Y, Liu Y, Wong W, Cheah K (2011) Biocompatible CdSe quantum dot-based photosensitizer under two-photon excitation for photodynamic therapy. J Mater Chem 8:2455–2458
69. Tsay JM, Trzoss M, Shi L, Kong X, Selke M, Jung ME, Weiss S (2007) Singlet oxygen production by peptide-coated quantum dot-photosensitizer conjugates. J Am Chem Soc 129:6865–6871
70. Starkey JR, Rebane AK, Drobizhev MA, Meng F, Gong A, Elliott A, McInnerney K, Spangler CW (2008) New two-photon activated photodynamic therapy sensitizers induce xenograft tumor regressions after near-IR laser treatment through the body of the host mouse. Clin Cancer Res 14:6564–6576
71. Khurana M, Karotki A, Collins H, Anderson HL, Wilson BC (2006) In vitro studies of the efficiency of two-photon activation of photodynamic therapy agents. Proc SPIE 6343, Photonics North 634306: doi:10.1117/12.706554
72. Liu Q, Guo B, Rao Z, Zhang B, Gong JR (2013) Strong two-photon-induced fluorescence from photostable, biocompatible nitrogen-doped graphene quantum dots for cellular and deep-tissue imaging. Nano Lett 13:2436–2441
73. Kratz F, Senter P, Steinhagen H (2011) Drug delivery in oncology. Wiley, Weinheim. doi:10.1002/9783527634057.index, Published Online: 16 DEC 2011
74. Samia AC, Chen X, Burda C (2003) Semiconductor quantum dots for photodynamic therapy. J Am Chem Soc 125:15736–15737
75. Fowley C, Nomikou N, McHale AP, McCaughan B, Callan JF (2013) Extending the tissue penetration capability of conventional photosensitisers: a carbon quantum dot-protoporphyrin IX conjugate for use in two-photon excited photodynamic therapy. Chem Commun 49:8934–8936
76. Ge J, Lan M, Zhou B, Liu W, Guo L, Wang H, Jia Q, Niu G, Huang X, Zhou H, Meng X, Wang P, Lee CS, Zhang W, Han X (2014) A graphene quantum dot photodynamic therapy agent with high singlet oxygen generation. Nat Commun 5:4596

77. Vaupel P, Kallinowski F, Okunief F (1989) Blood flow, oxygen and nutrient supply and metabolic microenvironment of human tumors: a review. Cancer Res 49:6449–6465
78. Moulder JE, Rockwell S (1984) Hypoxic fractions of solid tumors: experimental techniques, methods of analysis, and a survey of existing data. Int J Radiat Oncol Biol Phys 10:695–712
79. Chen Q, Huang Z, Chen H, Shapiro H, Beckers J, Hetzel FW (2002) Improvement of tumor response by manipulation of tumor oxygenation during photodynamic therapy. Photochem Photobiol 76:197–203
80. Caruso EB, Petralia S, Conoci S, Giuffrida S, Sortino S (2006) Photodelivery of nitric oxide from water-soluble platinum nanoparticles. J Am Chem Soc 129:480–481
81. Ignarro LJ (2010) Nitric oxide: biology and pathobiology, 2nd edn. Elsevier, Burlington
82. Wilson BC, Patterson MS, Burns DM (1986) Effect of photosensitizer concentration in tissue on the penetration depth of photoactivating light. Lasers Med Sci 1:235–244
83. Fowley C, McHale AP, McCaughan B, Fraix A, Sortino S, Callan JF (2014) Carbon quantum dot–NO photoreleaser nanohybrids for two-photon phototherapy of hypoxic tumors. Chem Commum 51:81–84

Top Curr Chem (2016) 370: 225–258
DOI: 10.1007/978-3-319-22942-3_8
© Springer International Publishing Switzerland 2016

Phototherapeutic Release of Nitric Oxide with Engineered Nanoconstructs

Aurore Fraix, Nino Marino, and Salvatore Sortino

Abstract The multiple role nitric oxide (NO) plays in a number of physiological and pathophysiological processes has, over the last few years, stimulated a massive interest in the development of new strategies and methods for generating NO in a controlled way, with the exciting prospect of tackling important diseases. Photochemical precursors of NO are particularly suited to this end because light triggering permits an exquisite control of location and timing of NO delivery. Integration of NO photodonors within the structure of appropriate materials represents a key step in the fabrication of functional devices for phototherapeutic applications. It also offers the advantage of concentrating a large number of chromophores in a restricted area with the result of significantly increasing the NO reservoir and the light harvesting properties. We present here an overview of the most significant advances made in the last 5 years in the fabrication of engineered nanoconstructs able to delivery NO under the exclusive control of light inputs, highlighting the logical design and their potential applications in battling cancer and bacterial infections.

Keywords Antibacterial therapy • Cancer therapy • Fluorescence imaging • Gel • Light • Nanomaterials • Nanoparticles • Nitric oxide • Polymers

Contents

A. Fraix, N. Marino, and S. Sortino (✉)
Laboratory of Photochemistry, Department of Drug Sciences, University of Catania, 95125 Catania, Italy
e-mail: ssortino@unict.it

Abbreviations

$^1O_2$	Singlet oxygen
CMC	Carboxymethyl chitosane
CQDs	Carbon quantum dots
Cys-NO	*S*-Nitroso cysteine
ETU	Energy transfer upconversion
FA	Folic acid
FRET	Förster resonance energy transfer
LbL	Layer-by-layer
LS	Langmuir Shaefer
MDR	Multidrug resistance
NIR	Near infrared
NO	Nitric oxide
NPs	Nanoparticles
PDT	Photodynamic therapy
PTT	Photothermal therapy
QDs	Quantum dots
ROS	Reactive oxygen species
SPE	Single photon excitation
TPE	Two-photon excitation
α-CD	α-Cyclodextrin
β-CD	β-Cyclodextrin
Δ	Heat

1 Therapeutic Roles of NO: Advantages and Limitations

Nitric oxide (NO) is an ephemeral inorganic free radical which has come into the limelight in the last two decades because of the fundamental role it plays in the bioregulation of vital functions, including neurotransmission, hormone secretion, and vasodilatation [1]. NO is also involved in a number of additional and important biological processes including platelet aggregation and adhesion [2], immune response to infections [3], reduction of radical-mediated oxidative pathways [4], wound repair [5], and cancer biology [6] (Fig. 1).

This multifaceted role and the difficulties to deliver gaseous NO have stimulated a tremendous interest in compounds able to store and release NO to selected targets

Fig. 1 Some of the most important bioregulatory and protective properties of NO

with the ambitious prospect of tackling important diseases, especially cancer and bacterial infections [7–10]. NO radical is able to attack biological substrates of different nature such as the plasma membrane [11], the mitochondria [12], and the cell nucleus [13], representing a multitarget cytotoxic agent and avoiding Multiple Drug Resistance (MDR) problems [14] encountered with target-specific "conventional" anticancer and antibacterial drugs. Furthermore, the short half-life in blood (<1 s), the lack of charge, the good lipophilicity, and the small size allow NO to diffuse easily in the cellular environment over short distances (<200 μm) [15], confining the region of action without inflicting systemic side effects common to several pharmaceuticals. An additional, peculiar advantage of NO is a low toxicity towards healthy cells, especially at concentrations toxic to cancer cells [6]. Furthermore, NO delivery to hypoxic regions, which are typical for solid tumors, increases significantly the local blood flow and enhances the susceptibility to radiotherapy [16].

NO has shown excellent broad-spectrum antibacterial activity towards both Gram-positive and Gram-negative bacteria, mainly because of its reactive by-products such as peroxynitrite and dinitrogen trioxide which affect the integrity of bacteria membrane and cell functions via oxidative and nitrosative stress [17]. Interestingly, the dose of NO required to kill bacteria is demonstrated to be non-toxic to human dermal fibroblasts. NO-based antimicrobials therefore represent good candidates to answer the alarmingly low turnover of new clinically approved antibiotic drugs [18].

Although many classes of low molecular weight NO donors have been developed to benefit a range of applications [19, 20], few have been translated to clinic. In fact, these NO releasers suffer two important limitations: they are cleared rapidly in the body and need to be tailored to reach pharmacological targets. Such drawbacks have inspired the development of NO-releasing macromolecular scaffolds [21, 22] and nanomaterials [23, 24] as potential therapeutics which better exploit NO biological roles, avoiding the above limitations. These molecular constructs represent suitable vehicles characterized by a number of advantages including enhanced NO storage, sustained NO release kinetics, easy modification with suitable targeting ligands, biocompatibility, improved therapeutic output, and reduced side-effects.

Another key point to be taken into account for developing useful therapeutics is that the therapeutic outcome of any NO-based drug is strictly dictated by three main parameters: concentration, delivery site, and dosage. In some cases, this creates a complex role for the molecule in opposing beneficial and deleterious events. For instance, it has been shown that, although micromolar concentrations of NO significantly inhibit tumor growth, picomolar concentrations of NO encourage tumor cells proliferation [6, 25, 26]. This dichotomy has made the development of new strategies and methods for generating NO with precise spatiotemporal control a *hot topic* in the burgeoning field of nanomedicine.

2 Light-Triggered NO Release: General Background

2.1 NO Photocages

In view of the easy manipulation, in terms of intensity, wavelength, duration, and localization, light represents a highly orthogonal, minimally invasive and finely tunable external stimulus to achieve controlled NO delivery with a superb spatiotemporal accuracy in biological systems. In addition, light triggering is biofriendly, provides fast reaction rates, and offers the great benefit of not affecting physiological parameters such as temperature, pH, and ionic strength, fundamental requisites for biomedical applications [27].

The working principle of an NO photochemical precursor is based on the concept of "photocaging". In general, this term refers to the temporary inactivation of a biologically active molecule by its covalent incorporation within a photoresponsive chromogenic center (the photocage) [28]. As illustrated in Fig. 2, the photosensitive moiety exploits the absorbed excitation energy to break a chemical bond, releasing the "caged" molecule, i.e., NO, in its active form. This enables the site of action of NO to be confined to the illuminated area with high spatial precision and its dosage to be controlled with great accuracy by tuning the light intensity and/or duration. These unique features make the NO photoreleasing compounds a powerful therapeutic arsenal much more appealing than those based

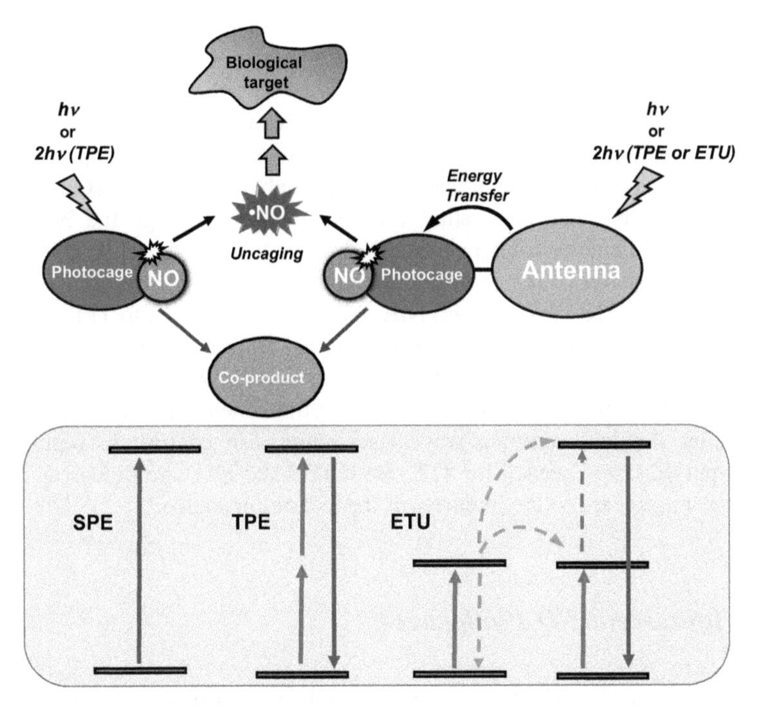

Fig. 2 *Top*. The photouncaging process: one or two-photon excitation of the photocage triggers cleavage of the covalent bond with the caged and inactive NO molecule, releasing it in its active form; alternatively, one or two-photon excitation of an antenna chromophore encourage the NO uncaging upon intramolecular energy transfer to the photocage. *Bottom*. Schematic representation for the SPE, TPE, and ETU processes which can be exploited to populate the excited state of NO photocages

on either thermal, enzymatic, or pH stimuli [29–33]. It should be noted that NO photouncaging is accompanied by the formation of a concomitant molecular fragment, which deserves specific attention. Indeed, physiological properties and toxicity of this product need to be carefully evaluated for an ideal functioning of the NO photodonor. Furthermore, such a by-product should not absorb excitation light in the same region as the NO photoprecursor so as to avoid undesired inner filter effects and unforeseen photochemical reactions.

Analogous to the photochemical uncaging of other bioactive molecules, effective use of NO photocages (especially those based on organic compounds) by single photon excitation (SPE) is limited by poor tissue transmission of UV and shorter visible wavelengths, the therapeutic window being ≈650–1350 nm where penetration depth in soft tissue is >1 cm [34]. This restricts, in principle, the application of many NO photoreleasers to topical uses. To overcome this drawback, NO photocages having high two-photon excitation (TPE) cross-section in the near infrared (NIR) region have been designed. TPE involves the simultaneous absorption of two less-energetic photons by using femtosecond laser pulses [35]. This process allows the excited state of the NO photoprecursor to be reached, with the additional advantage of a high spatial resolution because TPE excitation occurs primarily at the focal point of the light source [36]. Alternatively, in the case of NO photocages having either very low molar absorptivity or low TPE cross-section in the vis-NIR region, effective photochemical activation can be accomplished by using antenna chromophores covalently linked to the NO photocage through a molecular spacer. The antennas are appropriately chosen to act as one- or two-photon suitable light-harvesting centers, which encourage the NO release through effective energy transfer to the NO photocage (Fig. 2) [30, 33]. Energy transfer upconversion (ETU) from the antenna to the NO photocage can also be exploited to trigger NO release with NIR excitation. In contrast to TPE, where the antenna simultaneously absorbs the two photons, ETU is obtained by sequential absorption processes of NIR photons resulting in emission at shorter wavelengths (Fig. 2) [37]. ETU is achieved by using rare earth elements (i.e., Ln^{3+}) and can be achieved by simple continuum wave laser sources in contrast to sophisticated ultrafast-pulsed lasers needed for TPE. Both TPE and ETU are processes that are non-linear with regard to the intensity of the excitation source.

2.2 *Fluorescent NO Photocages*

The visualization of a NO photodonor in a cellular environment through fluorescence techniques represents an indispensable requisite in view of image-guided NO-based phototherapies. In such a way, excitation of the fluorophoric unit with low intensity light generates fluorescence emission, which allows localizing the NO photodonor in cells. Afterwards, excitation with high intensity light can provide a highly localized "burst" of NO precisely at the desired sites. Unlike photosensitizers used in photodynamic therapy (PDT) such as porphyrins and phthalocyanines [38], which are intrinsically fluorescent, the combination of fluorescent imaging and photosensitizing capacity in a single structure is less common in the case of NO photodonors. In the case of NO photocages lacking emissive properties, fluorescent characteristics can be imposed by their covalent or non-covalent combination with suitable flurogenic centers (Fig. 3a). In both cases, the most critical phase is the design. Inter-chromophoric interactions (i.e., energy or electron transfer) between the fluorophoric unit and the NO photocage need to be intentionally avoided with a careful choice of chromophoric groups by taking into account the energy of the lowest excited states of the components and their redox potential. Under these conditions, distinct photoresponsive moieties can be operated in parallel, despite their close proximity, to impose multifunctional characteristics of the resulting molecular construct. In particular, the possibility of exploiting the fluorescence of one component for imaging together with the ability of the other component to release NO enables diagnosis and therapy in tandem and, therefore, is especially valuable for phototheranostic applications [39].

As outlined in Sect. 1, the biological effects of NO are strictly dependent on its concentration. This feature makes the quantification of the NO delivery upon

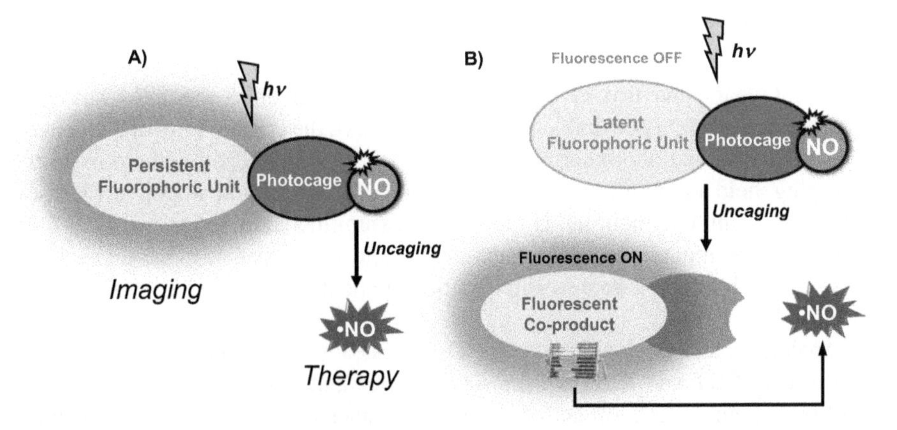

Fig. 3 (**A**) NO photocages can be joined with persistent fluorogenic units which do not change their emissive properties neither before nor after NO photorelease. (**B**) The release and report working principle: excitation of the photocage releases NO and concomitantly a fluorescent co-product, which acts as an optical counter for NO concentration

photouncaging a very important issue to be faced, especially when one is interested in reaching a critical molecule concentration to induce a specific effect. It should be considered that in some cases the knowledge of the quantum yield for the NO release and the intensity of the light irradiation cannot be sufficient to warrant the delivery of a precise amount of NO at the target site. In fact, in a living organism, the efficiency of NO release can dramatically change and the calibration of light power can be made difficult because of the uncertainties associated with many unknown medium characteristics regarding volume, viscosity, optical homogeneity, etc. A suitable way to address this quantification task is based on the use a *fluorescent reporter*. The principle of this simple and elegant strategy, also adopted for the quantification of other photocaged biologically active molecules [40], is sketched in Fig. 3b. The photocage is covalently linked to a "latent" fluorophoric center where fluorescence is temporarily quenched by photoinduced intermolecular processes (i.e., energy or electron transfer). Photoexcitation of the photocage triggers the release of the desired caged bioactive species which is accompanied by the simultaneous release of a co-product (the reporter) whose fluorescence is no longer quenched. In such a way, the uncaging process can be easily quantified by monitoring the fluorescence emission of the reporter which acts exactly as an optical counter for the NO released. Besides, fluorescence techniques allow the spatial distribution of the released species to be followed in a biological medium in real time. Photouncaging quantification of NO using fluorescent reporters has only recently attracted attention. Indeed, in several cases, the caged substrate and the co-product from the photolabile moiety after uncaging exhibit similar or even less intense brightness [41, 42].

2.3 NO and Bimodal Phototherapy

In recent years, the use of NO in conjunction with other therapies has been suggested to be highly desirable [43]. In this context, the combination of NO release with PDT [44] and photothermal therapy (PTT) [45] may open new doors in the emerging field of multimodal therapies for both cancer and infective diseases. These emerging treatment modalities aim at exploiting additive/synergistic effects caused by the simultaneous generation of multiple active species in the same region of space, with the final goal to amplify the therapeutic action and minimizing side effects [46].

Analogous to NO, singlet oxygen ($^1O_2$) and heat (Δ), the main cytotoxic agents involved in PDT and PTT, are transient species and thus offer the advantage of confining their region of action over short distances inside the cells without inflicting systemic side effects. Moreover, as with NO and in contrast to many conventional drugs, $^1O_2$ and Δ are multitarget agents capable of attacking biological substrates of different natures and do not suffer the MDR drawbacks which are mainly responsible for the failure of diverse therapeutic modalities. Finally, as both NO and Δ photorelease is independent of the O_2 availability, it complements PDT very well at the onset of hypoxic conditions, typical for some tumors, where PDT may fail.

Nanoplatforms integrating NO photodelivery properties with either PDT or PTT effects can be achieved by appropriate assembling of suitable photoactive components. In this case, inter-chromophoric interactions also need to be avoided intentionally with a careful choice of the chromogenic groups. As illustrated in several examples in the following sections, this allows the distinct photoresponsive components to be operated in parallel under light inputs within the same nanoconstruct, associating the photodelivery of NO with the simultaneous photogeneration of either $^1O_2$ or Δ (Fig. 4). A specific point to be addressed in the fabrication of such bimodal phototherapeutic systems should be focusing on the relative amounts of cytotoxic species photogenerated. In fact, although generation of $^1O_2$ and Δ is based on photophysical processes which, in principle, do not consume the photosensitizer, photogeneration of NO implies a neat photochemical reaction with

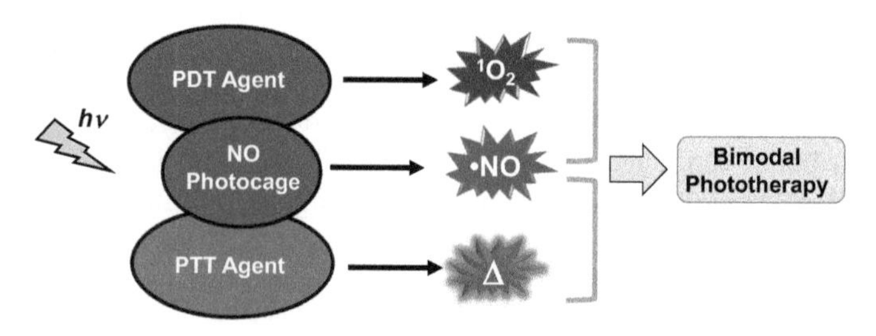

Fig. 4 NO photocages can be combined with photosensitizers for either PDT or PTT in close proximity. The parallel activation of the two chromogenic centers allows bimodal photoaction based on the combination of NO with the simultaneous photogeneration of either $^1O_2$ or Δ

consequent degradation of the NO photocage. As a result, to achieve effective bimodal performances, the multichromophoric nanoconstructs should take into account an appropriate balance between the potential reservoir of the NO available, the absorption coefficient of the NO photocage, and its photodegradation quantum yield in comparison with the photophysical features of the PDT and PTT agents.

3 Nanomaterials for NO Photodelivery

The assembling of photoactivable NO donors within appropriate materials is of fundamental importance for the development of NO delivering devices for therapeutic purposes [21–24, 31]. In fact, the assembling process (1) offers the advantage of concentrating a large number of chromophore in a restricted area/volume, increasing both the light harvesting properties and the reservoir of NO available, (2) facilitates the delivery process by means of tailored drug carriers, and (3) allows an easy integration with optical fiber light guides. Furthermore, materials science allows the combining in the same nanoplatform of NO photodonors with additional photoresponsive components performing functions useful for biomedical purposes, such as fluorescence imaging and photoregulated release of other therapeutic species. This represents a significant step forward for the achievement of multitasking NO-based nanodrugs with optimal phototheranostic activity. To this end, either covalent or non-covalent approaches can be used. In all cases, the fabrication of these multi-photoresponsive systems implies collective cross-disciplinary efforts because of the need for synthetic, assembling methodologies and physical characterization techniques. However, photochemistry plays a dominant role. In fact, the challenge in the design of these multifunctional photoactive constructs rests on how to assemble different units into a single architecture with specific characteristics, such as size, shape, and coordination environment, and to predict the response of the final systems to light excitation. It should be noted that, in contrast to the assembly of non-photoresponsive compounds, this is not a trivial task. In most cases, the response to light of photoactive centers located in a confined space of a specific material or close to other photoactive components can be considerably influenced in nature, efficiency, or both by the occurrence of competitive and undesired processes which preclude the final goal.

3.1 Inorganic Nanoparticles

Noble metal nanoparticles (NPs) stabilized with appropriate ligands are very suited for bio-applications. The large number of preparation protocols available allows different size ranges with narrow size distribution to be obtained. Besides, the easy chemical modification of their surface by the introduction of tailored functional units leads to nanohybrids which are evincing widespread interest in drug delivery

[47]. In the case of photoactive compounds, these three-dimensional nanoarchitectures can offer the advantages of (1) reduced quenching effects on the excited states of the photoactive units caused by the metal curvature, if compared to the bidimensional self-assembled monolayers counterparts [48] and (2) enhanced photochemical response produced by plasmonic effects [49].

AuNPs as a core scaffold offer the advantages of good biocompatibility with cells or tissues, excellent optical properties in the vis-NIR region, and superb capability to convert the excitation light into heat [50]. NO photoreleasing agents anchored at the surface of AuNPs generate photoactivable nanohybrids of great potential in bimodal phototherapy. In principle, such systems may combine the excellent optical properties of the dense Au core for PTT with a photoactivable shell for controlled release of NO. AuNPs exhibiting photoregulated NO release have been achieved based on a supramolecular approach involving the self-assembly of a host/guest complex between the water insoluble thiol derivative **1** and α-cyclodextrin (α-CD) on the surface of pre-formed AuNPs (Fig. 5A) [51]. Thiol **1** integrates a nitroaniline derivative in its molecular skeleton as an NO photocage which satisfies several biological prerequisites for photocages, such as dark stability, adequate absorption coefficient in the visible region, photodecaging of NO under the exclusive control of light inputs, and generation of non-toxic and nonabsorbing photoproducts [52]. The NO photorelease takes place through a nitro-to-nitrite photorearrangement followed by the rupture of the O–NO bond with formation of NO and a phenoxy radical as a key intermediate. The role of the α-CD is manifold. In fact, it conveys the photoactive unit to the metal surface, allows the solubilization of the resulting modified AuNPs in polar solvents, and, in view of its use in biocompatible materials, it is also expected to improve the biocompatibility of the resulting AuNPs. The coated AuNPs demonstrate different sensitivity to visible light excitation. In fact, irradiation at 400 nm (absorption maximum of the NO photodonor) leads to the typical bleaching of the nitroaniline chromophore associated with the light-controlled release of NO. In contrast, no spectral modifications are observed when the light excitation was centered in the plasmon absorption band at 532 nm. As a result, the distinct absorption spectral region of the Au scaffold and the NO photodonor units allows in principle the selective excitation of the core (triggering photothermia), the shell (triggering NO release), or both (triggering bimodal effects) by tuning the visible light energy.

Water dispersible NO photoreleasing AuNPs have recently been obtained by coating of the metal surface with 2-mercapto-5-nitro benzimidazole **2** (Fig. 5B) [53]. Interestingly, although the NO photorelease is insignificant for the free chromophore, remarkable NO release is observed after grafting the chromophore on the Au surface. This behavior was attributed to the inter-chromophoric interactions between the densely packed molecules and chromophore-plasmon electron interaction. These interactions force the nitro group to adopt a twisted conformation which is fundamental to observe the photorelease of NO through a mechanism similar to that observed for the photocage **1** (see Fig. 5A). The nanohybrid exhibited remarkable tumor mortality against cervical cancer cell lines via the apoptotic

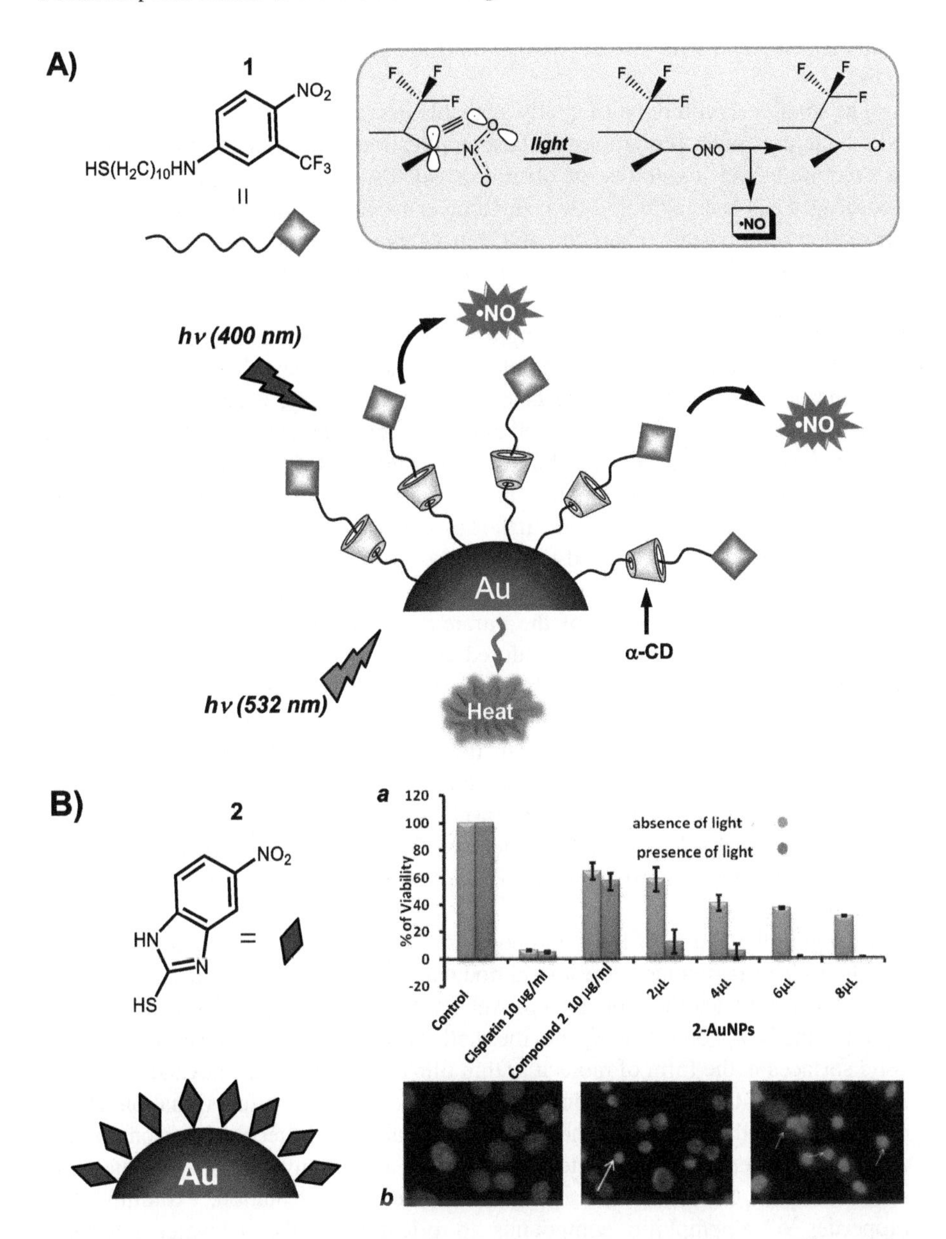

Fig. 5 (**A**) The non-hydrosoluble compound **1** can be first encapsulated in α-CD and the subsequent hydrosoluble host-guest complex can be grafted at the surface of AuNPs leading to bifunctional hybrids. The *inset* shows a sketch of the mechanism of NO release from the nitroaniline-based photocage. (**B**) Cytotoxic effects of AuNPs coated with **2** at various concentration compared with Cisplatin, free **2**, and control sample in the presence and in the absence of visible light (***a***). Hoechest 33342 staining of Hela nucleus, *white arrows* indicate the apoptotic cells. (***b***). (Adapted with permission from [51] and [53])

route, if compared with free **2**, clearly confirming the role of NO in killing cancer cells.

The small reservoir of NO actually available in monolayer-protected noble metal NPs remains a main limitation for those applications where large amounts of NO are required. The possibility of obtaining AuNPs with tunable payloads of NO photocages has been explored by using a layer-by-layer (LbL) approach. The LbL deposition of oppositely charged polyelectrolytes is an established method for the fabrication of thin films on flat solid surfaces, microparticles, and only years later on NPs [54–57]. This is because of problems associated with the wrapping of polyelectrolytes around NPs with high curvature such as aggregation caused by cross-linking of the particles by the polyelectrolyte chains and the separation of the unbound polyelectrolyte from the coated NPs [58]. Ideally, when the layer materials have been appropriately selected and the layer order carefully designed, LbL NPs can be rapidly fabricated in a build-to-order fashion [59], making these systems an extremely attractive option for the scale-up of industrial applications. LbL assembly has been used to obtain nanocapsules even from single NPs [60] and several studies have highlighted the viability of obtaining NPs for drug delivery [61]. In this framework, water soluble, colloidal platforms for NO release have been obtained by alternate coating of the citrate stabilized AuNPs template with the cationic polyelectrolyte **3** and the tailored anionic, photoactivable polyelectrolyte **4** which incorporates the nitroaniline-based photocage (Fig. 6) [62]. The LbL strategy was found to be very suitable for regulating the payload of the NO precursor around the Au core through the adjustment of the number of layers. The layered AuNPs preserve their nanodimensional character well and combine several advantages of an ideal NO-releasing vehicle such as (1) water solubility, (2) regulated payload of the NO precursor, (3) absence of significant clustering under physiological conditions, and (4) remote-controlled release of NO by visible light inputs.

Changing the nature of the metal core from Au to Ag generates photoactivable nanohybrids of particular interest in bimodal antibacterial therapy. In this case, the well-known antibacterial properties of AgNPs [63] can be coupled with bactericidal action of the NO photoreleased from the shell. The assembling of these hybrids onto solid surfaces in the form of molecular thin films is of particular relevance in view, for example, of topical applications and the achievement of bifunctional bioactive coatings for medical devices. The Langmuir–Schaefer (LS) technique is one of the most elegant methods to fabricate multilayers on almost any kind of substrate with precise control of thickness, packing, and molecular orientation, exploiting the properties of amphiphilic compounds to orientate at the air-water interface [64]. Hybrid multilayer films incorporating an NO photodonor and AgNPs were easily fabricated by the LS approach [65]. The strategy used was to dissolve naked and water soluble AgNPs, prepared with a green procedure [66], in the water subphase, and to spread the amphiphilic, tailored NO photodonor **6** having an amino-termination on the water surface (Fig. 7) [65].

The floating monolayers at the water–air interface are able to interact with the AgNPs through the amino functionality, encouraging their transfer onto quartz

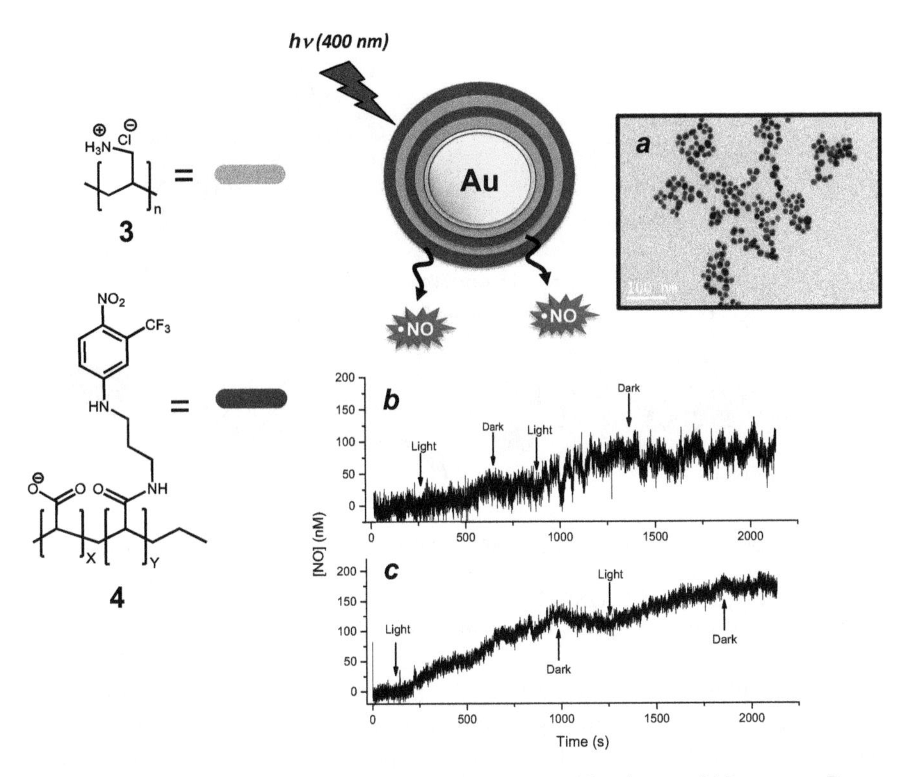

Fig. 6 Schematic for the NO photoreleasing AuNPs prepared by alternate LbL strategy. Representative transmission electron microscopy image of AuNPs coated with two layers of the photoactive polymer **4** (***a***) and NO release profiles observed upon light irradiation coated with one (***b***) and two (***c***) layers of the photoactive polymer **4** at pH 7.4 and $T = 25$ °C (Adapted with permission from [62])

slides by LS deposition, leading to hybrid multilayer films containing two potential antibacterial agents, AgNPs, and the NO photodispenser. The approach used allows the preservation of the nanodimensional character of the AgNPs, an indispensable requisite for the antibacterial action. The NO photodonor also retains its photochemical properties within the hybrid nanostructure, as confirmed by its ability to photorelease NO and effectively transfer this antibacterial species to a protein such as Myoglobin under visible light. The reservoir of AgNPs and NO available in the films can, of course, be adjusted by changing the number of layers, making these multilayers promising bifunctional model systems to be tested in biomedical research studies.

Semiconductor quantum dots (QDs) [67] and carbon quantum dots (CQDs) [68] are another fascinating class of inorganic NPs, the latter being superior in terms of ultralow toxicity and excellent biocompatibility. The huge one- and two-photon absorption cross section, the tunable emission bands, the superb photostability, and the simplicity of modification of their surface make QDs and CQDs very intriguing scaffolds for applications in bioimaging and drug delivery [67, 68]. NO photocages

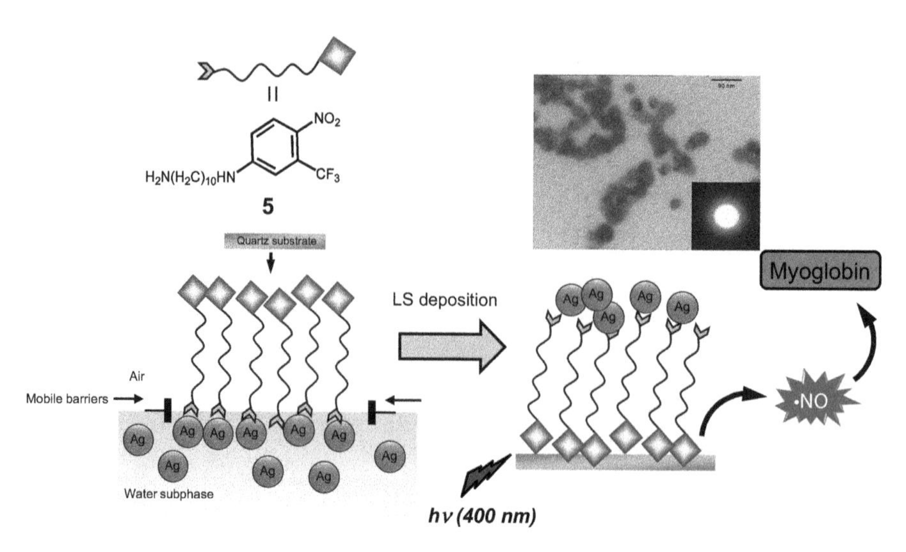

Fig. 7 Schematic view of the arrangement of **5** and the AgNPs at the air/water interface and after their transfer on a quartz substrate by LS deposition. Representative transmission microscopy image of a 15-layer hybrid **5**/AgNPs LS film. The *inset* reports the electron diffraction of the AgNPs in the film. (Adapted with permission from [65])

can be assembled at the surface of these inorganic nanoscaffolds to obtain nanohybrids in which (1) the core communicates with the shell through either energy or electron transfer mechanism and (2) the core and the shell do not interact with each other, working in parallel under light stimulation. Some of the most recent examples are illustrated in the following.

Ford and coworkers have developed an elegant approach to fabricate water soluble CdSe/ZnS core/shell QDs which enable the delivery of NO with high quantum efficiency from the Cr-complex **6**, having low molar absorptivity in the visible region (Fig. 8a) [69]. QDs surface was at first modified with dihydrolipoic acid to give aqueous solubility. Afterwards, the NO photocage was easily assembled at the QDs exterior by electrostatic interactions with the oppositely charged QDs surface. In this case the QDs core acts as effective light harvesting antenna and transfers the absorbed photons to the peripheral NO photocage, encouraging the NO release via the Förster Resonance Energy Transfer (FRET) mechanism. A detailed investigation carried out with Cr-complexes having slightly different absorption spectra demonstrated the importance of the spectral overlap between the emission of the QDs and the absorption of the complexes as a guideline essential to developing nanostructures that effectively utilize QDs as NO photosensitizers [70]. An inorganic NO photoprecursor electrostatically bound to Mn(II)-doped ZnS QDs has recently been reported by Tan et al. [71]. In this case, an ETU mechanism enables one to collect energy in the second NIR window at 1160 nm and convert it into a visible emission that triggers the NO release from the inorganic precursor-bound QDs surface. Robust CQD nanohybrids were recently prepared by covalent grafting of a nitroaniline-based NO photodonor [72]. In this case, this

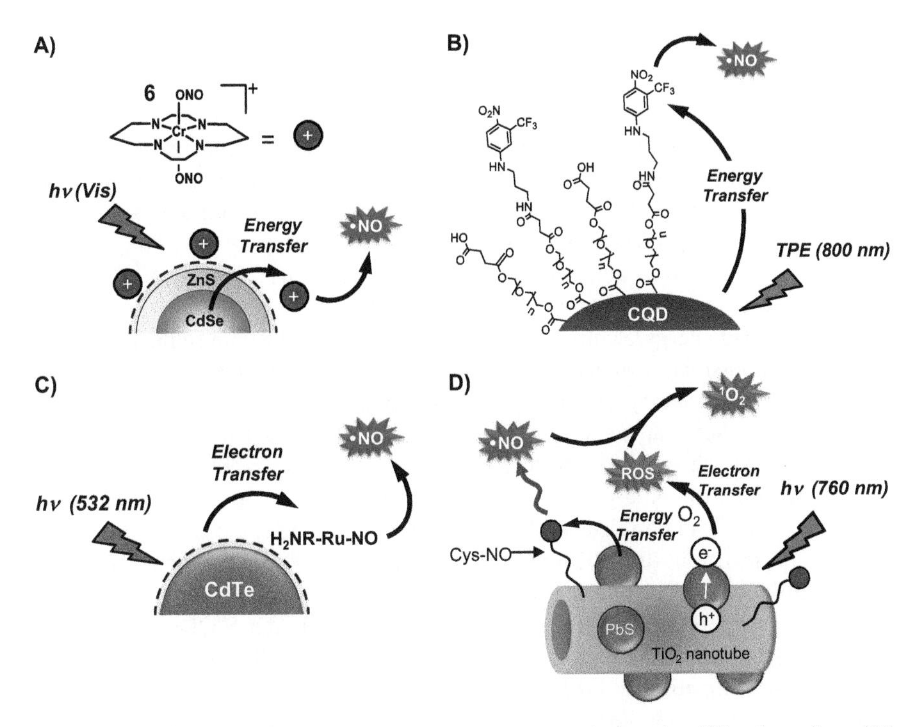

Fig. 8 Schematic for QDs and CQDs acting as antenna and triggering NO release from NO photocages electrostatically or covalently bound at their surface by energy transfer (**A**, **B**) or electron transfer (**C**). Both mechanisms operating in parallel allow enhanced $^1O_2$ generation mediated by NO (**D**). (Adapted with permission from [69, 72, 73] and [74])

nanoconstruct was able to photogenerate the NO radical via FRET mechanism activated by TPE with NIR light (Fig. 8b), allowing one to overcome the limitation of this organic photocage because of its absorption below 500 nm. The nanohybrid proved toxic to cancer cells in vitro and significantly reduced the size of human pancreatic BxPc-3 mice xenograft, which is relatively hypoxic, upon the highly biocompatible 800-nm light, demonstrating the potential of NO as a therapeutic alternative to PDT under hypoxic conditions.

An alternative to photoinduced energy transfer mechanism, the low energy absorbed by the QDs light-harvesting centers, can be exploited to trigger NO release from photocages via photoinduced electron transfer. This operating principle has been clearly demonstrated for CdTe QDs electrostatically coated with an Ru-complex NO photocage (Fig. 8c) [73]. The inorganic complex is photoreduced by the QDs, leading to the detachment of the NO radical under green light excitation. The cooperation of both energy and electron transfer processes was at the basis of the ingenious QDs-nanomaterial developed by Balkus and coworkers addressed to the enhanced generation of $^1O_2$ via NO photoproduction (Fig. 8d) [74]. The rationale behind this work was that NO is known to react with reactive oxygen species (ROS) to generate $^1O_2$ [75]. Photoactivable hybrid QDs were achieved by assembling the *S*-nitroso-cysteine (Cys-NO) at the surface of TiO_2 nanotubes doped with PbS QDs. Excitation of the QDs with 760-nm light, which falls in the therapeutic window, promotes NO release by energy transfer from cysteine-NO and simultaneously generates ROS by an electron transfer mechanism. The intermolecular reaction between NO and ROS was demonstrated to be responsible for the generation of $^1O_2$ with high efficiency. This nanomaterial is envisioned to have potential as a next generation photodynamic therapy agent. In this regard, the synthesis of less toxic quantum dots for this system such as CuS, FeS_2, and Ag_2S is a route which deserves attention in the near future.

Photoinduced energy/electron transfer processes from the core to the surface of QDs-based constructs, reflect on the luminescence of the QDs, limiting their performances as imaging agents. In contrast to the above systems, QDs coated with NO photocages can be designed in such a way as to avoid any energy/electron transfer process. This permits the core and the shell to be operated in parallel under light excitation without any light-driven communication. In this context, an interesting class of NO photoreleasing QDs for in vivo imaging is those able to display photoluminescence in the NIR region. These QDs display superior properties over those emitting in the visible region because of reduced auto-fluorescence, negligible tissue scattering, and great tissue-penetration depth. Ag_2S QDs proved very suited at this regard because of a perfect combination of excellent optical properties and good biocompatibility [76]. Intriguing bifunctional assemblies can be easily obtained by exploiting non-covalent interactions of Ag_2S scaffolds with both organic and inorganic NO photoreleasers [77, 78]. For example, carboxy-terminated Ag_2S QDs were entangled within a network of a chitosane-based *S*-nitrosothiol whereas amino-terminated Ag_2S were coated with $F_4S_3(NO)_7^-$ Roussin's black salt anion. The resulting hybrid NPs can be imaged thanks to their easily detectable NIR emission and are able to release NO rapidly upon excitation with visible light [77, 78]. Alternatively, *S*-nitrosothiol **7** (Fig. 9) can be covalently attached at the surface of the Ag_2S QDs, leading to more robust, biocompatible, and water dispersible nanohybrids [79]. The NIR signals from these modified QDs were clearly detected in the liver, stomach, and spleen of mice after intravenous injection. In all the Ag_2S-based systems developed, the two functions can be performed without interference from each other, because excitation light for NO photorelease and NIR imaging occurs in different wavelength regions, opening new prospects for the application of these nanoconstructs as phototheranostics.

NO releasing QDs with good selectivity towards cancer cells have been achieved by incorporation of specific ligands able to bind specific receptors overexpressed by tumor cells. Gui et al. have developed an intriguing nanoconstruct by the self-assembling of three independent components designed for targeting, imaging, and controlled NO release [80]. Carboxymethyl chitosan (CMC), a biocompatible hydrophilic polymer, was functionalized with folic acid (FA), a high affinity ligand for folate receptor overexpressed by several tumor cell lines, via carboxy-amine coupling reaction between the –COOH of FA and $–NH_2$ of CMC to give a CMC–FA conjugate. This polymeric scaffold was suitable for the facile introduction of $F_4S_3(NO)_7^-$ salt as NO precursor and polyethyleneimine-stabilized ZnS-Mn QDs

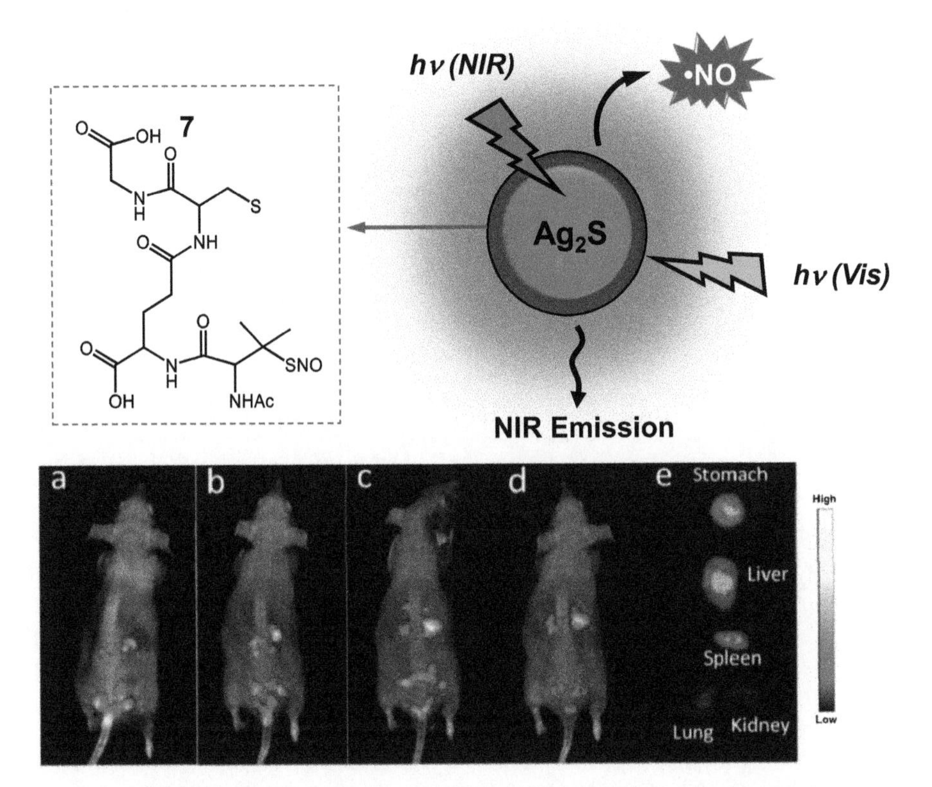

Fig. 9 Ag_2S QDs coated with *S*-nitrosothiols shell release NO upon excitation with visible light and emit in the NIR when excited with NIR light. In vivo NIR images of nude mice at 1 h (**a**), 6 h (**b**), 12 h (**c**), and 24 h (**d**) after intravenous injection of the QDs. (**e**) Ex vivo NIR images of relevant organs from a sacrificed mouse at 12 h post-injection. (Adapted with permission from [79])

as effective green photoemitters via electrostatic adsorption interactions. The hybrid nanospheres enable effective light-triggered release of NO and targeted fluorescence imaging in tumor cells with over-expression of folate receptors.

A significant step forward in this field has been achieved by Liu and coworkers who fabricated a multifunctional nanoplatform in which a folate targeting ligand and the Ru-complex **8** were covalently grafted at the surface of TiO_2 QDs (Fig. 10) [81]. The Ru-complex is fluorescent and acts simultaneously as NO photocage and as $^1O_2$ photosensitizer, being very suited for bimodal phototherapy. The nano-construct exhibits an excellent selectivity towards cancer cells overexpressing the folate-receptor, can be tracked because of its inherent fluorescence, and induces synergistic cytotoxic effects in cancer cells under the control of visible light.

An intriguing nanoconstruct based on all inorganic materials has been developed by Ford and coworkers. This systems is based on core-shell upconverting NPs coated with a silica layer surface modified with cationic amino groups to bind electrostatically the $F_4S_3(NO)_7^-$ salt as NO photoreleaser. Excitation of the NPs

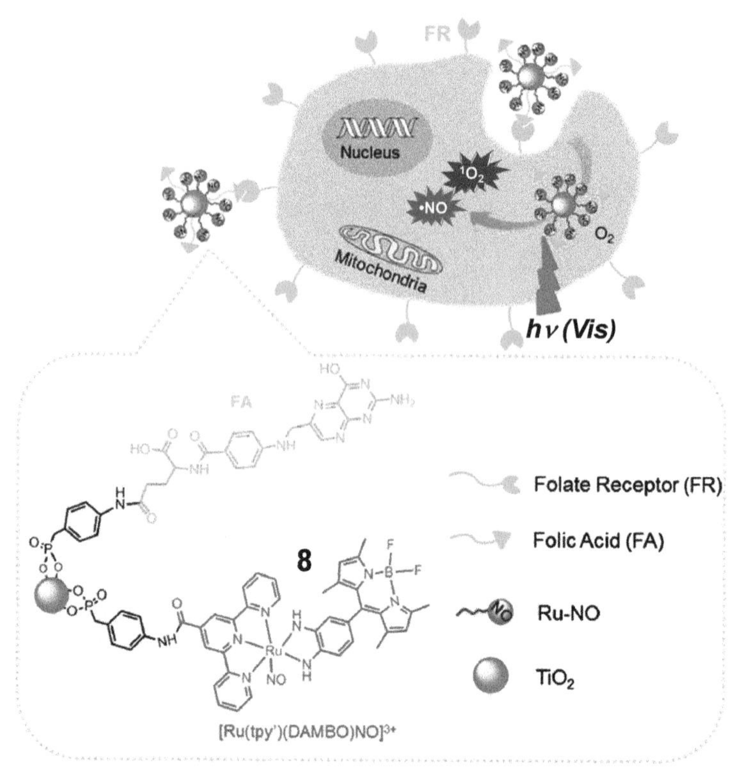

Fig. 10 Schematic of the TiO_2-based nanoplatform for the targeted delivery of NO and $^1O_2$ to specific cancer cells. (Adapted with permission from [81])

with 980-nm NIR light generates an emission centered at 550 nm which overlaps the absorption spectrum of the NO photocage, encouraging the NO release by ETU mechanism [82].

3.2 Polymer Nanoparticles

Polymer NPs are promising vehicles for the delivery of drugs to target different locations in living organisms [83]. They can represent versatile nanocontainers to reduce toxicity of photoresponsive molecules [84] as well as to enhance their photoreactivity [85]. For example, Gomes et al. have recently demonstrated that NPs of poly(D,L-lactic-*co*-glycolic acid), prepared by the double emulsification process, are able to entrap effectively Ru-complexes as NO photoprecursors. These NPs reduced significantly the intrinsic toxicity of the inorganic complexes and allowed diffusion of NO photogenerated out of the matrix, reaching the cell membranes and killing the tumor cell [84]. The photochemical performances, in terms of efficiency of NO photorelease, from an NO photodonor based on the same nitroaniline-based photocage integrated in compound **1** (see Fig. 5A) [85] were significantly improved after its encapsulation within the interior of core-shell micelles of Pluronic F127. This was caused by the presence of abstractable hydrogen from the polymer which is consistent with the photodecomposition route associated with the nitro-to-nitrite photorearrangement of this NO photocage.

Interestingly, polymer NPs can also represent fascinating scaffolds for the achievement of photoactivable nanoconstructs with distinct photofunctionalities [86]. Multiple photoresponsive character in a confined region of space can be easily imposed on polymer NPs by the non-covalent co-encapsulation of distinct and non-interacting photoresponsive species. The non-covalent approach requires minimal, if any, synthetic efforts, permitting the relative amounts of the co-encapsulated guests to be easily regulated with adjustments in the composition of the mixture of self-assembling components. In most instances, the photoresponsive species in their native form are sufficiently hydrophobic to enter spontaneously the interior of the host and can be encapsulated within the same nanosized container even in a single step. Alternatively, hydrophilic photoresponsive guests can be entangled at the surface of the polymer NPs by Coulombic interactions.

Cationic core-shell polymeric NPs composed of a hydrophobic inner core of polyacrylate surrounded by a hydrophilic shell of quaternary ammonium salts have been used as a suitable scaffold to obtain bichromophoric NPs for imaging and simultaneous photodelivery of NO and $^1O_2$ [87]. This nanoconstruct was achieved by the electrostatic entangling of two anionic components, the porphyrin **9** and NO photocage **10**, within the cationic shell of the NPs (Fig. 11). As demonstrated in pioneering work using the same photoactive components [88], the two chromophores preserve their photochemical behavior quite well despite the fact that they are located in close proximity. Accordingly, they can be activated independently exclusively by light control, as demonstrated by their capability to photogenerate the cytotoxic $^1O_2$ and NO effectively and exhibit remarkable red fluorescence from the porphyrin moiety, supporting the validity of the logical design. The co-loaded NPs transport their cargo within melanoma cancer cells without suffering displacement despite the non-covalent binding, representing one of the first examples in this regard. The NPs were well tolerated by cancer cells in the dark and exhibited strongly amplified cell mortality under visible light excitation, most likely because of the combined action of $^1O_2$ and NO.

Focusing on solubility and aggregation issues in aqueous medium, cyclodextrin (CD) polymers offer the possibility of guest interactions with diverse binding sites, i.e., within the 3D macromolecular network and the CD cavities, thereby enhancing the apparent solubility and regulating the self-association tendency of guests [89]. Polymer **11** (Fig. 12) consists of β-CD units interconnected by epichlorohydrin spacers to form glyceryl cross-linked β-CD polymer. This polymer is well tolerated in vivo [90] and highly soluble in water where it exists under the form of neutral NPs of ca. 25 nm diameter [91]. Because of the presence of different hydrophobic nanodomains, these NPs are able to entrap a variety of guests [92–94] with enhanced stability constants and payloads as compared with the

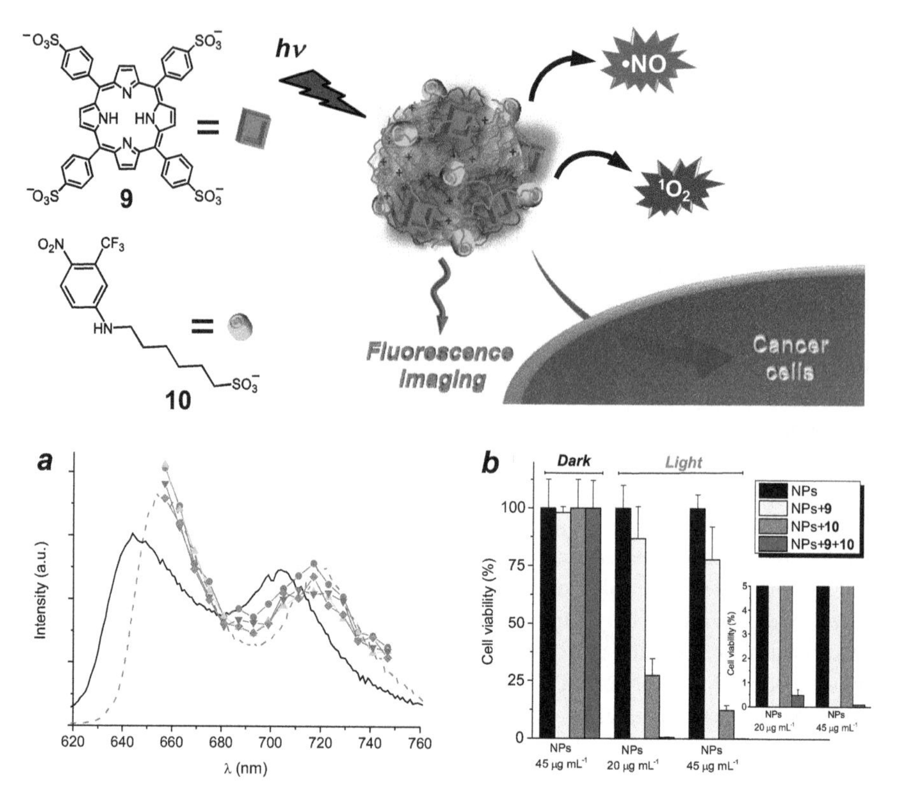

Fig. 11 Cationic core-shell NPs co-entangle the anionic components **9** and **10**, leading to bichromophoric NPs able to release simultaneously NO and $^1O_2$ under visible light excitation. Confocal spectra ($\lambda_{exc} = 640$ nm) for different region of interest of cells incubated with NPs loaded with **9** + **10** (lines + symbols) and fluorescence emission spectra of **9** unloaded (*solid*) and loaded (*dotted*) on the NPs in water (***a***). Cell viability of B78H1 murine melanoma cells incubated for 1 h with polymeric NPs and either kept in the dark or irradiated for 5 min with white light (100 mW cm^{-2}) (***b***). The *inset* shows a zoom of the irradiated samples. (Adapted with permission from [87])

unmodified β-CD. A number of nanoassemblies for NO photodelivery with multiple photofunctionalities based on these polymer NPs have been obtained by the appropriate self-assembling of the ad hoc chosen molecular components reported in Fig. 12.

The photochemical independence of the porphyrin center and the nitroaniline derivative inspired the design of the molecular hybrid **12** in which a pyridyl porphyrin and the NO photodonor are covalently connected through an alkyl spacer [95]. This molecular conjugate would offer in principle the great advantage of a much more precise control of timing, location, and dosage of the cytotoxic species because its covalent nature ensures that the photodelivery events occur exactly in the "very same region of space" of the cell component. The molecular hybrid is sparingly soluble in water medium where it forms aggregates which are

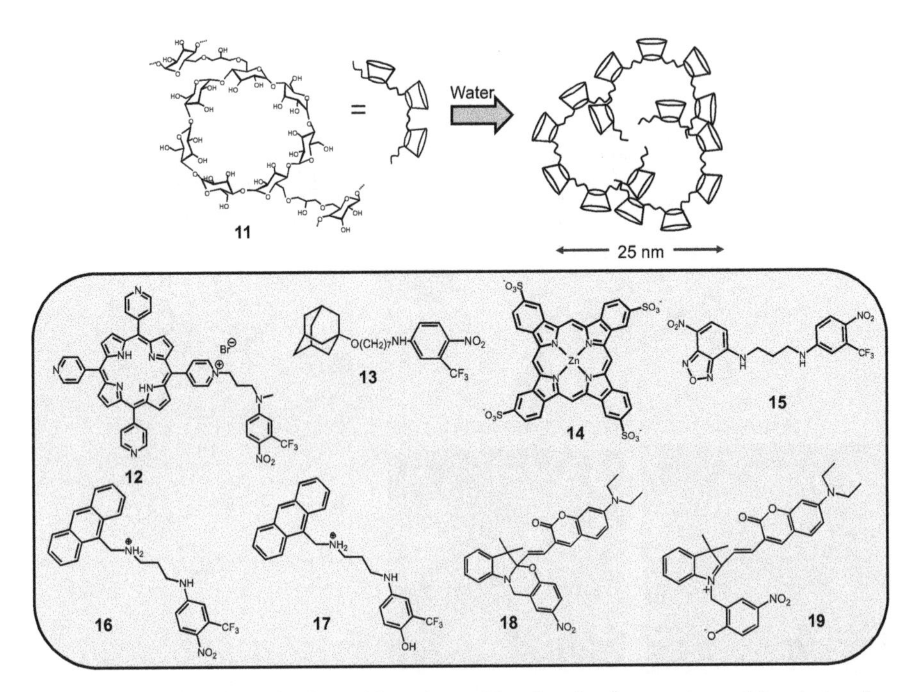

Fig. 12 Schematic of the NPs formed by polymer **11** and molecular structures of the photoactive guests that can be encapsulated therein

non-responsive to light whereas it can be effectively encapsulated within NPs of polymer **11**. The host-guest system photogenerates $^1O_2$ and NO simultaneously under visible light stimuli, exhibits satisfactory red fluorescence allowing the visualization of its localization in melanoma cancer cells, and induces amplified cell mortality by bimodal photoaction.

Bichromophoric NPs have been obtained by the non-covalent co-encapsulation of the NO photodonor **13** and the zinc phthalocyanine **14** [96]. The phthalocyanine chromophore is an excellent $^1O_2$ photosensitizer and offers the advantage to exploit TPE fluorescence for imaging because of the high two-photon cross section [97]. However, it is completely aggregated and non-photoresponsive in water medium. In this case, the polymer NPs disrupt significantly the aggregation of **14**, entrapping this chromophore mainly under its photoactive monomeric form. Besides, the CD cavities tightly accommodate the tailored NO photodonor **13** because of its adamantane appendage, a perfect guest for the β-CD host. The co-entrapment of the two guests in the very same host generate photoresponsive NPs very suited for TPE fluorescence imaging and NO photorelease. The macromolecular host delivers its photoresponsive cargo of active molecules not only within the cytoplasm but also in human skin as exemplified in ex vivo experiments [96].

A major limitation of the above system is represented by the fact that the NO photodonor used is not fluorescent and cannot be imaged in cells. A step forward in this direction has been made with the supramolecular engineered nanoplatform

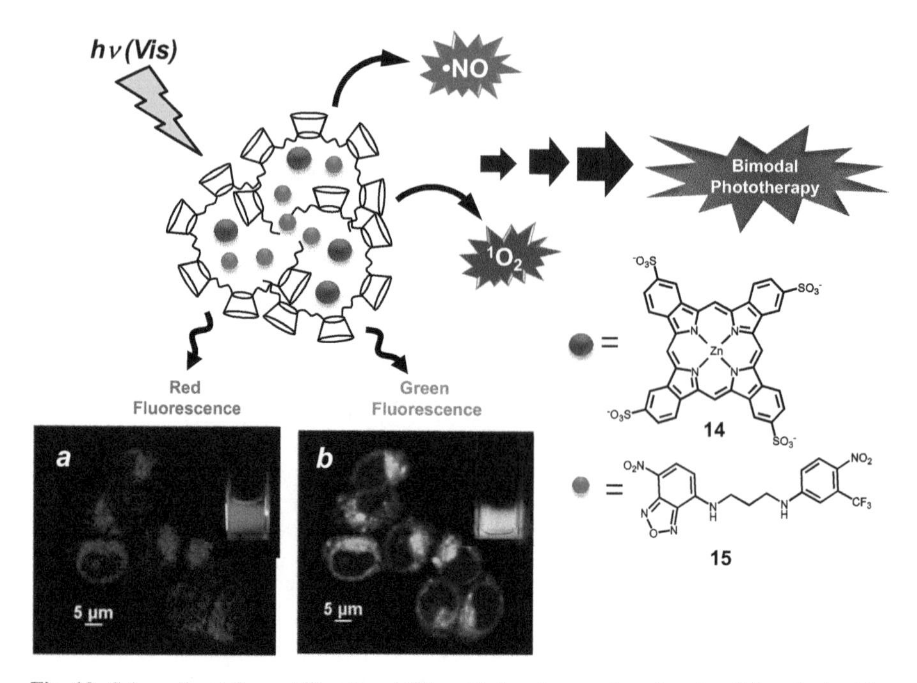

Fig. 13 Schematic of the multifunctional NPs with four-in-one photofunctionalities obtained by the co-encapsulation of components **14** and **15** within NPs of polymer **11**. Illumination with visible light produces simultaneously $^1O_2$ and NO together with the red and fluorescence of the photoprecursors. Confocal images of melanoma cells incubated for 4 h with the NPs and obtained with excitation at 640 and 457 nm collecting fluorescence in the range 660–730 (***a***) and 500–550 (***b***). The *insets* show the actual images of the NPs dispersed in water solution at the same excitation wavelength. (Adapted with permission from [98])

obtained by the co-encapsulation of the zinc phthalocyanine **14** with the NO photodonor **15** within NPs of **11** [98]. Compound **15** contains an aminonitro benzofurazane appendage which emits strongly in the green region without precluding the NO photoreleasing properties of the NO photocage. The whole construct has been designed in such a way that the excited states responsible for the emission and the photogeneration of the cytotoxic species of each compound cannot be quenched by the other and vice versa. Appropriate tuning of the relative concentration of the two chromophores in the NPs allows comparable absorption in the visible region to be reached. This nanoplatform shows the convergence of four-in-one photoresponsive functionalities (Fig. 13). In fact, illumination with visible light generates both green and red fluorescence as well as NO and $^1O_2$. Remarkably, this nanoconstruct (1) effectively internalizes compounds **14** and **15** in cancer cells, where they can be easily mapped because of their dual-color fluorescence and (2) induces amplified level of cell mortality because of the concomitant photoproduction of two cytotoxic agents in the same region of space.

As outlined in Sect. 2.2, the design of NO photocages with a fluorescent reporting function is an elegant strategy for the quantification of the concentration of NO in a biological environment. A multifunctional biocompatible nanoconstruct

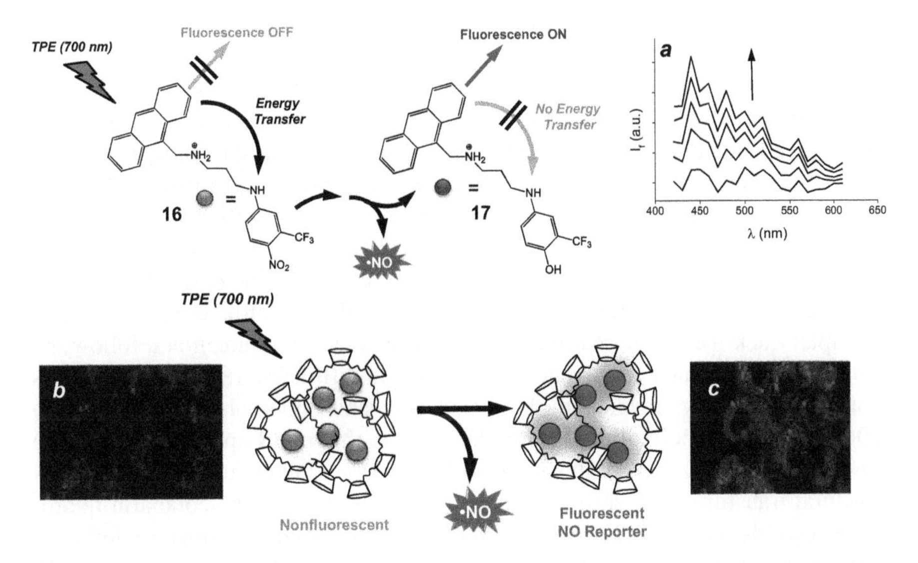

Fig. 14 TPE of **16** encourages the NO release by FRET, leading to the concomitant generation of the fluorescent reporter **17**. Evolution of the TPE fluorescence emission spectra of aqueous dispersions of polymer NPs loaded with **16** upon irradiation from 0 to 1500 s (***a***). Representative TPE fluorescence microscopy images observed before (***b***) and after (***c***) TPE of the same NPs incubated with A431 cancer cells. (Adapted with permission from [99])

able to photorelease NO with a fluorescent reporter has recently been achieved, based on polymeric NPs of **11** encapsulating the ad hoc designed molecular conjugate **16** (Fig. 14) [99]. The conjugate integrates two chromogenic centers within the same covalent skeleton, an anthracene moiety and the nitroaniline-based NO photodonor. The typical emission of the anthracene fluorophore is completely suppressed by FRET mechanism because of the considerable overlap between the emission of the anthracene with the absorption of the NO photocage, making the conjugate **16** intrinsically nonfluorescent. However, photouncaging of NO leads to the formation of the strongly fluorescent co-product **17** (in which FRET can no longer occur) which can be used as *a fluorescent reporter* for the delivery in living cells [100]. This guest is poorly soluble in water but can be encapsulated in NPs of **1**, leading to latent-fluorescent NPs [99]. TPE using NIR 700-nm laser light can be used for monitoring and triggering the photorelease of NO, wherein the uncaging of a strongly fluorescent co-product acts as TPE fluorescent reporter for the concomitant NO release from the nanoassembly (Fig. 14a). This system is then capable of the near instantaneous quantification of such an important bioactive agent by two-photon fluorescence in living cells (Fig. 14b, c) [99].

Although fluorescence imaging represents a powerful tool for the mapping of therapeutic agents in cells, the diffraction barrier limits the spatial resolution of biological samples to hundreds of nanometers [101]. These physical dimensions are much greater than the sizes of most molecules and cellular components. It follows that conventional fluorescence microscopes cannot provide spatial information at the molecular level. The ability to photoactivate fluorescence, however, offers the opportunity to overcome diffraction and acquire images with resolution at the nanometer level [102]. Specifically, fluorophores co-localized within the same subdiffraction volume can be resolved in time, if their fluorescence is designed to switch independently at different intervals of time. Therefore, the ability to activate fluorescence under optical control offers the opportunity to monitor dynamic processes in real time and reconstruct fluorescence images with nanoscaled precision [103]. On these bases, the assembling of photoactivable fluorophores with NO photoreleasers in very close proximity can permit the activation of the fluorescence and release of NO in parallel under optical control. In such a way, one can, in principle, track the NO photocage in the intracellular environment and follow the biological effects induced therein by NO, with subdiffraction resolution. Raymo's group has recently developed the photoswitchable fluorophore **18** (Fig. 15) [104]. Under ultraviolet illumination, the oxazine ring of **18** opens and brings the coumarin appendage in conjugation with the 3*H*-indolium cation of **19**. This structural transformation shifts the main absorption band of the coumarin fluorophore towards the red region. Irradiation at a wavelength positioned within the photogenerated absorption band can, therefore, be exploited to excite selectively **19** with concomitant fluorescence. This species, however, reverts spontaneously back to **18** on a microsecond timescale. As a result, the fluorescence of this system can be switched reversibly on the basis of the photoinduced opening and thermal closing of the oxazine ring. It was shown that these molecular switches can be encapsulated within the hydrophobic interior of polymer micelles [105] and the resulting supramolecular assemblies can be dispersed in aqueous media and imaged with subdiffraction resolution [106]. The photoactivable fluorophore **18** can be co-encapsulated with the NO photodonor **13** within the same NPs of the polymer **11** (Fig. 15) [107]. The two guests can be operated together under optical control without interference. Indeed, NO bursts (Fig. 15a) can be produced on the basis of the irreversible transformation of **13**, although the emission intensity (Fig. 15b) can be switched on and off, relying on the reversible interconversion of **18** and **19**. Furthermore, these polymer nanocarriers can cross the membrane of A375 melanoma cells and transport their photoresponsive and multifunctional cargo to the cytosol [107]. Under these conditions, the fluorescence of one component allows the visualization of the labeled cells, and its switchable character could, in principle, be used to acquire superresolution images, and the release of NO from the other can be exploited to photoinduce cell mortality. These results open intriguing prospects in this direction. In fact, the combination of NO photocages, photoswitchable fluorophores, and specific targeting ligands in a single nanoarchitecture may evolve into the realization of valuable multifunctional photoresponsive assemblies for a variety of NO-based biomedical applications.

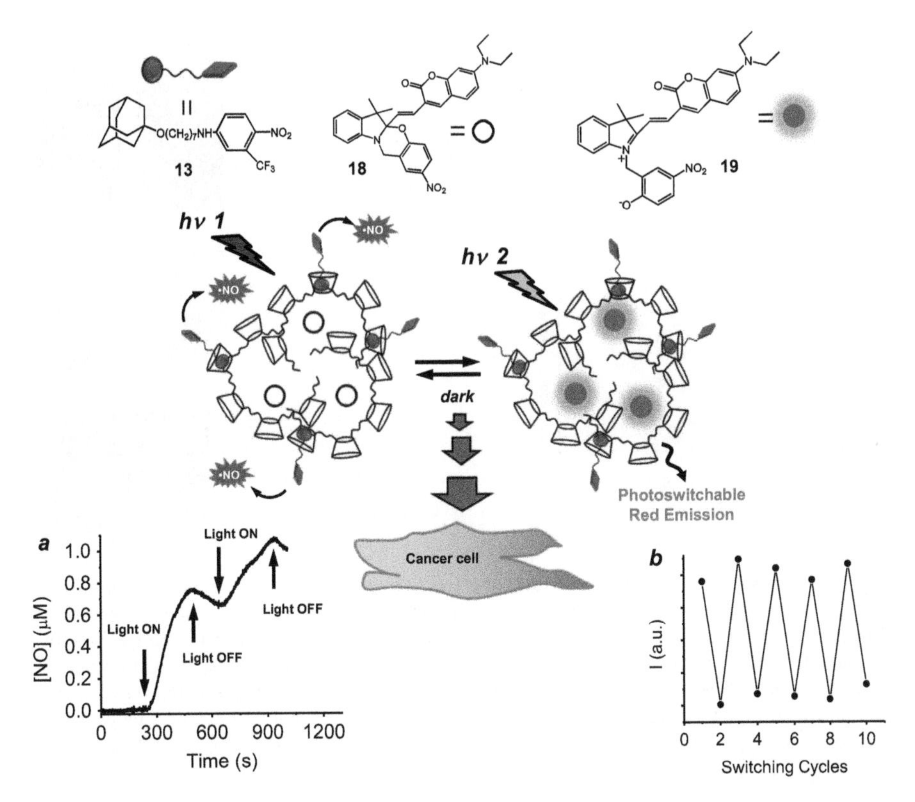

Fig. 15 Nanoparticles of **11** co-encapsulate **13** and **18** in their hydrophobic interior and permit the photoinduced control of both guests in parallel. The photoinduced and irreversible transformation of **13** produces NO revealed by an amperometric signal under illumination (***a***) although the photochemical and reversible interconversion of **18** and **19** translates into the modulation of the red emission intensity (***b***). (Adapted with permission from [107])

3.3 Gels and Matrices

Assembling NO photodonors into biocompatible polymeric gel and optically transparent matrices is promising, especially in view of applications such as devices coating. The resulting materials give the possibility of high loading capacity and provide controlled NO flux at device/target interface because of their high NO-permeability. A simple approach involves the spontaneous, non-covalent incorporation of the photoactive units into a suitable scaffold which can then be shaped and cut as needed. Alternatively, NO functionalities can be covalently introduced in the scaffold molecular skeleton.

A very simple "green" procedure has been developed to obtain supramolecular hydrogels in the absence of any toxic solvents or reagents by the self-assembling of two hydrosoluble polymers: the poly-β-CD polymer **11** and a dextran modified with alkyl side chains **20** (Fig. 16) [108]. The gel formation is based on a "lock-and-key"

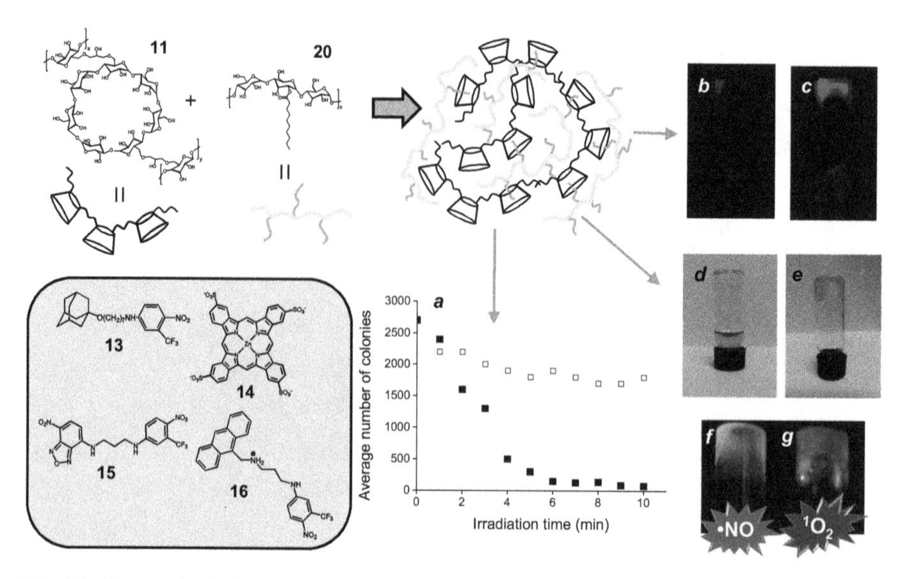

Fig. 16 Photoactive hydrogels can be formed at room temperature by adding aqueous solutions of **20** to aqueous polymer NPs of **11** encapsulating guests **13–16**. Bactericidal action of hydrogel of **13** against *E. coli* in the dark (*open symbols*) and upon visible light irradiation (*filled symbols*) (**a**). Actual fluorescence of the hydrogel of **16** before (**b**) and after (**c**) 45 min of visible light irradiation. Pictures taken before (**d**) and after (**e**) adding the aqueous solution of **20** to the polymer NPs co-encapsulating **14** and **15** and removing the supernatant, and actual fluorescence images of the bichromophoric gel observed upon excitation at 470 (**f**) and 650 nm (**g**). (Adapted with permission from [109, 110] and [111])

mechanism in which practically all the dextran alkyl side chains are included into some β-CD cavities of polymer **11**, leaving most of them still available for further complexation with additional guests. Hydrogel with different photofunctionalities can be obtained by using as a common strategy the pre-encapsulation of the desired photoactive guests within polymer NPs of **11** and then by adding aqueous solutions of the modified dextran **20**, observing the instantaneous formation of the gel at room temperature. A hydrogel embedding the adamantane-terminated NO photocage **13** has been prepared [108] and shown to be stable in the dark towards NO release. In contrast, NO delivery can be conveniently remote-controlled with exposure to visible light stimuli. Once photogenerated, the NO radical can promptly diffuse out of the gel matrix to reach biological targets as demonstrated by the NO binding to a supernatant solution of myoglobin. This NO-delivery gel platform shows excellent and strictly light-dependent bactericidal activity against the Gram-negative antibiotic resistant *Escherichia coli* DH5α bacterial strains (Fig. 16a) [109]. NO photoreleasing gel with a fluorescence reporting function was obtained by encapsulating within the gel network the molecular hybrid **16** discussed above. The preservation of the photochemical properties of this photoactive guests once embedded in the polymeric matrix enables one to photoactivate the blue fluorescence of the reporter concomitantly to the photorelease of NO, and allows in principle an easy optical quantification of its dosage (Fig. 16b, c) [110].

This gel matrix also proved to be suitable for the co-entrapping of multiple guests with excellent preservation of their photochemical properties. The polymer NPs entrapping the green emitting NO photodonor **15** and the red emitter $^1O_2$ photogenerator **14** illustrated in the previous section can be instantaneously turned into gel exhibiting similar response to light excitation [111]. The unchanged color of the polymer NPs before and after the gel formation (Fig. 16d, e) unequivocally demonstrates the successful encapsulation of the photoactive guest components in the gel matrix. Moreover, the very similar absorption spectra of the two chromophores before and after the gel formation account for the absence of any significant molecular rearrangement (i.e., displacement/aggregation) caused by the absence of relevant inter-chromophoric interactions in the ground state, even in the gel matrix. The multi-photoresponsive character of the hydrogel is demonstrated by the combination of fluorescence and photochemical experiments. The two fluorogenic guests retain their luminescence behavior when entangled in the hydrogel, making it dual-color fluorescent. In fact, the selective excitation at 680 and 480 nm results in the characteristic red and green fluorescence arising from **14** and **15** (Fig. 16f, g). The capability of the hydrogel to release simultaneously NO and $^1O_2$ under visible light stimuli is demonstrated by the direct and real-time monitoring of these transient species. In this view, it represents an intriguing model system for potential image-guided phototherapeutic applications. In principle, irradiation with appropriate light sources allows the precise localization of the hydrogel implanted in a bioenvironment by means of its double emission and to activate controlled release of $^1O_2$, NO, or both these active species. All these findings indicate that the role of the modified dextran **20** is exclusively to "freeze" the photophysical behavior observed in solution and to translate it into the gel phase, without altering the photochemical and photophysical features of the individual guests.

An important point to be stressed is related to the stability of these hydrogels under physiological conditions. The multivalent character of the interactions between all the components ensures the stability of the hydrogel and the negligible leaching of the photoactive components from the gel network under physiological conditions.

NO photoreleasing gel containing organic NO photocages can also be prepared by a covalent approach. Schoenfisch's group has recently reported on xerogels with enhanced NO storage stability at physiological temperature based on the condensation of a novel a tertiary thiol-bearing silane precursor with alkylsilanes or alkylalkoxysilanes [112]. The resulting xerogels were finally nitrosated to obtain RSNO photofunctionalities. The sterical hindrance near the nitroso moieties significantly enhances their usually low stability at room temperature, making light strictly necessary to release the NO radical. Biomedical utility of these materials was demonstrated by their capability to reduce *Pseudomonas aeruginosa* adhesion under light excitation. These xerogels offer great potential for commercialization because of the inexpensive reagents, mild synthetic conditions, ease of application to a large number of substrates, and the straightforward extension to scaled-up preparation.

NO photocages based on inorganic complexes of transition metals have been extensively used as photoactive agents in gels and matrices. In these cases, incorporation of the NO-releasing complex into the scaffold is also intended to avoid toxicity associated with the heavy metal ion. Mascharak's group has carried out extensive work in this regard by using the Mn-complex **21** as NO photoprecursors under visible light (Fig. 17). An engineered NO-delivery platform has been fabricated by LbL assembling [113]. The Mn-complex was embedded in a pluronic F127 gel which, in turn, was sandwiched between two layers of polydimethylsiloxane and polydimethoxysiloxane, respectively. The polymeric layers ensure complete sealing of the NO photocage and its photoproducts while allowing NO diffusion (Fig. 17A). This device leads to a drastic reduction of microbial loads of *Acinetobacter baumannii* and *P. aeruginosa* via NO released in 1 h of illumination, making this flexible patch of potential interest as a bandage material for chronic infections of antibiotic resistant strains of bacteria. The same group has developed a prototype fiber-optic NO delivery system based on the same inorganic photoprecursor [114]. This functional construct was made from a polymethylmethacrylate optical fiber where the cone-tipped end was equipped with the NO photoprecursor **21** embedded in a sol–gel (Fig. 17B). Following unmolding, the

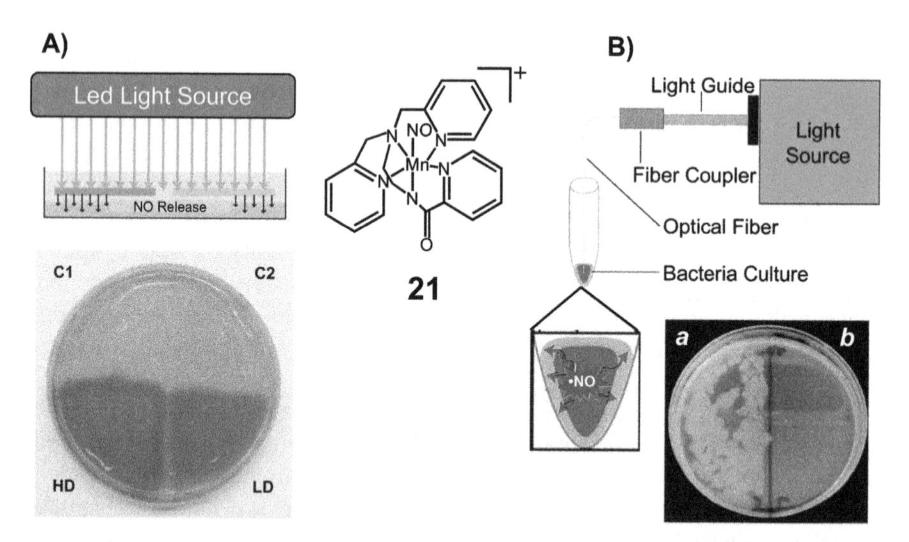

Fig. 17 (**A**) Illumination scheme to test NO delivery from a patch to an agar plate that contains the Griss reagent, which generates a pink dye when reacted with NO_2^-, the reaction product of NO. Photograph of agar after 1 h illumination and removing the polymeric patches. The quadrants C1 and C2 are regions containing the gel empty and loaded with **21** pre-irradiated so all NO has been released, respectively. HD and LD quadrants are regions containing the gel loaded with high and low dose of **21**. (**B**) Schematic setup for the antibiotic experiments using the fiber-optic-based NO delivery systems containing the photoprecursor **21** in a sol–gel material. *P. aeruginosa* incubated with the fiber-optic system in the dark (***a***) and after 10 min illumination (***b***). (Adapted with permission from [113] and [114])

modified fiber tip was coated with a polyurethane wall to prevent leaching. In this case, much lower doses of NO delivered via light triggering resulted in the drastic reduction of various bacterial loads, making this prototype endoscopic delivery system worthy of consideration for future applications. Finally, the complex **21** has also been loaded into the columnar pores of silicate-based porous host scaffolds [115]. In the case of an aluminosilicate-based host, strong electrostatic interactions between the negatively charged host and the cationic photoprecursor ensured an enhanced loading capacity compared with the neutral Si-based host. In any case, leaching of the nitrosyl inorganic complex from the two different hosts was minimal under physiological conditions. Interestingly, light exposure resulted in the rapid release of NO from the entrapped precursors whereas the photoproducts are retained within the structure of the biocompatible host. Experiments carried out with the typical light flux value for a sunny day (100 mW/cm^2) revealed the capability of the material to reduce the *A. baumannii* load. Consequently, powders of this kind could be employed as a first line of treatment for infections in battlefield wounds

As described in Sect. 2.1, ETU is a very well suited strategy to obtain NO photorelease by NIR excitation. Ford and coworkers have recently fabricated a novel implantable solid material encapsulating separate components in a biocompatible matrix [116]. This material consists of polymethoxysiloxane composite with NIR upconverting NPs which are cast into a biocompatible disk impregnated with the Roussin's salt as NO photoprecursor. The encapsulating strategy also isolates the components from the external medium, thereby reducing toxicity. Irradiation with 980-nm light results in the release of physiologically relevant NO concentration from the material which, in principle, can be implanted in a desired location and triggered on demand using tissue-penetrating light (Fig. 18). A similar approach based on the same polymeric composite has been adopted to trigger NO release from the Cr-complex **6** (see Fig. 8) [117], which typically shows very little photochemistry in the solid state. The polymeric composites obtained show long-term and controlled release of NO after direct visible light excitation of the complex or indirect 980-nm NIR excitation by means of upconverting NPs. The NO release from the polymer composites was largely dependent on gas permeability of the polymer which is critical to maintaining high flux of NO release during irradiation. A remarkable point of interest in this study was the demonstration that it is the polymer composition that ultimately affects release of NO from the material the most and not the inclusion of more NO photoprecursor. Careful choice of polymer is therefore needed when identifying potential polymers to serve as platforms for optimal controlled NO release. These findings provided useful guidelines for the preparation of composite materials for light-controlled NO release and basic research necessary to move toward testing of the materials on actual disease models.

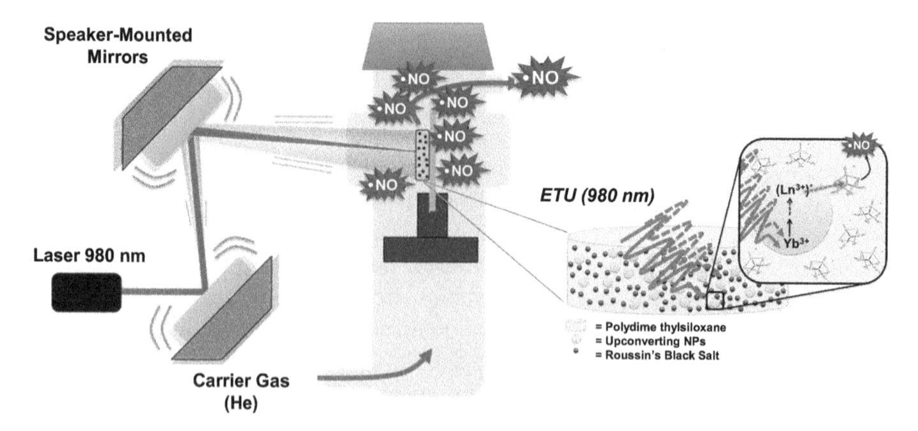

Fig. 18 Schematic of the optical train for the photolysis of the biocompatible disk based on the ETU principle. NIR irradiation of the upconverting NPs generates visible emission, which triggers NO release from the Roussin's salt co-encapsulated in the disk. (Adapted with permission from [116])

4 Conclusions and Outlook

A number of nanomaterials releasing NO under the exclusive control of vis-NIR light for potential and minimally invasive therapeutic applications in cancer and bacterial infection can be obtained by both covalent and non-covalent approaches, paying specific attention to the choice of the photoactive components. NO photoreleasing nanoconstructs with multiple photofunctionalities are achieved by combining ad hoc chosen chromophoric units that can be designed to interact upon light excitation and exchange either energy or electrons. Such photoinduced processes permit the engineering of properties into the final constructs which would be impossible to replicate with the separate components. As an alternative, non-interacting guests with distinct functions can be entrapped in the same host scaffold to ensure multifunctional character in a confined region of space. In contrast to PDT, which has already found clinical application, therapies based on light-triggered NO release are still confined to the research environment because of the lack of converging research activity from different backgrounds, even though they have very promising features. However, the massive progress in nanomaterial science, the impressive technological advances in lasers and fiber-optic tools, and the ability to exploit TPE and ETU with high tissue-penetrating NIR light strongly drive progress in NO-based therapeutics and are expected to facilitate an entirely new category of clinical solutions.

Acknowledgements S.S. is very grateful to his former students, the external collaborators, and all the scientists who have been contributing to this field and whose names are listed in the references for their inestimable contribution to the work described herein.

We thank AIRC (Project IG-12834) and the Marie Curie Program (FP7-PEOPLE-ITN-2013, CYCLON-HIT 608407) for financial support.

References

1. Ignarro LJ (ed) (2010) Nitric oxide: biology and pathobiology. Elsevier, Burlington
2. Walford G, Loscalzo J (2003) J Thromb Haemost 1:2112–2118
3. Fang FC (1997) J Clin Invest 99:2818–2825
4. Packer L (ed) (1999) Nitric oxide, part C: biological and antioxidant activities, vol 301, Methods in enzymology. Academic, San Diego
5. Luo JD, Chen AF (2005) Acta Pharmacol Sin 26:259–264
6. Fukumura D, Kashiwagi S, Jain RK (2006) Nat Rev Cancer 6:521–534
7. Wang PG, Xian M, Tang X, Wu X, Wen Z, Cai T, Janczuk AJ (2002) Chem Rev 102:1091–1134
8. Wang PG, Cai TB, Taniguchi N (eds) (2005) Nitric oxide donors for pharmaceutical and biological applications. Wiley, Weinheim
9. Halpenny GM, Mascharak PK (2010) Anti Infect Agents Med Chem 9:187–197
10. Carpenter AW, Schoenfisch MH (2012) Chem Soc Rev 41:3742–3752
11. Gupta S, McArthur C, Grady C, Ruderman NB (1994) Am J Physiol 266:H2146–H2151
12. Nishikawa M, Sato EF, Utsumi K, Inoue M (1996) Cancer Res 56:4535–4540
13. Routledge MN, Wink DA, Keefer LK, Dipple A (1994) Chem Res Toxicol 7:628–632
14. Szakács G, Paterson JK, Ludwig AJ, Booth-Gent C, Gottesman MM (2006) Nat Rev Drug Discov 5:219–234
15. Ignarro LJ (2009) Arch Pharm Res 32:1099–1101
16. Cook T, Wang Z, Alber S, Liu K, Watkins SC, Vodovotz Y, Billiar TR, Blumberg D (2004) Cancer Res 64:8015–8021
17. Fang FG (1999) Nitric oxide and infections. Kluwer Academic/Plenum, New York
18. Taubes G (2008) Science 321:356–361
19. Lehamann J (2000) Expert Opin Ther Pat 10:559–574
20. Wimalawansa SJ (2008) Expert Opin Pharmacother 9:1935–1954
21. Riccio DA, Schoenfisch MH (2012) Chem Soc Rev 41:3731–3741
22. Jen MC, Serrano MC, Van Lith R, Ameer GA (2012) Adv Funct Mater 22:239–260
23. Seabra AB, Duran N (2010) J Mater Chem 20:1624–1637
24. Kim J, Saravanakumar G, Choi HW, Park D, Kim WJ (2014) J Mater Chem B 2:341–356
25. Hickok JR, Thomas DD (2010) Curr Pharm Des 16:381–391
26. Chang CF, Diers AR, Hogg N (2015) Free Radic Biol Med 79:324–336
27. Sortino S (2012) J Mater Chem 22:301–318
28. Klan P, Solomek T, Bochet CG, Blanc A, Givens R, Rubina M, Popik V, Kostikov A, Wirz J (2013) Chem Rev 113:119–191
29. Ostrowski AD, Ford PC (2009) Dalton Trans 10660–10669
30. Ford PC (2008) Acc Chem Res 41:190–200
31. Sortino S (2010) Chem Soc Rev 39:2903–2913
32. Fry NL, Mascharak PK (2011) Acc Chem Res 44:289–298
33. Ford PC (2013) Nitric Oxide 34:56–64
34. Smith AM, Mancini MC, Nie S (2009) Nat Nanotechnol 4:710–771
35. Prasad PN (2003) Introduction to biophotonics. Wiley, New York
36. Zipfel WR, Williams RM, Webb WW (2003) Nat Biotechnol 21:1369–1377
37. Dong H, Sun LD, Yan CH (2015) Chem Soc Rev 44:1608–1634
38. Celli JP, Spring BQ, Rizvi I, Evans CL, Samkoe KS, Verma S, Pogue BW, Hasan T (2010) Chem Rev 12:2795–2838
39. Rai P, Mallidi S, Zheng X, Rahmanzadeh R, Mir Y, Elrington S, Khurshid A, Hasan T (2010) Adv Drug Deliv Rev 62:1094–1124
40. Veldhuyzen WF, Nguyen Q, McMaster G, Lawrence DS (2003) J Am Chem Soc 125:13358–13359
41. Rose MJ, Fry NL, Marlow R, Hinck L, Mascharak PK (2008) J Am Chem Soc 130:8834–8846

42. Rose MJ, Mascharak PK (2008) Chem Commun 33:3933–3935
43. Jia Q, Janczuk A, Cai T, Xian M, Wen Z, Wang PG (2002) Expert Opin Ther Pat 12:819–826
44. Hasan T, Moor ACE, Ortel B (2000) Cancer medicine, 5th edn. Decker BC Inc, Hamilton
45. Jain PK, Huang X, El-Sayed IH, El-Sayed MA (2008) Acc Chem Res 41:1578–1586
46. Komarova NL, Boland CR (2013) Nature 499:291–292
47. Deldheim DL, Foss CA (eds) (2000) Metal nanoparticles-synthesis, characterization, and applications. Marcel Dekker, New York
48. Callari F, Petralia S, Sortino S (2006) Chem Commun 9:1009–1011
49. Scaiano JC, Stamplecoskie K (2013) J Phys Chem Lett 4:1177–1188
50. Hu M, Chen J, Li ZY, Au L, Hartland GV, Li X, Marquez M, Xia Y (2006) Chem Soc Rev 35:1084–1094
51. Kandoth N, Vittorino E, Sortino S (2011) New J Chem 35:52–56
52. Caruso EB, Petralia S, Conoci S, Giuffrida S, Sortino S (2007) J Am Chem Soc 129:480–481
53. Sudesh P, Tamilarasa K, Arumugam P, Bervhmans S (2013) ACS Appl Mater Interfaces 5:8263–8266
54. Decher G, Schlenoff J (2002) Multilayer thin films. Wiley, Weinheim
55. Tang Z, Wang Y, Podsiadlo P, Kotov NA (2006) Adv Mater 18:3203–3224
56. Ung T, Liz-Marzan LM, Mulvaney P (1998) Langmuir 14:3740–3748
57. Pastoriza-Santos I, Pérez-Juste J, Liz-Marzán LM (2006) Chem Mater 18:2465–2467
58. Schneider G, Decher G (2008) Langmuir 24:1778–1789
59. Schneider G, Decher G (2004) Nano Lett 4:1833–1839
60. Moncada S, Palmer RMJ, Higgs EA (1991) Pharmacol Rev 43:109–142
61. Elbakry A, Zaky A, Liebl R, Rachel R, Goepferich A, Breuning M (2009) Nano Lett 9:2059–2064
62. Taladriz-Blanco P, Pérez-Juste J, Kandoth N, Hervés P, Sortino S (2013) J Colloid Interface Sci 407:524–528
63. Pal S, Tak YK, Song JM (2007) Appl Environ Microbiol 73:1712–1720
64. Ulman A (ed) (1991) An introduction to ultrathin organic films: from Langmuir-Blodgett to self-assembly. Academic, New York
65. Vittorino E, Giancane G, Manno D, Serra A, Valli L, Sortino S (2012) J Colloid Interface Sci 368:191–196
66. Giuffrida S, Ventimiglia G, Sortino S (2009) Chem Commun 48:4055–4057
67. Zrazhevskiy P, Sena M, Gao X (2010) Chem Soc Rev 39:4326–4354
68. Cao L, Wang X, Meziani MJ, Lu F, Wang H, Luo PG, Lin Y, Harruff BA, Veca LM, Murray D, Xie DY, Sun YP (2007) J Am Chem Soc 129:11318–11319
69. Neuman D, Ostrowski AD, Mikhailovsky AA, Absalonson RO, Strouse GF, Ford PC (2008) J Am Chem Soc 130:168–175
70. Burks PT, Ostrowski AF, Mikhailovsky AA, Chan EM, Wagenknecht PS, Ford PC (2012) J Am Chem Soc 134:13266–13275
71. Tan L, Wan A, Zhu X, Li H (2014) Chem Commun 50:5725–5728
72. Fowley C, McHale AP, McCaughan B, Fraix A, Sortino S, Callan JF (2015) Chem Commun 51:81–84
73. Franco LP, Cicillini SA, Biazzotto JC, Schiavon MA, Mikhailovsky AM, Burks P, Ford PC, Santana da Silva R (2014) J Phys Chem A 118:12184–12191
74. Ratanatawanate C, Chyao A, Balkus KJ Jr (2011) J Am Chem Soc 133:3492–3497
75. Khan AU (1995) J Biolumin Chemilumin 10:329–333
76. Jiang P, Zhu CN, Zhang ZL, Tian ZQ, Pang DW (2012) Biomaterials 33:5130–5135
77. Tan A, Wan A, Li H (2013) Langmuir 29:15032–15042
78. Tan L, Wan A, Zhu X, Li H (2014) Analyst 139:3398–3406
79. Tan A, Wan A, Li H (2013) ACS Appl Mater Interfaces 5:11163–11171
80. Gui R, Wan A, Zhang Y, Li H, Zhao T (2014) RSC Adv 4:30129–30136
81. Xiang HJ, An L, Tang WW, Yang SP, Liu JG (2015) Chem Commun 51:2555–2558
82. Garcia JV, Yang J, Shen D, Yao C, Li X, Stucky GD, Zhao DY, Ford PC, Zhang F (2012) Small 8:3800–3805
83. Husseini AG, Pitt WG (2008) Adv Drug Deliv Rev 60:1137–1152

84. Gomes AJ, Espreafico EM, Tfouni E (2013) Mol Pharmaceutics 10:3544–3554
85. Taladriz-Blanco P, de Oliveira MG (2014) J Photochem Photobiol A Chem 293:65–71
86. Swaminathan S, Garcia-Amoròs J, Fraix A, Kandoth N, Sortino S, Raymo FM (2014) Chem Soc Rev 43:4167–4178
87. Fraix A, Manet I, Balestri M, Guerrini A, Dambruoso P, Sotgiu G, Varchi G, Camerin M, Coppellotti O, Sortino S (2015) J Mater Chem B 3:3001–3010
88. Caruso EB, Cicciarella E, Sortino S (2007) Chem Commun 47:5028–5030
89. Anand R, Manoli F, Manet I, Daoud-Mahammed S, Agostoni V, Gref R, Monti S (2012) Photochem Photobiol Sci 11:1285–1292
90. Daoud-Mahammed S, Grossiord JL, Bergua T, Amiel C, Couvreur P, Gref R (2008) J Biomed Mater Res Part A 86:736–748
91. Othman M, Bouchemal K, Couvreur P, Desmaële D, Morvan E, Pouget T, Gref R (2011) J Colloid Interface Sci 354:517–527
92. Renard E, Deratani A, Volet G, Sebille B (1997) Eur Polym J 33:49–57
93. Gref R, Amiel C, Molinard K, Daoud-Mahammed S, Sebille B, Gillet B, Beloeil JC, Ringard C, Rosilio V, Poupaert J, Couvreur P (2006) J Control Release 111:316–324
94. Battistini E, Gianolio E, Gref R, Couvreur P, Fuzerova S, Othman M, Aime S, Badet B, Durand P (2008) Chem Eur J 14:4551–4561
95. Fraix A, Guglielmo S, Cardile V, Graziano ACE, Gref R, Rolando B, Fruttero R, Gasco A, Sortino S (2014) RSC Adv 4:44827–44836
96. Kandoth N, Kirejev V, Monti S, Gref R, Ericson MB, Sortino S (2014) Biomacromolecules 15:1768–1776
97. Mir Y, van Lier JE, Allard JF, Morris D, Houd D (2009) Photochem Photobiol Sci 8:391–395
98. Fraix A, Kandoth N, Manet I, Cardile V, Graziano ACE, Gref R, Sortino S (2013) Chem Commun 49:4459–4461
99. Kirejev V, Kandoth N, Gref R, Ericson MB, Sortino S (2014) J Mater Chem B 2:1190–1195
100. Vittorino E, Sciortino MT, Siracusano G, Sortino S (2011) ChemMedChem 6:1551–1554
101. Murphy DB (ed) (2001) Fundamentals of light microscopy and electronic imaging. Wiley-Liss, New York
102. Heintzmann R, Gustafsson MGL (2009) Nat Photonics 3:362–364
103. Fernandez-Suarez M, Ting AY (2008) Nat Rev Mol Cell Biol 12:929–943
104. Deniz E, Sortino S, Raymo FM (2010) J Phys Chem Lett 1:3506–3509
105. Cusido J, Battal M, Deniz E, Yildiz I, Sortino S, Raymo FM (2012) Chem Eur J 18:10399–10407
106. Deniz E, Tomasulo M, Cusido J, Yildiz I, Petriella M, Bossi ML, Sortino S, Raymo FM (2012) J Phys Chem C 116:6058–6068
107. Deniz E, Kandoth N, Fraix A, Cardile V, Graziano ACE, Lo Furno D, Gref R, Raymo FM, Sortino S (2012) Chem Eur J 18:15782–15787
108. Daoud-Mahammed S, Couvreur P, Bouchemal K, Cheron M, Lebas G, Amiel C, Gref R (2009) Biomacromolecules 10:547–554
109. Kandoth N, Mosinger J, Gref R, Sortino S (2013) J Mater Chem B 1:3458–3463
110. Fraix A, Kandoth N, Gref R, Sortino S (2015) Asian J Org Chem 4:256–261
111. Fraix A, Gref R, Sortino S. (2014) J Mater Chem B 2:3443–3449
112. Riccio DA, Coneski PN, Nichols SP, Brodnax AD, Schoenfisch MH (2012) ACS Appl Mater Interfaces 4:796–804
113. Halpenny GM, Heilman B, Mascharak PK (2012) Chem Biodivers 9:1829–1839
114. Halpenny GM, Gandhi KR, Mascharak PK (2010) ACS Med Chem Lett 1:180–183
115. Heilman BJ, St John J, Oliver SRJ, Mascharak PK (2012) J Am Chem Soc 134:11573–11582
116. Burks PT, Garcia JV, GonzalezIrias R, Tillman JT, Niu M, Mikhailovsky AA, Zhang J, Ford PC (2013) J Am Chem Soc 135:18145–18152
117. Mase JD, Razgoniaev AO, Tschirhart MK, Ostrowski AD (2015) Photochem Photobiol Sci 14:775–785

Index

CPSIA information can be obtained
at www.ICGtesting.com
Printed in the USA
LVHW02*1443040318
568593LV00001B/264/P